Lithium-Ion and Next-Generation Batteries Recycling

Lithium-Ion and Next-Generation Batteries Recycling

Guest Editor

Sascha Nowak

MDPI

Basel • Beijing • Wuhan • Barcelona • Belgrade • Novi Sad • Cluj • Manchester

Guest Editor
Sascha Nowak
MEET Battery
Research Center
University of Münster
Münster
Germany

Editorial Office
MDPI AG
Grosspeteranlage 5
4052 Basel, Switzerland

This is a reprint of the Special Issue, published open access by the journal *Recycling* (ISSN 2313-4321), freely accessible at: https://www.mdpi.com/journal/recycling/special_issues/C43BQ90B3N.

For citation purposes, cite each article independently as indicated on the article page online and as indicated below:

Lastname, A.A.; Lastname, B.B. Article Title. *Journal Name* **Year**, *Volume Number*, Page Range.

ISBN 978-3-7258-6268-9 (Hbk)
ISBN 978-3-7258-6269-6 (PDF)
https://doi.org/10.3390/books978-3-7258-6269-6

Contents

About the Editor

Sascha Nowak

Sascha Nowak holds a scientific staff position at the MEET Battery Research Center at Münster University as the head of the division Analytics and Environment. He studied chemistry at the University of Münster and received his PhD in Analytical Chemistry. After his doctorate, he joined the working group of Prof. Winter at the MEET Battery Research Center in 2009 as a postdoctoral researcher where he established the analytical department. From 2010 to 2012 he was the head of the competence areas Analytics and Recycling, and since 2012 he has held a scientific staff position at the MEET Battery Research Center.

recycling

MDPI

Article

Investigation of Aqueous Delamination Processes for Lithium-Ion Battery Anodes

Eric Trebeck [1,*], Anting Grams [2], Jan Talkenberger [2], Sricharana Prakash [1], Julius Eik Grimmenstein [1], Thomas Krampitz [1], Holger Lieberwirth [1] and Adrian Valenas [1]

1 Institute of Mineral Processing Machines and Recycling Systems Technology, TU Bergakademie Freiberg, 09599 Freiberg, Germany; sricharana.prakash@student.tu-freiberg.de (S.P.)
2 Weber Entec GmbH, 76275 Ettlingen, Germany
* Correspondence: eric.trebeck@iart.tu-freiberg.de

Abstract

Recycling of lithium-ion batteries (LIBs) requires efficient separation of active material from current collectors to enable high-quality recovery of both the coating and the metal foil. In this study, a water-based delamination process for anode foils was systematically investigated under variations in temperature, particle size, ultrasonic power, and prior mechanical stressing of the particles. Mechanically cut and pre-folded foil pieces were treated in a batch setup at different temperatures (room temperature to 100 °C) and ultrasonic power levels (50 and 100%). Results show that higher temperatures strongly promote delamination, with 100% removal of the active layer achieved on the smooth foil side at 80 °C without ultrasonic treatment. Ultrasonic treatment at moderate power (50%) yielded greater delamination than at full power (100%), likely due to more effective cavitation dynamics at moderate intensity. Mechanical pre-stressing by folding significantly reduced delamination, with three folds effectively preventing separation. In comparison, mechanically comminuted particles from a granulator achieved similar delamination to three-folded particles after 5 min treatment, and higher delamination after 30 min. These findings highlight the importance of process parameters in achieving efficient aqueous delamination, providing insights for scaling low-energy recycling processes for LIB production scrap.

Keywords: lithium-ion battery recycling; copper foil recovery; thermal decoating; aqueous delamination; circular economy; anode recycling

1. Introduction

1.1. Background and Relevance

Global electric vehicle sales have increased to 6.6 million in 2021, with electric vehicles accounting for 10% of the worldwide car market in 2021 and the global electric vehicles fleet expected to account for 30% of all vehicles sold by 2030 [1]. In the EV sales outlook, Bloomberg New Energy Finance forecasts a substantial rise in lithium-ion battery demand, reaching 408 GWh by 2025 and 1293 GWh by 2030 [2].

However, this growth presents challenges related to resource consumption, as LIB production requires large quantities of metals such as lithium, cobalt, nickel, and manganese. With rising demand driven by modernization and electrification, the implementation of efficient disposal and recycling strategies is essential to minimize environmental impacts and promote sustainable waste management [3].

Moreover, production scrap can constitute up to 30% of the electrode mass during manufacturing [4]. With a specific capacity of 300 Wh/kg, a production scrap rate of 300,000 metric tons can therefore be expected for 2030 [5].

Graphite, designated as a critical raw material by the EU [6], is the dominant anode material in lithium-ion batteries due to its high electrical conductivity, layered carbon structure enabling efficient ion intercalation, and relatively low cost. By 2030, the demand for battery-grade graphite is projected to be four times higher than in 2023, when it reached 1000 kt for electric vehicles alone [7]. Copper, likewise, has extensive uses in defense applications such as in the aerospace, naval, space, and electronics industries [8], and offers high recovery value [9]. Therefore, efficient separation and recovery of anode materials, particularly graphite and copper from spent lithium-ion batteries, is vital for maximizing resource utilization and reducing environmental impacts.

Consequently, achieving the goal of improving the recyclability of these anode components and reintegrating them into the supply chain is essential in reducing the overall CO_2 footprint, ideally through approaches that minimize chemical and thermal processing.

1.2. Direct Recycling as an Approach

Direct recycling is a process that reuses battery materials without converting them back to their raw chemical forms [9]. Unlike conventional methods such as pyrometallurgy or hydrometallurgy, it recovers functional cathode particles without decomposition into their constituent elements or dissolution and precipitation of the whole particle [10]. This preserves the original crystal structure of the material, enabling direct reuse in battery manufacturing and avoiding many process steps such as hydrometallurgy and resynthesis of the active material [11].

By preventing the destruction of spent battery materials and directly restoring degraded electrode materials, direct recycling retains a significant portion of the energy invested during the original manufacturing process [9,11]. This results in lower energy consumption, shorter processing times, and reduced CO_2 emissions compared to conventional recycling methods [11]. In fact, if recycled products from each process are reused in LIB manufacturing, GHG emissions can be reduced by 2.85% for pyrometallurgy, 10.24% for hydrometallurgy, and 34.52% for direct recycling [12].

However, reconditioning of the recovered active materials may still be necessary before reuse [11].

The success of this approach relies heavily on the ability to separate electrode coatings cleanly and without structural damage, ensuring the recovery of pure, undamaged material fractions. Current research initiatives, such as the U.S. Department of Energy's ReCell Center [13], are actively working to optimize these processes and promote the development of a closed-loop battery recycling system.

1.3. Current Research Landscape

Lithium-ion battery recycling typically begins with discharge, disassembly, and separation. Pyrometallurgy recovers metals such as cobalt and nickel through high-temperature treatment but entails considerable energy consumption and environmental impacts from combustion and calcination processes. Hydrometallurgy achieves higher recovery rates using chemical leaching solutions, but it demands substantial reagent input and subsequent wastewater treatment [14].

Among solvent-based delamination methods, N-methyl-2-pyrrolidone is widely used due to its high polarity and thermal stability. However, increasing regulatory restrictions and toxicity are driving efforts to replace it with safer alternatives [15]. Alternative reaction media, such as $AlCl_3$–NaCl molten salt and chloride–glycerol deep eutectic solvents, were explored for detaching cathode materials from aluminum foils, but both the morphology and composition of the recovered cathode materials were altered [16].

Recently, the use of power ultrasound (10–1000 W/cm^2, <100 kHz) has gained attention as a green and energy-efficient alternative for LIB recycling. Ultrasound has demonstrated potential in enhancing the delamination, separation, and regeneration of electrode materials, offering significant advantages over conventional solvent or heat-based methods [17].

1.4. Research Needs and Objectives

The rising demand for lithium-ion batteries calls for efficient and environmentally sustainable recycling methods, particularly for the recovery of critical anode materials such as graphite and copper. Existing delamination techniques typically involve high energy consumption, use of toxic acids, and significant wastewater generation, which can result in secondary pollution [18].

Although water-based delamination approaches have been explored, the quantitative evaluation of the effects of comminution on the separation efficiency between graphite and copper remains insufficient. This represents a significant research gap in the development of optimized, scalable recycling processes.

To address this, the present study investigates a simple, chemical-free delamination process using only water under systematically varied temperature conditions and particle sizes. Supported by fundamental investigations on the influence of temperature and particle size, the focus is on the effect of stressed, crushed particles. To further enhance delamination efficiency, ultrasound treatment is explored as a complementary process.

The primary objective of this work is the investigation of water-based delamination methods for recovering high-purity anode materials with minimal environmental impact. The findings contribute to enhancing the overall recyclability of lithium-ion batteries while ensuring high separation efficiency.

2. Results

2.1. Temperature

In Figure 1 the delamination progress is shown as a function of the temperature. There is no delamination till T = 60 °C. At 80 °C the delamination jumps to 50%. Obviously, there is a big difference in the delamination progress of the two sides of the foil. As you can also see in Figure 2, the particles are delaminated just on the side with the smooth surface. Even at T = 100 °C, the rough side remains coated.

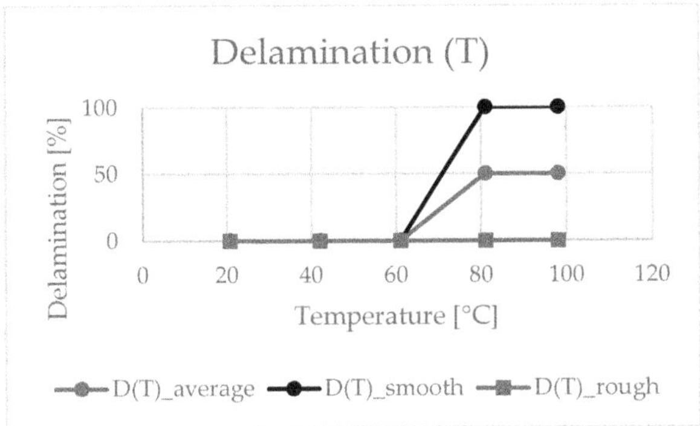

Figure 1. Delamination as a function of the temperature.

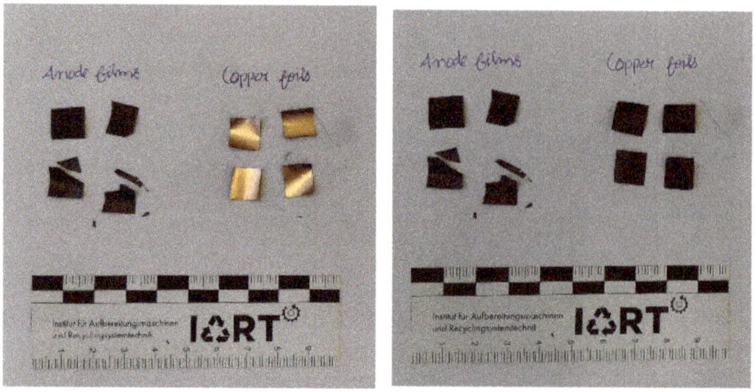

Figure 2. Particles after delamination at 80 °C: the front side (**left**) shows the smooth, delaminated surface, while the back side with the rough surface (**right**) remains coated.

2.2. Particle Size

Based on the findings in Section 2.1, the experiments were carried out at 80 °C. In Figure 3 the delamination is shown as a function of the particle size. Across all particle sizes, an overall delamination of 50% was observed. As before, delamination occurred exclusively on the smooth side (100%), while the rough side remained completely intact (0%) (see Figure 4).

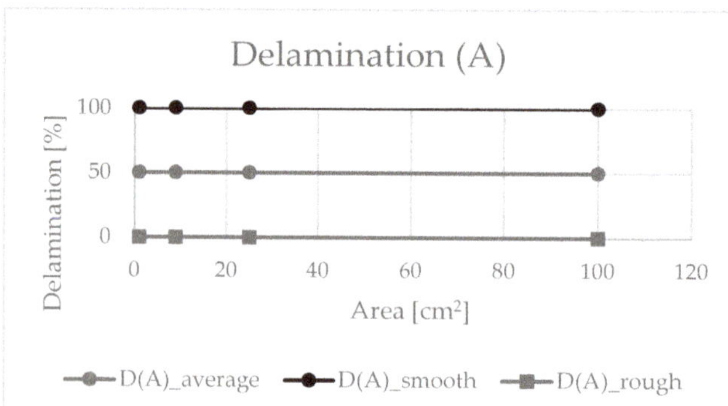

Figure 3. Delamination as a function of the particle size.

Figure 4. The 10 × 10 cm particles after delamination at 80 °C: the front side (**left**) shows the smooth, delaminated surface, while the back side with the rough surface (**right**) remains coated.

2.3. *Ultrasonic Treatment*

The results from the ultrasonic treatment are shown in Table 1. With an addition of ultrasonic power of 100%, an increase in delamination at T = 20 °C from 0 to 29% was observed. When reducing the ultrasonic power to 50%, the delamination increased to 65%. An increase in temperature (T = 60 °C) led to further improvements in delamination (92.65%). Even after 3 min experimental time, a delamination of about 72% was observed. In comparison to the initial experiments, there was better delamination behavior when using ultrasonic stress than without it under otherwise identical conditions. You can see the particles from the different experiments in Figures 5–8.

Table 1. Delamination results with ultrasonic treatment.

Nr.	T [°C]	t [min]	P [%]	D [%]
V1	20	5	100	29.04
V2	20	5	50	64.69
V5	60	5	50	92.65
V6	60	3	50	71.69

Figure 5. Particles after treatment in experiment V1.

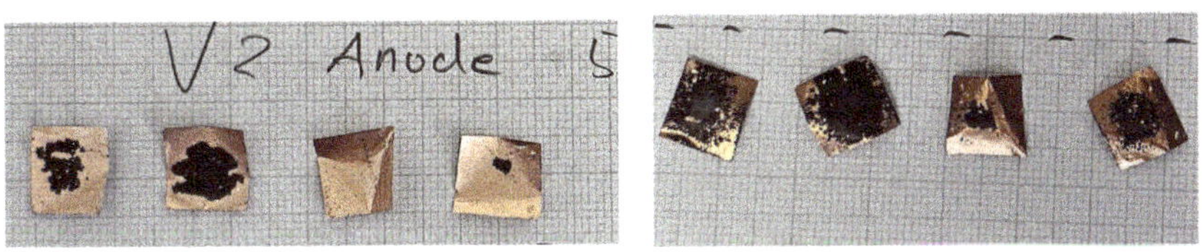

Figure 6. Particles after treatment in experiment V2.

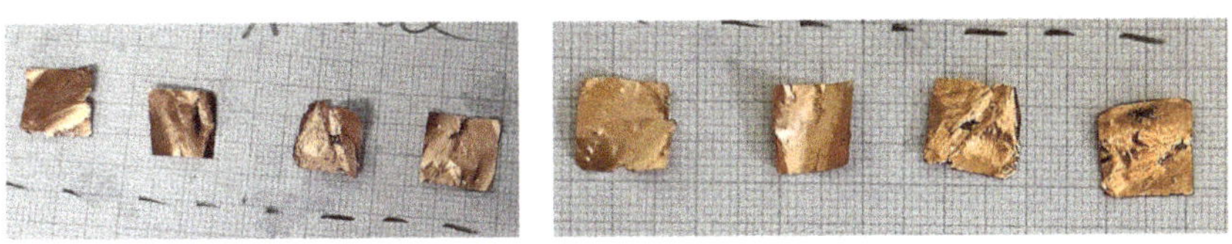

Figure 7. Particles after treatment in experiment V5.

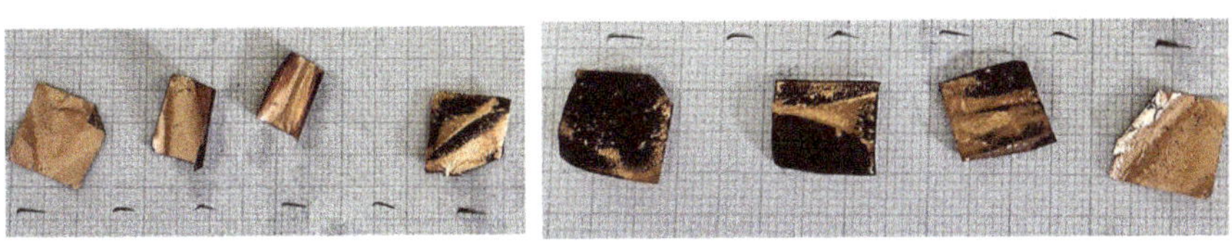

Figure 8. Particles after treatment in experiment V6.

2.4. *Precrushing and Deformation*

In Figures 9 and 10, the delamination as a function of the folds is shown. The difference is that in the experiments shown in Figure 9, the side with the smooth surface was folded inward, whereas in the experiments shown in Figure 10, it was folded outward. When folding the smooth surface to the inside, you can see one fold is enough to inhibit any delamination. The slight increase after 3 folds is solely related to the stress caused by the folding itself. When the smooth surface is folded outward, a different pattern emerges. After a single fold, 100% of the smooth surface remains on the outside, resulting in 100% delamination on this side. When the smooth surface is halved by a second fold, the delaminated area is also reduced by half. From the third fold onward, no delamination occurs. The rough side remained laminated in all cases. Figure 11 shows, on the one hand, the delamination of folded but subsequently unfolded pieces. It can be seen that with an increasing number of mechanical stresses—in this case, folds—the degree of delamination decreases. The mechanically produced particles (granulator) were treated for 5 min in one case and for 30 min in the other. The particles treated for 5 min exhibit a similar degree of delamination (12.38%) to those folded three times. However, the delamination degree could be further increased (36%) by extending the treatment duration. In Figures 12–15 the particles from the experiments are shown for visualization of the results.

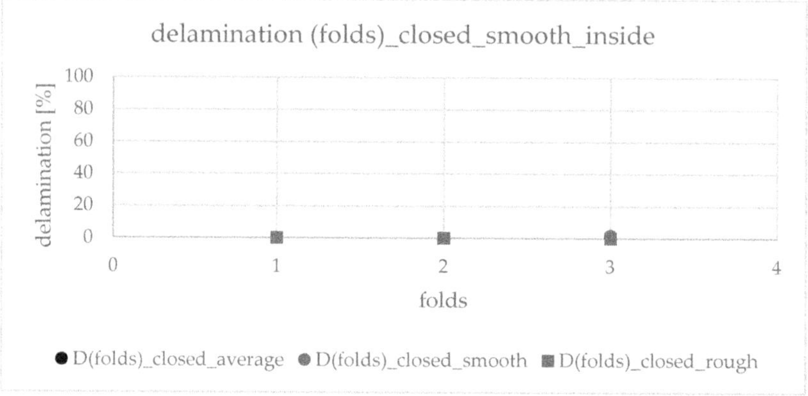

Figure 9. Delamination as a function of the folds. Folding configuration with the smooth (active) surface folded inward.

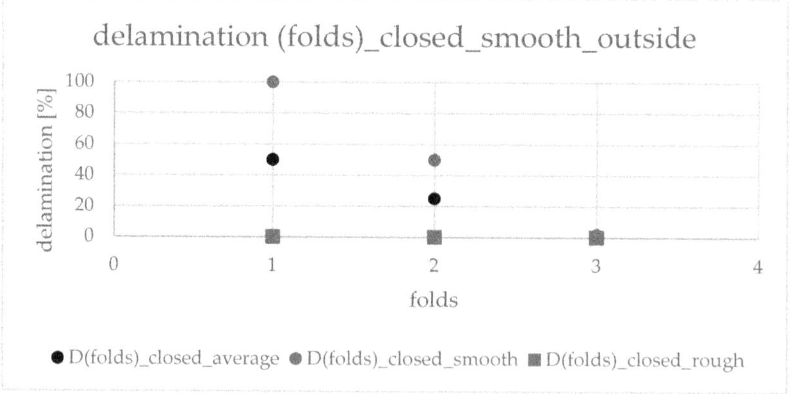

Figure 10. Delamination as a function of the folds. Folding configuration with the smooth (active) surface folded outward.

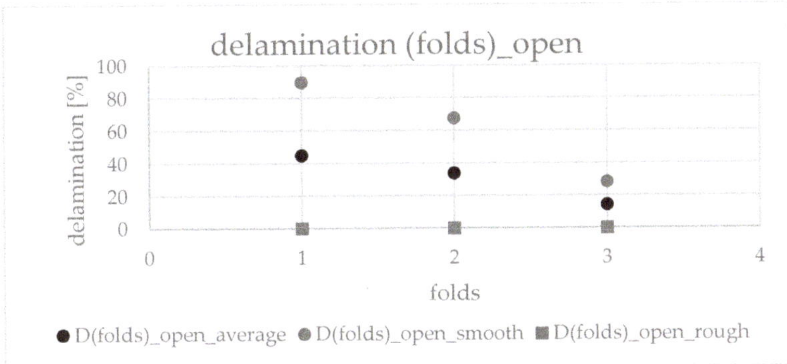

Figure 11. Delamination for folded and subsequently unfolded pieces.

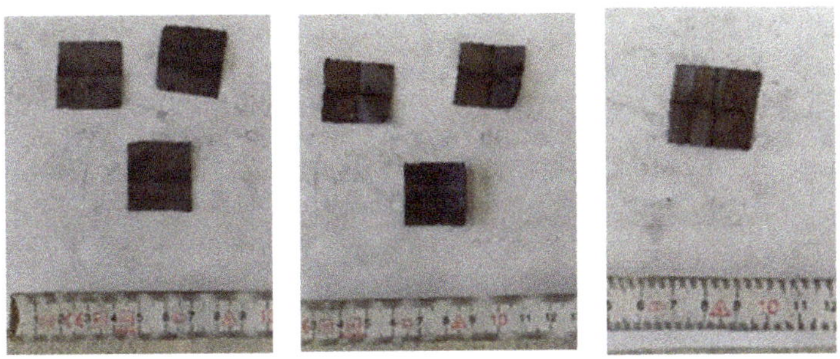

Figure 12. Particles after delamination test with the smooth (active) surface folded inward.

Figure 13. Particles after delamination test with the smooth (active) surface folded outward.

Figure 14. Particles after delamination tests on folded and subsequently unfolded pieces.

Figure 15. Mechanically produced particles from a granulator after delamination treatment.

3. Discussion

This study demonstrates that aqueous delamination of LIB anode foils can be effectively achieved through adjustment of process parameters, particularly temperature, ultrasonic power, and particle pre-treatment. Across all particle sizes, the smooth foil side exhibited complete delamination under favorable conditions, whereas the dendritic, rough side remained coated in all cases without additional processing steps. Temperature proved to be the most influential factor, with elevated temperatures (80 °C) enabling 100% delamination on the smooth surface even without ultrasonic support. This effect is likely not caused by thermal softening of the binder (which remains thermally stable up to around 150 °C) but rather by a reduction in adhesive forces at the foil–coating interface induced by enhanced capillary action. At higher temperatures, the viscosity and surface tension of water decrease, facilitating deeper and faster penetration of the liquid into interfacial gaps and pores. This penetration may further weaken the adhesion between the coating and the foil, promoting more effective delamination. Ultrasonic treatment further enhanced delamination, but only when applied at moderate intensity (50%). A possible explanation is that moderate power promotes more stable cavitation behavior, generating effective shear forces at the interface, whereas excessive power may lead to uncontrolled bubble collapse and turbulent flow that reduce delamination efficiency. Mechanical pre-stressing by folding reduced delamination efficiency, which can be attributed to a decrease in the accessible active surface area and potential local compaction of the coating in the fold region. Three successive folds prevented delamination entirely. Similarly, particles mechanically comminuted by a granulator showed reduced initial delamination after 5 min treatment. An extended treatment time (30 min) overcame these effects, suggesting that prolonged exposure compensates for reduced surface accessibility. These results showed that efficient aqueous delamination depends on a combination of electrode surface roughness, accessible surface area, thermal effects on binder adhesion, and cavitation behavior. Moderate ultrasonic intensity at elevated temperatures appears to offer a favorable balance, while excessive mechanical pre-stressing during comminution or fold formation should be avoided. Future work should focus on scaling these findings to continuous systems, optimizing residence time, and developing selective delamination strategies for both foil sides to further improve process efficiency in LIB recycling.

4. Materials and Methods

A typical lithium-ion battery anode consists of graphite, conductive carbon, and a binder, such as polyvinylidene difluoride or a combination of carboxymethyl cellulose and styrene-butadiene rubber, coated onto a current copper collector [19].

The anode foils used in this study (Figure 16) consist of copper foil, produced by electrolysis, coated with graphite using a polyvinylidene difluoride binder.

Figure 16. Anode foil used in the experiments.

For characterization of the morphology of the foil, the uncoated part of the anode was evaluated via microscope. As you can see in Figure 17, the surfaces of the two sides differ. One side is smooth while the other side exhibits a more rough and dendritic morphology. The reason for that lies in the production method of the foil. Compared to rolled foils, foils produced by electrolysis exhibit two different surface structures.

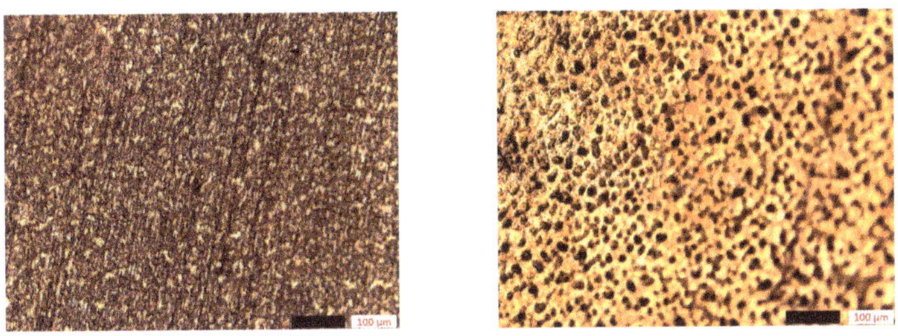

Figure 17. Anode foil with smooth surface (**left**) and dendritic, rough surface (**right**).

Experiments were performed using water baths heated with a standard laboratory heating device and continuous temperature monitoring. Mechanical stirring was applied during all tests. Ultrasound-assisted delamination trials were carried out at Weber Entec GmbH (76275 Ettlingen, Germany) using specialized equipment in the form of a Biopush flow cell (Figure 18).

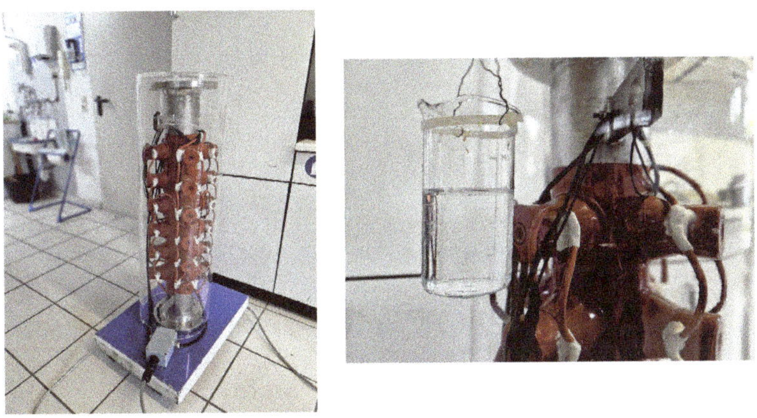

Figure 18. BioPush flow cell setup by Weber Entec GmbH.

To simulate real recycling conditions, the foils were comminuted using two devices. One was a rotary shear and the other one a granulator, resulting in fragments with various sizes and deformations (e.g., bends and folded edges).

The degree of delamination is defined as the percentage of the copper foil surface from which the graphite coating has been removed. It is calculated using image analysis in ImageJ software (Version 1.54p), as in Equation (1). Therefore, every particle was photographed separately under flat conditions to avoid inaccuracies.

$$Delamination[\%] = \frac{Area\ decoated}{Area\ total} * 100 \tag{1}$$

4.1. Experimental Series

4.1.1. Temperature

A systematic series of delamination experiments was performed at temperatures between 20 °C and 100 °C (20 °C, 40 °C, 60 °C, 80 °C, 100 °C), using a fixed stirring time of 5 min, which was defined based on the investigations by Bai et al. [19] to study the influence of temperature on coating removal. For this purpose, the anode foil was cut into pieces measuring 1 × 1 cm.

4.1.2. Particle Size

To evaluate the effect of particle size on delamination, foil fragments measuring 1 × 1 cm, 3 × 3 cm, 5 × 5 cm, and 10 × 10 cm were prepared and treated at 80 °C for 5 min under constant stirring.

4.1.3. Ultrasonic Treatment

For the ultrasonic stress tests 1 × 1 cm particles were prepared. Both the influence of ultrasonic power (50 and 100%) and that of temperature (20 °C and 60 °C) were examined.

4.1.4. Precrushing and Deformation

To systematically investigate the influence of mechanical deformation, the particles were intentionally folded in a defined manner. The effect of one to three consecutive foldings was examined. For this purpose, 3 × 3 cm particles were folded once in the middle. In one scenario, the particles were unfolded again, simulating the fragmentation process in a granulator or rotary shear. In the second scenario, the folded particles were treated in their closed state.

In addition, the delamination behavior of the mechanically crushed particles was investigated.

Author Contributions: Conceptualization, E.T.; methodology, E.T.; formal analysis, E.T. and S.P.; investigation, A.G., J.T., S.P., A.V. and J.E.G.; data curation, S.P. and E.T.; writing—original draft preparation, S.P. and E.T.; writing—review and editing, T.K. and H.L.; visualization, E.T.; supervision, T.K. and H.L.; project administration, E.T.; funding acquisition, E.T., T.K. and H.L. All authors have read and agreed to the published version of the manuscript.

Funding: This research was sponsored by Deutsche Bundesstiftung Umwelt, grant number 39311/01-21/2.

Data Availability Statement: The original contributions presented in this study are included in the article. Further inquiries can be directed to the corresponding author.

Conflicts of Interest: Anting Grams and Jan Talkenberger are employees in Company Weber Entec GmbH. The remaining authors have no conflicts of interest to declare.

References

1. Connelly, E.; Aryanpur, V.; Paoli, L.; Teter, J. Global Electric Vehicle Outlook 2022. Securing Supplies for an Electric Future. 2022. Available online: https://iea.blob.core.windows.net/assets/ad8fb04c-4f75-42fc-973a-6e54c8a4449a/GlobalElectricVehicleOutlook2022.pdf (accessed on 6 October 2025).
2. Curry, C. Lithium-ion Battery Costs and Market. Squeezed Margins Seek Technology Improvements & New Business Models. 2017. Available online: https://roar-assets-auto.rbl.ms/documents/36549/BNEF-Lithium-ion-battery-costs-and-market.pdf (accessed on 6 October 2025).
3. Ren, X.; Tong, Z.; Dai, Y.; Ma, G.; Lv, Z.; Bu, X.; Bilal, M.; Vakylabad, A.B.; Hassanzadeh, A. Effects of Mechanical Stirring and Ultrasound Treatment on the Separation of Graphite Electrode Materials from Copper Foils of Spent LIBs: A Comparative Study. *Separations* **2023**, *10*, 246. [CrossRef]
4. Gaines, L.; Zhang, J.; He, X.; Bouchard, J.; Melin, H.E. Tracking Flows of End-of-Life Battery Materials and Manufacturing Scrap. *Batteries* **2023**, *9*, 360. [CrossRef]
5. Wiechers, P.; Hermann, A.; Koob, S.; Glaum, F.; Gleiß, M. Development of a Process for Direct Recycling of Negative Electrode Scrap from Lithium-Ion Battery Production on a Technical Scale and Its Influence on the Material Quality. *Batteries* **2024**, *10*, 218. [CrossRef]
6. Jiang, M.; Li, B.; Liu, X. Global value chain restructuring driven by critical material trade: A case study of graphite. *Resour. Conserv. Recycl.* **2025**, *222*, 108489. [CrossRef]
7. Park, J.; Cho, S.-J.; Shin, S.; Kim, R.; Shin, D.; Shin, Y. Overview of graphite supply chain and its challenges. *Geosci. J.* **2025**, *29*, 329–341. [CrossRef]
8. Bobba, S.; Carrara, S.; Huisman, J.; Mathieux, F.; Pavel, C. *Critical Raw Materials for Strategic Technologies and Sectors in the EU: A Foresight Study*; European Union, Publications Office of the European Union: Luxembourg, 2020.
9. Xu, P.; Tan, D.H.S.; Jiao, B.; Gao, H.; Yu, X.; Chen, Z. A Materials Perspective on Direct Recycling of Lithium-Ion Batteries: Principles, Challenges and Opportunities. *Adv. Funct. Mater.* **2023**, *33*, 2213168. [CrossRef]
10. Sloop, S.; Crandon, L.; Allen, M.; Koetje, K.; Reed, L.; Gaines, L.; Sirisaksoontorn, W.; Lerner, M. A direct recycling case study from a lithium-ion battery recall. *Sustain. Mater. Technol.* **2020**, *25*, e00152. [CrossRef]
11. Lefherz, H.; Ahuis, M.; Dilger, N.; Kwade, A.; Zellmer, S. Sustainability of electrode scrap recycling—Quantifying the environmental benefit of direct recycling routes for battery production scrap. *J. Energy Storage* **2025**, *132*, 117688. [CrossRef]
12. Shin, Y.; Kim, S.; Park, S.; Lee, J.; Bae, J.; Kim, D.; Joo, H.; Ban, S.; Lee, H.; Kim, Y.; et al. A comprehensive review on the recovery of cathode active materials via direct recycling from spent Li-ion batteries. *Renew. Sustain. Energy Rev.* **2023**, *187*, 113693. [CrossRef]
13. Recell Center. Direct Recycling of Materials. Available online: https://recellcenter.org/research/direct-recycling-of-materials/ (accessed on 6 October 2025).
14. Baum, Z.J.; Bird, R.E.; Yu, X.; Ma, J. Lithium-Ion Battery Recycling—Overview of Techniques and Trends. *ACS Energy Lett.* **2022**, *7*, 712–719. [CrossRef]
15. Sliz, R.; Valikangas, J.; Vilmi, P.; Hu, T.; Lassi, U.; Fabritius, T. Replacement of NMP solvent for more sustainable, high-capacity, printed Li-ion battery cathodes. In Proceedings of the IEEE 16th Nanotechnology Materials and Devices Conference (NMDC), Vancouver, BC, Canada, 12–15 December 2021.
16. Bai, Y.; Muralidharan, N.; Li, J.; Essehli, R.; Belharouak, I. Sustainable Direct Recycling of Lithium-Ion Batteries via Solvent Recovery of Electrode Materials. *ChemSusChem* **2020**, *13*, 5664–5670. [CrossRef]
17. Tong, Z.; Ren, X.; Ni, M.; Bu, X.; Dong, L. Review of Ultrasound-Assisted Recycling and Utilization of Cathode Materials from Spent Lithium-Ion Batteries: State-of-the-Art and Outlook. *Energy Fuels* **2023**, *37*, 14574–14588. [CrossRef]
18. Gupta, V.; Appleberry, M.; Li, W.; Chen, Z. Direct recycling industrialization of Li-ion batteries: The pre-processing barricade. *Next Energy* **2024**, *2*, 100091. [CrossRef]
19. Bai, Y.; Li, M.; Jafta, C.J.; Dai, Q.; Essehli, R.; Polzin, B.J.; Belharouak, I. Direct recycling and remanufacturing of anode scraps. *Sustain. Mater. Technol.* **2023**, *35*, e00542. [CrossRef]

recycling

MDPI

Article

Graphene Recovery in Both Dispersed and Decanted Fractions from Lithium-Ion Battery Graphite via Sonication

Erasmo Arriola-Villaseñor [1,2], Alba Nelly Ardila Arias [1,*], Santiago Bedoya Betancour [1,2], Luz Marina Ocampo-Carmona [2], Trino Armano Zepeda Partida [3], Sergio A. Gómez Torres [4] and Gustavo Ariel Fuentes Zurita [4]

1 Research Group on Environmental Catalysis and Renewable Energies—CAMER, Politécnico Colombiano Jaime Isaza Cadavid, Medellín 050010, Antioquia, Colombia
2 Facultad de Minas, Universidad Nacional de Colombia Sede Medellín—UNAL, Medellín 050034, Antioquia, Colombia; earriola@unal.edu.co (E.A.-V.); sabedoyab@unal.edu.co (S.B.B.); lmocampo@unal.edu.co (L.M.O.-C.)
3 Centro de Nanociencias y Nanotecnología CNyN-UNAM, Ensenada 22800, Baja California, Mexico; trino@ens.cnyn.unam.mx
4 Departamento Ingeniería de Procesos e Hidráulica, Universidad A. Metropolitana-Iztapalapa—UAMI, Iztapalapa, Mexico City 09310, Mexico
* Correspondence: anardila@elpoli.edu.co

Abstract: In this study, graphene production via liquid-phase exfoliation assisted by sonication was evaluated using deionized water as a solvent and two graphite sources: one recovered from spent lithium-ion batteries (LIBs) and a commercial counterpart. A 750 W, 20 kHz ultrasonic processor was used, with sonication amplitudes ranging from 50% to 80% for two hours while maintaining a constant temperature of 45 °C. The resulting dispersions were left undisturbed for 24 h at ambient temperature to allow natural phase separation between decanted and dispersed fractions. These fractions were subsequently dried and weighed to determine exfoliation yield. High-quality graphene was successfully obtained via direct liquid-phase exfoliation of graphite recovered from LIBs, assisted by sonication in deionized water. Graphene formation was confirmed in both suspended and decanted fractions after two hours of sonication at 80% amplitude through complementary characterization techniques, including UV-Vis, Raman spectroscopy, HRTEM, and XRD. Comparative experiments using thermally pretreated battery graphite and commercial graphite revealed that graphene dispersions derived from untreated LIB-derived graphite exhibited greater long-term stability than those obtained from commercial or thermally pretreated battery graphite before sonication.

Keywords: graphite; spent; lithium-ion batteries; direct liquid phase exfoliation; graphene

1. Introduction

Graphene production has been the subject of extensive research due to its remarkable physical, chemical, and electronic properties, including high electrical conductivity, mechanical flexibility, and large surface area [1–3]. Graphene can be synthesized by several methods, with the specific method significantly impacting the resulting graphene type, as synthesis procedures can induce structural defects in accordance with the second law of thermodynamics [4]. The top-down method uses graphite as precursor. This method relies on the principle of exfoliating graphite, which serves as the initial material. Graphite, essentially a stack of graphene layers bound by Van der Waals forces, undergoes exfoliation through mechanical or micro-mechanical means, graphite intercalation, nanotube

https://doi.org/10.3390/recycling10030119

slicing utilizing carbon nanotubes as precursors, pyrolysis, reduction of graphite oxide, electrochemical exfoliation, sonication, and ball milling [5].

In recent years, innovative methods by liquid phase exfoliation (LPE) have emerged focused on the use of solvents and ultrasound, with the aim of reducing production costs and using more environmentally friendly raw materials [5–10]. The research development about LPE involves flaking natural graphite through sonication (cavitation) or high shear mixing under mild operating conditions [5,9–12]. This technique has enabled the production of graphene from various raw materials, including natural graphite, expanded graphite, and, more recently, graphite recovered from the anodes of spent lithium-ion batteries. However, the synthesis of pure natural graphite for graphene production typically involves high energy input and the use of toxic organic solvents, rendering such production methods either costly or potentially environmentally hazardous [9,13–15]. Additionally, natural graphite may not always be an optimal choice for graphene synthesis. Furthermore, natural graphite has been designated as a critical resource by the European Union (EU), indicating a high risk of supply shortages that could significantly impact the economy [16].

The production of graphene from waste materials holds great promises for reducing its long-term cost. The global demand for lithium-ion batteries (LIBs) is projected to surge from 185 GWh in 2020 to over 2000 GWh by 2030, largely driven by emerging applications such as electric vehicle (EV) technology [17]. However, this rapid increase in demand is expected to lead to significant waste disposal challenges as these batteries reach the end of their lifespan. Consequently, battery reuse and recycling have garnered considerable attention as measures to mitigate environmental impacts. Researchers have recently begun focusing on anode recycling, which constitutes 12–21% of the total mass of LIBs composition [4,9]. Utilization of substantial quantities of waste graphite from LIBs presents a significant opportunity for graphene production to contribute to waste management and the circular economy. In recent years, researchers have increasingly focused their efforts on producing graphene from LIBs graphite, recognizing its potential to provide a balance between scalability and quality in a cost-effective manner [6,18–23]. However, in most studies conducted to date, the focus has primarily been on characterizing the dispersed graphene fraction, while the sedimented fraction has received little attention. This asption is largely based on the premise that most graphene is obtained in the dispersed fraction.

Previous research has demonstrated that graphite recovered from LIBs has the potential to serve as a viable source of graphene, contributing not only to the sustainable management of hazardous waste but also to the production of high-value material [17,20,24]. However, a key limitation of these studies is the lack of evaluation of the sedimented fraction of the processed graphite, which may contain particles with properties similar or complementary to those of the dispersed graphene fraction. This gap in literature underscores the need for a more comprehensive approach that includes the full characterization of the exfoliated material, encompassing both the dispersed and sedimented fractions, to gain a complete understanding of the distribution and quality of the graphene obtained.

Additionally, in liquid-phase graphite exfoliation for graphene production, sonication plays a crucial role, as the applied ultrasound amplitude directly influences the degree of exfoliation and, consequently, the yield and quality of the resulting graphene [6,25,26]. Despite this, there is a notable lack of studies investigating the impact of sonication amplitude on the properties of graphene derived from graphite recovered from LIBs. Most research has focused on fixed amplitudes or optimizing other parameters, such as solvent type or sonication duration, without thoroughly examining the relationship between sonication amplitudes and process efficiency. This gap highlights the need for a systematic investigation of different sonication amplitudes to gain a deeper understanding of their influence on graphite exfoliation and the quality of the resulting graphene.

For these reasons, the present study focuses on the production of graphene from the graphite contained in the anodes of LIBs via ultrasound-assisted liquid-phase exfoliation, using water as the solvent. Water was selected due to its low cost, availability, and environmental safety, offering a more sustainable approach to graphene synthesis. The primary objective was to investigate the effect of sonication amplitude on both graphene yield and its structural and composition. Additionally, a comprehensive physicochemical characterization of both the sedimented and dispersed fractions was conducted, enabling a holistic evaluation of the material properties in both phases.

This innovative approach not only optimizes the graphene production process but also complements existing research by providing critical insights into the influence of sonication amplitude on the exfoliation of recycled graphite. Furthermore, it expands the current understanding of the sedimented fraction, which has remained largely unexplored. The findings of this study may serve as a valuable reference for future research on graphene recovery from electronic waste and the development of more sustainable and cost-effective production techniques.

2. Materials and Methods

The pretreatment of LIBs involved multiple steps, including safe discharge, manual dismantling, crushing, screening, and basic cleaning. Initially, different types of spent LIBs were placed in separate discharge tanks and submerged in a 5% sodium chloride solution for 24 h to ensure complete and safe discharge and washed with deionized water. Afterward, the discharged LIBs manually disassembled, and the graphite-based negative electrode strips were extracted. These graphite strips, sourced from various LIBs, were then fed into a universal crusher and pulverized for 5 min, followed by sieving through an 80-mesh screen to achieve a uniform particle size.

Graphene production was carried out via liquid-phase exfoliation assisted by ultrasonic homogenization, using deionized water as the solvent. A Cole-Parmer 750-Watt, 20 kHz ultrasonic processor was employed for this purpose. In each experiment, 100 mg of graphite recovered from spent LIBs was dispersed in 150 mL of deionized water in a 200 mL of 50-mm-diameter glass beaker. The suspension was then subjected to ultrasonic homogenization at varying amplitudes for 2 h while maintaining the temperature at 45 ± 2 °C using an external cooling system with continuous water circulation around the reaction vessel. The specific sonication conditions (amplitude and graphene raw) investigated for aqueous liquid-phase exfoliation of graphite are detailed in Table 1. Sonication was performed using amplitudes of 50% and 80%. These values were selected based on reports indicating that an amplitude of approximately 70% enhances graphite exfoliation in similar systems [27]. Thus, 50% and 80% were chosen as reference points below and above this threshold, respectively, to evaluate their effect on the exfoliation process.

After the sonication process, a thin surface layer with a silver-gray color was observed, corresponding to unexfoliated material. This fraction was carefully removed using a glass Pasteur pipette (mass < 1% of the initial graphite) and excluded from the yield analysis. Subsequently, the dispersions were centrifuged at 6000 rpm for 30 min. The resulting fractions—supernatant (dispersed fraction) and sediment (decanted fraction)—were carefully separated and dried in an oven at 80 °C until constant weight was reached. The approximate graphene yield in each fraction was estimated using a gravimetric method by dividing the dry mass of each fraction by the sum of the total dry masses of both fractions, multiplied by 100%. This approach was adopted as a relative approximation of the exfoliated material distribution, following similar methodologies reported in the literature [28–32]. Although this method does not provide a fully precise quantification of pure graphene content, the structural analyses included in this study indicate that both fractions contain material with graphene-like characteristics.

Table 1. Codes and experimental conditions for evaluated material types.

Material	Graphene Obtention Conditions
Gf-C	Commercial graphite, (Flakes, Sigma Aldrich).
GPi	Graphite from LIBs.
Gn-C	Graphene commercial (Graphene nanoplatelets, Sigma Aldrich).
Gc-S5-De	Graphene from commercial graphite under sonication at 50% amplitude for 2 h followed by decantation.
Gc-S5-Di	Graphene from commercial graphite by sonication at 50% amplitude for 2 h and contained in the dispersed fraction.
Gc-S8-De	Graphene from commercial graphite under sonication at 80% amplitude for 2 h followed by decantation.
Gc-S8-Di	Graphene obtained from commercial graphite by sonication at 80% amplitude for 2 h and contained in the dispersed fraction.
Gn-S5-De	Graphene from LIBs graphite under sonication at 50% amplitude for 2 h followed by decantation.
Gn-S5-Di	Graphene from LIBs graphite by sonication at 50% amplitude for 2 h and contained in the dispersed fraction.
Gn-S8-De	Graphene from LIBs graphite under sonication at 80% amplitude for 2 h followed by decantation.
Gn-S8-Di	Graphene from LIBs graphite by sonication at 80% amplitude for 2 h and contained in the dispersed fraction.

The tests were conducted using commercial graphite and graphite extracted from spent LIBs to investigate the differences between the graphene obtained from each source. Additionally, commercial graphene and graphene oxide were used as reference materials for characterization purposes, allowing for a comparative analysis of their properties with those of the materials obtained in this study.

3. Material Characterization

An initial characterization was performed to determine the basic physicochemical properties and chemical composition of both the dispersed and sedimented fractions. The analyzed properties included density by volumetric displacement (pycnometer method) using water as the reference fluid (USP-699 Density of Solids), Brunauer-Emmett-Teller (BET) surface area, average pore volume, average pore diameter, dispersion pH, zeta potential, dispersion conductivity, and contact angle. graphene and characterize its structural and physicochemical properties. Additionally, more advanced analytical techniques were employed to confirm the formation of graphene and characterize its structural and physicochemical properties.

The metal composition was determined using Atomic Absorption Spectroscopy (AAS) on an Agilent 240 FS atomic absorption spectrophotometer, employing an air-acetylene flame.

For the contact angle analysis, the different materials were compacted onto a glass sample holder using axial pressure to form a flat and uniform surface. Contact angles were measured for two water volumes. Initially, a single water (~50 μL) droplet was deposited on the surface of each material, and high-resolution photographs were taken at an initial time (1 s) and monitored over a period of 10 s. Subsequently, four additional droplets (~200 μL) were added to assess the material's stability and permeability. Each image was analyzed using the LB-ADSA (Liquid Bridge—Axisymmetric Drop Shape Analysis) method, implemented in the ImageJ V1.54k software with the Drop Shape Analysis (DSA) plugin [33]. This method numerically solves the Laplace-Young equation, which relates the axisymmetric shape of the droplet to surface tension and gravitational forces. This approach compensated for roughness effects, ensuring reliable measurements on the compacted material surfaces.

The specific surface area of the samples was determined by nitrogen (N_2) physisorption at 77 K using the Brunauer–Emmett–Teller (BET) method on a Quantachrome Autosorb Automated Gas Sorption System. Prior to measurement, the samples were degassed under vacuum at 380 °C for 12 h.

Zeta potential measurements were performed on a Malvern ZEN 2600 equipment, Malvern, UK.

XPS analyses were carried out on samples using an AES-XPS PHI-548 spectrometer, Amsterdam, The Netherlands in which the samples were excited by an unmonochromatized Al Kα line at 1486.6 eV. The working pressure was $<1 \times 10^{-10}$ Pa. Survey scans were obtained in the range from 1205 to (-10) eV energy interval at 1.0 eV per step and pass energy of 100 eV. Additionally, the high-resolution XPS scans were completed at 0.2 eV energy steps and pass energy of 50 eV (the constant pass energy mode).

High-resolution transmission electron microscopy (HRTEM) and energy-dispersive X-ray spectroscopy (EDS) were conducted using an INCA Oxford Si (Li) detector integrated into a JEOL 2100F electron microscope, Tokyo, Japan. The microscope operated at an accelerating voltage of 200 kV, providing a resolution of 0.19 nm.

Ultraviolet–Visible (UV–Vis) spectroscopy, recorded with a Shimadzu UV–Vis spectrophotometer, Kyoto, Japan to identify the formation of graphene in the samples. The dispersed fraction was analyzed directly, while the decanted fraction required the dried samples to be dispersed in distilled water by agitation with a magnetic stirrer. Immediately after, all dispersions were transferred into 10 × 10 mm vials and measured to prevent partial agglomeration of the flakes and their sedimentation in the aqueous solution. Each UV–Vis spectrum was collected with an acquisition time of 20 s within the spectral range of 200 to 800 nm, covering the region where graphene-related peaks are expected.

X-ray diffraction (XRD) analysis was performed using a Rigaku, Tokyo, Japan, D-Max 2200 Series diffractometer equipped with Cu-Kα radiation (λ = 1.54 Å). The scanning rate was set at 3° per minute, with an operating tube voltage and current of 40 kV and 40 mA, respectively. Diffraction patterns were recorded over a 2θ angular range of 2–90°.

Raman analyses were performed at room temperature using a Horiba Jobin Yvon, Stow, MA, USA, Labram HR confocal Raman spectrometer, equipped with a 632.81 nm laser. The spectra were recorded in the range of 600 to 4000 cm^{-1}.

H_2-Temperature programmed reduction (H_2-TPR) of the materials was carried out at 10 °C/min from 25 °C up to 900 °C, in the presence of 10% H_2/Ar (Infra). The gas flow rate was 25 mL/min and it was kept constant using a mass flow controller. The apparatus was calibrated by reduction of 30 mg of CuO powder (99%, Merck, Rahway, NJ, USA) under the same experimental conditions.

Attenuated total reflectance-Fourier transform spectroscopy (ATR-FTIR) analyses were carried out over the wave number range of 4000–500 cm^{-1} using a Shimadzu IRAffinity-1S.

4. Results and Discussion

4.1. Metal Composition

The analysis was conducted using AAS identified trace metallic impurities in the selected materials (Table 2). Specifically, elements such as cobalt (Co), nickel (Ni), manganese (Mn), lithium (Li), and copper (Cu) were detected. However, these impurities were not found in commercial graphite (Gf-C), indicating a high purity of the source material. In contrast, recycled graphite from LIBs (GPi) exhibited significant amounts of these elements, reflecting the composition of the internal components of the batteries, particularly those used in cathodes and current collectors [34–36].

Table 2. Composition metal trace in some materials.

Material	Metal Composition (% wt.)					
	Co	**Ni**	**Mn**	**Li**	**Cu**	**Al**
Gf-C	0.000	0.000	0.000	0.000	0.000	0.000
GPi	0.231	0.004	0.136	0.115	1.272	0.000
Gn-S8-Di	0.217	0.004	0.018	0.000	0.465	0.000
Gn-S8-De	0.231	0.006	0.085	0.000	1.031	0.000

In the graphene obtained via liquid-phase exfoliation, both in the dispersed fraction (Gn-S8-Di) and the sedimented fraction (Gn-S8-De), metallic impurities were detected at varying concentrations, with Co, Ni, Mn, and Cu being the most prominent. However, lithium was not identified in either graphene fraction, suggesting that the ultrasound-assisted exfoliation process facilitated the removal of this metal. This result may be attributed to the solubility of the electrolyte and the deintercalation of Li during treatment, a phenomenon previously reported in studies on anode material recovery from LIBs [17].

The absence of metallic impurities in commercial graphite aligns with its use in applications requiring high purity standards. On the other hand, the detected impurities in recycled graphite originate directly from materials used in LIBs manufacturing. Co and Ni are commonly found in cathodes, while Cu is used in current collectors. During the recycling process, residual traces of these materials often remain in the recovered graphite, affecting its chemical composition. The results obtained are consistent with findings reported in the scientific literature. Previous studies have identified metallic impurities in graphene produced from the exfoliation of graphite recovered from LIBs, highlighting the persistence of elements such as Co, Al, Cu, and Mn, which confirm the presence of residues from cathode materials and other internal battery components [17].

On the other hand, X-ray photoelectron spectroscopy (XPS) analysis confirmed the presence of metallic impurities and complemented the observations made by AAS. Figures 1 and 2 show the XPS spectra of the sonicated materials at 80% amplitude for the sedimented fraction of both commercial graphite (Gc-S8-De) and graphite recovered from LIBs (Gn-S8-De). This analysis is crucial for evaluating the purity of the obtained graphene and the presence of residual metallic impurities, which are key factors for its potential applications, particularly in catalytic systems [37].

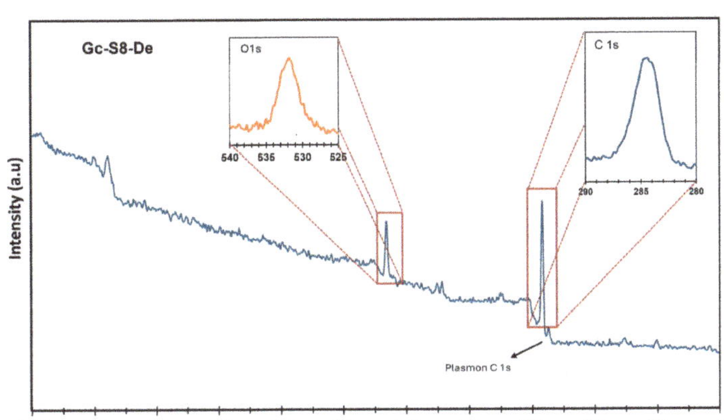

Figure 1. XPS spectra of the Gc-S8-De material.

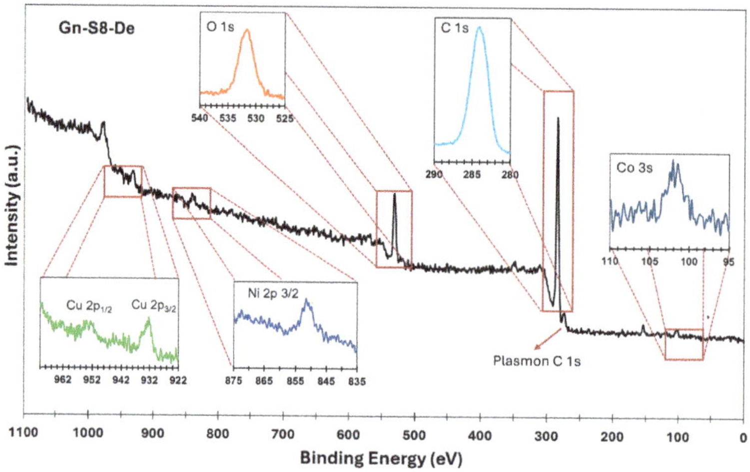

Figure 2. XPS spectra of the Gn-S8-De material.

Figure 1, corresponding to the material derived from commercial graphite (Gc-S8-De), presents an XPS spectrum dominated by the C 1s bands at 284.1 eV, characteristic of C–C and C=C bonds associated with graphene structures along with its π–π* plasmon, confirming the presence of graphene structure after sonication.

In contrast, the spectrum shown in Figure 2, corresponding to the material derived from graphite recovered from LIBs (Gn-S8-De), reveals a more complex surface composition. In addition to the C 1s (284.1 eV) and O 1s (532.1 eV) bands, contributions at 932.5 eV (Cu $2p_3/_2$) and 953.5 eV (Cu $2p_1/_2$), corresponding to copper, as well as signals at 852.6 eV for nickel (Ni $2p_3/_2$) and 101.4 eV associated with cobalt (Co 2p), were identified.

The presence of these transition metals, which are typical of LIB cathodes, indicates that some residual metallic impurities persist after the exfoliation process. This partial incorporation of metals into the carbonaceous matrix could significantly modify the electronic properties of graphene, creating opportunities for its application in heterogeneous catalysis, where Cu, Co, Mn, and Ni could act as active sites or promoters, or in energy storage devices, where these metallic species may enhance charge storage capacity and transfer [38,39]. If high-purity graphene is required, these findings highlight the need for additional purification processes. However, the controlled presence of these metals could be strategically leveraged to develop functional materials with properties tailored to specific applications.

4.2. Zeta Potential

The zeta potential (ζ) analysis is a measure of the electrical charge on the surface of dispersed particles in a liquid and is used to assess their colloidal stability in water. High absolute values (greater than $\pm$30 mV) indicate electrostatic repulsion between particles, which prevents aggregation and promotes stable dispersions [25,40,41]. The results obtained in this study, presented in Table 3, show predominantly negative values, consistent with Lewis-type charge transfer at the particle-solvent interface. Highly negative zeta potential values ($\zeta < -40$ mV) in aqueous graphene dispersions are primarily attributed to the ionization of oxygen-containing functional groups (e.g., -COOH, -OH) on the material surface. These groups undergo deprotonation in neutral or alkaline media, generating negatively charged species ($-COO^-$, $-O^-$) that induce electrostatic repulsion between particles [41,42]. Additionally, the adsorption of hydroxyl ions (OH^-) onto the graphene surface, which is favored at high pH, further increases the negative charge density. These mechanisms align with the DLVO theory, wherein electrostatic repulsion predominates over van der Waals forces, thereby stabilizing the colloidal dispersion [43]. Although Lewis acid-base interactions may occur at defective sites on graphene (e.g., edges with electron-deficient sp^2 carbon), their contribution to surface charge is secondary in aqueous systems, where ionic and pH-dependent effects dominate [41,43].

Table 3. Physicochemical properties for different materials. Experimental data for the materials characterization. Data generated by the authors.

Material	Density Solid (g/mL)	BET Surface Area (m^2/g)	Average Pore Volume BJH (cm^3/g)	Average Pore Diameter (nm)	PH Dispersion	Zeta Potential (mV)	Dispersion Conductivity (μS/cm)	Contact Angle (°)
Gf-C	0.366	2.11	0.008	15.45	8.48	-	-	148.1
GPi	0.760	5.13	0.036	22.35	8.15	-	-	143.9
Gn-C	0.065	49.05	0.096	11.97	8.91	−47.8	-	19.4
Gc-S5-De	0.293	3.18	0.010	24.75	8.81	−45.9	50.2	149.3
Gc-S5-Di	0.217	6.97	0.030	38.00	8.57	−48.5	70.6	138.4
Gc-S8-De	0.254	2.06	0.005	39.73	9.11	−43.7	91.2	143.5
Gc-S8-Di	0.202	4.39	0.020	37.56	9.33	−50.8	104.3	22.1
Gn-S5-De	0.654	5.01	0.021	22.53	7.35	−43.4	143	147.2
Gn-S5-Di	0.409	14.39	0.060	22.85	9.41	−58.6	265	135.4
Gn-S8-De	0.575	4.81	0.020	23.09	7.73	−42.4	164	21.5
Gn-S8-Di	0.249	14.75	0.060	17.63	9.85	−58.0	352	13.9

A difference in stability is observed between the dispersed (Di) and decanted (De) fractions of each material. In all cases, for each type of raw material, the dispersed fractions exhibit higher zeta potential magnitudes than the decanted fraction. For example, for graphene derived from commercial graphite at 50% sonication, the dispersed fraction (Gc-S5-Di) has a zeta potential of −58.6 mV, while the decanted fraction (Gc-S5-De) shows −45.9 mV. This trend is consistent in materials derived from LIBs, indicating that dispersed fractions are more stable due to their higher surface charge, which enhances electrostatic repulsion and prevents sedimentation. This behavior is attributed to the sonication process, which increases the surface charge of graphene by introducing defects, exposing a greater surface area of graphene layers, and consequently leading to a higher exposure of ionized functional groups.

Regarding the influence of sonication amplitude, the data show that graphene obtained at 80% and 50% amplitude tends to have very similar zeta potential values. This suggests that stability is comparable among materials derived from the same raw source, which is consistent with the stability images (Table 4), where suspensions of materials sonicated at 80% exhibited greater stability.

Moreover, graphene derived from LIBs exhibits zeta potential values further from zero in its dispersed fractions compared to that from commercial graphite. For example, the material (Gn-S5-Di) obtained at 50% sonication amplitude shows a zeta potential of −58.6 mV, while Gc-S5-Di has −48.5 mV. Similarly, at 80% amplitude, Gn-S8-Di reaches −58.0 mV, whereas Gc-S8-Di shows −50.8 mV. This indicates that materials derived from LIBs (dispersed fraction) are more stable in dispersion. In contrast, the zeta potential of the decanted fractions showed similar values regardless of the graphite source, which is consistent given that this fraction settled immediately after the exfoliation process.

Table 4. Dispersion stabilization time and yields for the decanted and suspended fractions.

Sample	Yield (%)		Dispersion Stabilization Time (Days)					
	Di	De	0	15	30	60	90	180
Gc-S5	4.46	95.54					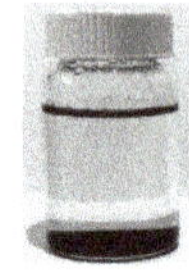	
Gc-S8	5.55	94.45				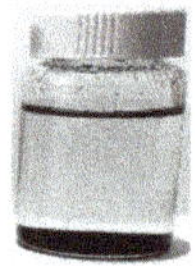		
Gn-S5	8.60	91.4						
Gn-S8	9.95	90.05						
Gn-C	-	-	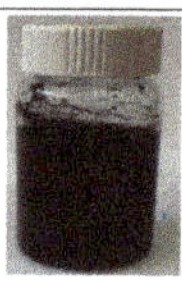					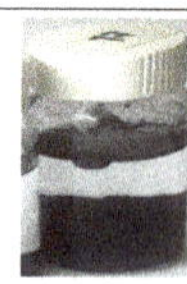

As explained in the analysis of the stability of different material dispersions, the greater stability observed in materials derived from LIBs may be related to the presence of residual metals. These metals modify the surface charge of the particles, specifically through the adsorption of metal ions onto the graphene surface, generating additional charges that increase the absolute value of the zeta potential. This modification in surface charge enhances electrostatic repulsion between particles of the same charge, reducing aggregation tendencies and promoting greater suspension stability [44]. Additionally, certain metals can serve as active sites for the formation of oxygenated functional groups during treatment processes, further contributing to an increase in surface charge. Together, these factors explain why LIB-derived materials, particularly Gn-S5-Di and Gn-S8-Di, exhibit greater stability compared to those obtained from commercial graphite.

On the other hand, if graphene has a low zeta potential value (close to 0 mV), it means that electrostatic repulsion forces between particles are weak or nonexistent. In this case, graphene particles tend to aggregate rapidly due to Van der Waals attraction forces. This aggregation leads to particle sedimentation, preventing effective dispersion in water and causing the formation of large flocs or aggregates [43]. Pure graphene without functionalization, or exfoliated graphite that has not been properly processed through sonication or chemically modified, typically exhibits low zeta potential values and, consequently, low dispersion capacity in water. Although this value is not listed in Table 3, this observation is consistent with the stability images in Table 4, where commercial graphene sedimented within a few days after dispersion preparation.

In other words, the analysis of the results reveals that materials derived from LIBs exhibit greater colloidal stability compared to those from commercial graphite. Moreover, while sonication at 80% amplitude enhances exfoliation, the zeta potential values suggest that optimal stability is achieved at 50% amplitude. Finally, dispersed fractions demonstrate greater stability than decanted fractions, highlighting the importance of controlling sonication conditions to obtain materials with optimal properties for specific applications.

The differences in zeta potential (Z) values between materials derived from commercial graphite and those recovered from lithium-ion batteries (LIBs) (ranging from −42.4 to −58.6 mV) demonstrate the influence of material origin and prior treatment on dispersion properties. Although both values fall within a narrow range associated with colloidal stability, the ~16 mV gap can be linked to enhanced aggregation resistance and long-term dispersion uniformity in materials with higher absolute zeta potential values, as reported in previous studies [44,45]. Such works indicate that even moderate differences in zeta potential may significantly influence dispersion stability and homogeneity, particularly in systems like composites or coatings there is a minimal difference between 50% and 80% amplitude, suggesting that graphene with these properties can be obtained at 50% amplitude while achieving energy savings compared to 80% amplitude sonication. The differences observed between graphene derived from commercial graphite and that recovered from LIBs emphasize the influence of material origin and prior treatment on the dispersion properties of graphene.

For a long time, graphite was considered hydrophobic.

4.3. Contact Angle

For a long time, graphite was considered hydrophobic, with a water contact angle of 90° or higher [46]. However, it has been found that this hydrophobicity may be related to the presence of active sites on the material [47]. Additionally, studies have shown that the wettability of monolayer and bilayer graphene is significantly higher than that of graphite, exhibiting a water contact angle of $\theta < 30°$ [6]. The rigorous determination of hydrophobicity in carbonaceous materials, such as graphite and graphene, requires the integration of precise contact angle (θ) measurements with theoretical models adapted to rough or porous surfaces [48,49].

Wetting is the phenomenon in which a liquid spreads over a solid surface upon contact. The contact angle is commonly used to describe the degree of wetting, where Young's equation (Equation (1)) defines the equilibrium state of a resting droplet on an ideal, smooth surface.

$$\gamma_{sv} = \gamma_{lv}\cos\theta_Y + \gamma_{sl} \quad (1)$$

where γ_{sv}, γ_{lv} y γ_{sl} correspond to the interfacial surface tensions of the solid–vapor, liquid–vapor, and solid–liquid interfaces, respectively, and θ_Y is the resulting contact angle [50]. A hydrophilic surface typically exhibits a contact angle in the range of $10° < \theta < 90°$, while hydrophobic surfaces fall within $90° < \theta < 150°$. Surfaces with are classified as ultrahydrophilic, whereas those with θ $150° < \theta < 180°$ are considered superhydrophobic [51–53]. However, this model assumes an ideal, perfectly smooth surface. For rough surfaces, such as compressed powder materials, wetting behavior can be understood through two principal models: the Wenzel model and the Cassie-Baxter model. The Wenzel model is particularly applicable to hydrophilic materials, as it describes a liquid making uniform contact with the surface, regardless of the roughness or porosity. In this case, liquid adsorption into surface pores or irregularities is a defining characteristic, and if the material's porosity allows, complete penetration of the liquid into surface asperities occurs, leading to total wetting [50]. This behavior is characteristic of hydrophilic surfaces, where the presence of polar functional groups (such as -OH, -COOH) in graphene or graphite enhances water interaction [44,51,54].

On the other hand, the Cassie-Baxter model describes heterogeneous wetting, in which the water droplet remains suspended on the peaks of a rough surface while air becomes trapped in the valleys. This condition is known as the Cassie-Baxter state Where γ_{sv}, γ_{lv} y γ_{sl} correspond to the interfacial surface tensions of the solid–vapor, liquid–vapor, and solid–liquid interfaces, respectively, and θ_Y is the resulting contact angle [50]. A hydrophilic surface typically exhibits a contact angle in the range of $10° < \theta < 90°$, while hydrophobic surfaces fall within $90° < \theta < 150°$. Surfaces with are classified as ultrahydrophilic, whereas those with θ $150° < \theta < 180°$ are considered superhydrophobic [51–53]. However, this model assumes an ideal, perfectly smooth surface. For rough surfaces, such as compressed powder materials, wetting behavior can be understood through two principal models: the Wenzel model and the Cassie-Baxter model. The Wenzel model is particularly applicable to hydrophilic materials, as it describes a liquid making uniform contact with the surface, regardless of the roughness or porosity. In this case, liquid adsorption into surface pores or irregularities is a defining characteristic, and if the material's porosity allows, complete penetration of the liquid into surface asperities occurs, leading to total wetting [50]. This behavior is characteristic of hydrophilic surfaces, where the presence of polar functional groups (such as -OH or -COOH and epoxides) in graphene or graphite enhances water interaction [44,51,54].

On the other hand, the Cassie-Baxter model describes heterogeneous wetting, in which the water droplet remains suspended on the peaks of a rough surface while air becomes trapped in the valleys. This condition is known as the Cassie-Baxter state [50,55]. In this regime, the formation of liquid marbles can be observed, where liquid droplets (typically water) are coated with hydrophobic particles that prevent wetting [56]. This behavior is characteristic of superhydrophobic materials [55], in which surface roughness traps air pockets, creating a liquid-air-solid interface. The presence of liquid marbles, which roll freely across the surface, is a key indicator of the Cassie-Baxter state. This phenomenon is commonly observed in materials with micro- or nanostructures that minimize contact between the liquid and the solid, such as graphene functionalized with nonpolar groups or surfaces with hydrophobic coatings [50].

A modified Young's model ($\theta > 90°$\theta > 90^\circθ > $90°$ without liquid marble formation) describes intrinsically hydrophobic surfaces, where the contact angle exceeds 90° but does not reach superhydrophobicity. This behavior is attributed to the presence of non-oxidized graphene domains (sp^2), which are inherently hydrophobic due to their low surface energy [48]. Unlike the Cassie-Baxter model, liquid marbles do not form because there is no significant surface roughness to trap air [53].

The combination of these data with qualitative criteria, such as the formation of liquid marbles or liquid absorption, allows for the classification of hydrophobicity according to these models. Table 3 presents the details contact angle of the materials. Based on these parameters, it was determined that the initial graphite samples Gpi ($\theta = 143.9°$) and Gf-C ($\theta = 148.1°$) exhibited superhydrophobicity (Cassie-Baxter model) with the formation of stable liquid marbles. In contrast, commercial graphene (Gn-C, $\theta = 19.4°$) displayed extreme hydrophilicity (Wenzel model), absorbing the droplet in less than 2 s.

For materials derived from commercial graphite subjected to sonication at 50% amplitude, the sedimented fraction (Gc-S5-De, $\theta = 149.3°$) retained the hydrophobicity of the original graphite, while the dispersed fraction (Gc-S5-Di, $\theta = 138.4°$) showed a slight decrease in contact angle, although it maintained the formation of liquid marbles. In the case of sonication at 80% amplitude, only the sedimented fraction (Gc-S8-De) preserved the hydrophobic properties of the starting graphite, with a moderate reduction in contact angle ($\theta = 143.5°$) and the formation of a partial liquid marble. Conversely, the dispersed fraction (Gc-S8-Di) exhibited a transition toward extreme hydrophilicity ($\theta = 22.1°$ within the first second), with complete absorption occurring in less than 5 s.

For materials obtained from graphite recovered from LIBs and sonicated at 50% amplitude, a similar trend was observed: the sedimented fraction (Gn-S5-De, $\theta = 147.2°$) and the dispersed fraction (Gn-S5-Di, $\theta = 135.4°$) retained high hydrophobicity with liquid marble formation. However, when sonication was increased to 80%, both fractions (Gn-S8-De and Gn-S8-Di, with $\theta = 21.5°$ and $\theta = 13.9°$ within the first second, respectively) transitioned toward extreme hydrophilicity, with complete absorption in less than 5 s.

Based on the literature, where contact angle increases of approximately 1° have been reported for a volume increment of 15 μL [57], it can be inferred that the contact angle variation obtained in this study (Δθ\Delta\thetaΔθ close to or below 10° for a volume increment of ~200 μL) suggests a uniform chemical hydrophobicity at the macroscopic scale.

The images in Figure A1 (Appendix A) confirm that, in materials sonicated at 50% amplitude, the sedimented fraction formed partial liquid marbles, whereas the dispersed fraction exhibited more extensive liquid coverage. This behavior suggests that these materials retain their hydrophobic properties and promote the formation of a partial liquid marble. Additionally, the low variation in Δθ implies that the observed hydrophobicity is homogeneous at the macroscopic level, although the incomplete formation of liquid marbles reveals potential heterogeneity in surface roughness. It is proposed that the limited mobility of sedimented particles in the decanted fraction restricts their ability to fully coat the air-liquid interface, whereas in materials processed at high amplitude (S8-Di), fragmentation into nanoparticles and surface oxidation promote localized liquid absorption.

On the other hand, sonication at 80% amplitude in the dispersed fraction homogenizes the extreme hydrophilic response, regardless of the starting graphite, possibly due to the generation of polar functional groups such as -OH and -COOH, and may also promote the formation of ether-type epoxide groups during sonication, which increases the solid-liquid interfacial energy and favors complete wetting. The fragmentation of graphite into nanoparticles creates a porous structure that facilitates liquid penetration via capillary action, in accordance with the Wenzel model. Furthermore, the presence of trace metals (Co, Cu, etc.) could act as active sites for oxidation reactions, promoting the hydrophilicity of the materials.

Overall, it is concluded that sonication at 50% amplitude favors the production of superhydrophobic materials, whereas amplitudes of 80% induce extreme hydrophilicity, except in the case of materials derived from commercial graphite. These findings suggest a multifaceted approach to designing materials with controlled wetting properties by adjusting ultrasonic treatment intensity and surface morphology.

In this regime, the formation of liquid marbles can be observed, where liquid droplets (typically water) are coated with hydrophobic particles that prevent wetting [56]. This behavior is characteristic of superhydrophobic materials [55], in which surface roughness traps air pockets, creating a liquid-air-solid interface. The presence of liquid marbles, which roll freely across the surface, is a key indicator of the Cassie-Baxter state. This phenomenon is commonly observed in materials with micro- or nanostructures that minimize contact between the liquid and the solid, such as graphene functionalized with nonpolar groups or surfaces with hydrophobic coatings [50].

4.4. Textural Properties

The textural analysis of the materials indicates that they fall within the mesoporous range (2–50 nm). Specific surface area (BET) measurements revealed that the decanted fraction did not exhibit significant changes compared to the starting graphite [58]. In contrast, the dispersed fraction of graphene showed substantial increases in specific surface area, with an approximate 100% increase for materials derived from commercial graphite and up to 200% for those obtained from LIB graphite. This significant increase in the dispersed fraction can be attributed to the greater exposure of internal surfaces following exfoliation. The enhancement in the surface area of LIB-derived graphene is likely due to a combination of exfoliation, pre-existing defects, and the formation of few-layer sheets, which promote the opening of new active surfaces [58]. Conversely, the decanted fraction exhibited minimal increases, likely due to its composition of thicker (~3-layer) and less exfoliated sheets, which restrict access to internal surfaces.

Regarding pore diameter, materials derived from commercial graphite exhibited a significant increase, ranging from 70% to 160%, from 15.45 nm in the original commercial graphite to 39.73 nm in the decanted fraction sonicated at 80% amplitude. In contrast, materials derived from LIB graphite maintained a nearly constant pore diameter of approximately 22.4 nm, with no clear correlation between pore size and recovered fraction (dispersed or decanted) or sonication amplitude (50% or 80%).

The observed expansion in pore diameter for materials derived from commercial graphite, particularly in the decanted fraction, can be attributed to the more ordered and less defective initial structure of this type of graphite [35]. During sonication, mechanical stresses and cavitation promote mesopore growth by widening pre-existing pores. This effect is more pronounced in the decanted fraction because the partially exfoliated, multi-layered sheets (approximately three layers, according to Raman analysis) can retain expanded porous structures. In contrast, the constant pore diameter in graphene derived from LIBs suggests that these materials already possess a more heterogeneous and defective porous structure due to charge-discharge cycles during their use. This is consistent with their slightly larger initial pore diameter compared to commercial graphite, making them less sensitive to sonication-induced modifications in pore diameter.

Regarding pore volume, the decanted fraction materials remained nearly unchanged compared to reference graphite, whereas the dispersed fraction materials exhibited a substantial increase in pore volume-approximately 100% for LIB-derived materials and up to 200% for those obtained from commercial graphite. The significant increase in pore volume in the dispersed fraction can be explained by the formation of thin graphene sheets (approximately two layers), which provide greater accessibility to internal pores. Sonication at 80% amplitude facilitates the near-complete exfoliation of these sheets, generating open structures with increased available pore volume. For LIB-derived materials, the moderate increase in pore volume may be related to residual impurities and structural defects that restrict pore growth.

4.5. Dispersion Conductivities

The electrical conductivity results of the material dispersions (Table 3) reveal significant correlations between the graphite source, sonication amplitude, and recovered fraction. In general, materials derived from graphite recovered from LIBs exhibit significantly higher conductivity than those obtained from commercial graphite. For instance, the dispersed fraction (Di) of LIB-derived graphene sonicated at 80% amplitude (Gn-S8-Di) reaches 352 mS/cm, approximately 3.4 times higher than its commercial counterpart (Gc-S8-Di: 104.3 mS/cm).

The graphite source plays a crucial role in determining the conductivity behavior of the different graphene dispersions. LIB-derived graphene (Gn) not only outperforms commercial graphene (Gc) in this property but even under lower sonication conditions (50% amplitude), Gn-S5-Di (265 mS/cm) surpasses Gc-S8-Di (104.3 mS/cm), which was processed at 80% amplitude. This advantage may be related to the electrochemical history of LIB graphite. During charge/discharge cycles, the repeated intercalation of various metals induces a controlled expansion of the layered structure, effectively "pre-conditioning" the material for more efficient exfoliation. Additionally, the presence of metallic impurities enhances the conductivity of the recovered graphite, which could be beneficial for electronic applications.

This pronounced difference suggests that LIB-derived graphite possesses intrinsic structural characteristics that favor the production of high-conductivity graphene. This finding aligns with the zeta potential values, where materials derived from recovered graphite exhibited higher absolute values compared to those obtained from commercial graphite.

The sonication amplitude proved to be a significant factor, regardless of the graphite source or recovered fraction. As the amplitude increased, the conductivity of all dispersions improved. For instance, in the decanted fraction of commercial graphite, Gc-S8-De (91.2 mS/cm) exhibited an 81% increase compared to Gc-S5-De (50.2 mS/cm). Similarly, in LIB-derived graphite (Gn), Gn-S8-De (164 mS/cm) surpassed Gn-S5-De (143 mS/cm) by 15%. This trend aligns with the observed zeta potential values.

This behavior may be attributed to the greater mechanical energy transfer at higher amplitudes, which effectively breaks the interlayer Van der Waals bonds, reducing the number of layers. However, this decrease in layer count is only evident in Raman spectroscopy for the dispersed fraction at 80% amplitude. Additionally, in all cases, the dispersed fractions (Di) exhibited significantly higher conductivities than the decanted fractions (De). For example, in LIB-derived graphite treated at 50% amplitude, Gn-S5-Di (265 mS/cm) exceeded Gn-S5-De (143 mS/cm) by 85%.

4.6. Density Solids

The density of all analyzed materials decreased compared to the original graphite, which is consistent with the exfoliation process. The most significant reduction in density was observed in the dispersed fraction, particularly in LIB-derived graphene subjected to sonication at 80% amplitude (Gn-S8-Di). This effect is attributed to the presence of thin layers, approximately two in number, which lead to less compact structures. The observed decrease in density across all cases aligns with the exfoliation process, where layer separation reduces material compaction.

4.7. Dispersion Stability and Exfoliation Yields

The results indicated that both the sonication amplitude and the type of graphite used significantly influenced the dispersion stability and exfoliation efficiency. Notably, graphite recovered from LIBs yielded more stable dispersions compared to commercial graphite. Furthermore, increasing the sonication amplitude from 50% to 80% enhanced graphene formation in the dispersed fraction, as evidenced by a more intense coloration of the suspensions, indicative of a higher concentration of exfoliated material.

As summarized in Table 4, the exfoliation yield varied depending on the graphite type and sonication conditions. Commercial graphite subjected to 50% sonication amplitude exhibited a limited exfoliation yield, with 95.54% of the material remaining in the decanted fraction and only 4.46% in the dispersed fraction. Increasing the sonication amplitude to 80% resulted in a slight improvement, with the dispersed fraction reaching 5.55% and the decanted fraction accounting for 94.45%. This suggests that while higher sonication energy may promote exfoliation, the effect is less pronounced for commercial graphite, likely due to its lower intrinsic exfoliation susceptibility in the absence of natural stabilizers.

In contrast, graphite recovered from LIBs demonstrated superior exfoliation efficiency under identical conditions. At 50% sonication amplitude, the dispersed fraction reached 8.60%, increasing to 9.95% at 80% amplitude. Although the absolute difference in exfoliated material was not drastic, the dispersions derived from LIB-derived graphite exhibited significantly greater long-term stability, indicating a reduced tendency for aggregation and sedimentation of exfoliated graphene.

This behavior can be attributed to the presence of trace metal impurities in the recycled graphite from LIBs, such as manganese, nickel, cobalt and copper (identified by both AAS and XPS), which may act as counter-cations within the dispersion. These metallic species could interact with the surface of exfoliated graphene, enhancing its stabilization in the aqueous medium and mitigating its propensity to reaggregate. Consequently, dispersions obtained from LIB-derived graphite exhibited superior stability and lower sedimentation rates compared to those derived from commercial graphite, even when the amount of exfoliated material was comparable.

These findings align with prior literature reports [6,7,9,59–61], which have demonstrated that liquid-phase exfoliation efficiency and graphene dispersion stability are influenced by factors such as sonication energy, precursor surface chemistry, and the presence of stabilizing species. In this context, the presence of metallic traces in recycled LIB graphite appears to play a critical role in enhancing dispersion stability, presenting a potential advantage for graphene production from battery waste.

On the other hand, sonication amplitude and graphite origin are key factors influencing graphene production via liquid-phase exfoliation. While higher sonication amplitudes moderately improve exfoliation efficiency, dispersion stability is significantly enhanced by the presence of metallic species in LIB-derived graphite. This finding highlights a potential advantage in leveraging battery waste for the sustainable synthesis of advanced nanomaterials, offering promising prospects for large-scale graphene production.

4.8. UV-Vis Analysis

A preliminary characterization was conducted to confirm the formation of graphene in the different samples, with Ultraviolet–Visible (UV-Vis) spectroscopy being the first technique used for this purpose. The UV-Vis spectra were analyzed to detect characteristic peaks associated with the presence of graphene in the dispersed and decanted fractions of the materials obtained from both commercial graphite and discarded battery graphite after ultrasonic exfoliation. The UV-Vis spectra are displayed in Figure 3. This technique allowed for the initial identification of graphene-related features and provided insights into the extent of graphene formation under the different experimental conditions.

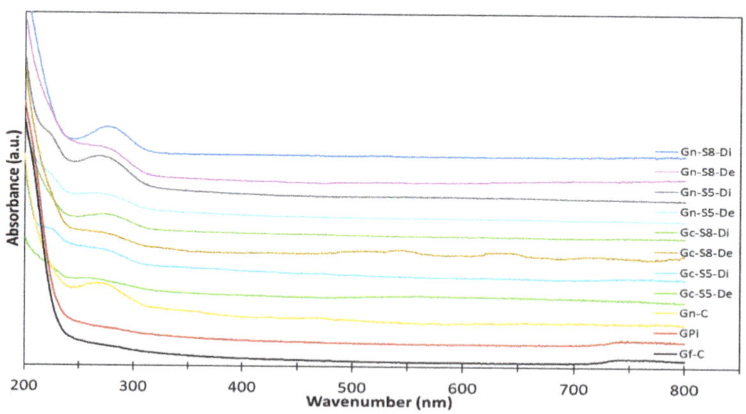

Figure 3. UV-Vis spectra of different materials.

In the analysis of these spectra, no band was identified in either the graphite from discarded batteries or the commercial graphite. It is important to note that the absence of specific bands in the UV-Vis spectra for both types of graphite is consistent with findings reported in the literature [12,15,62]. According to studies, graphite itself does not exhibit distinct UV-Vis absorption bands in the 200–800 nm range, as the conjugated π-electron system of graphite typically absorbs in a different region, primarily in the infrared or visible range. Therefore, the lack of observable bands in this region for both commercial and discarded battery graphite aligns with the expected behavior for unmodified graphite, which does not possess the specific electronic transitions seen in graphene oxide or reduced graphene oxide. This result highlights the need for further processing or exfoliation to induce the characteristic features of graphene, such as the π-π^* transition, which would appear in the UV-Vis spectra.

However, for the commercial graphene, a broad band was identified with a center around 265 nm [12,62], which is characteristic of graphene and is associated with the sp^2 hybridization of the C=C bonds in graphene [14,63]. Specifically, this band was more pronounced in materials sonicated from LIB graphite compared to those from commercial graphite, when compared at the same sonication percentage. Similarly, it was found that materials sonicated at a higher amplitude percentage (80%) exhibited a more pronounced band compared to those sonicated at 50%. Additionally, the materials from the dispersed fraction showed a higher intensity of this band, suggesting a greater amount of graphene is present in this fraction, although it cannot be ruled out that some graphene may also be present in the decanted fraction. Thus, based on this technique, it is initially evident that materials obtained via sonication at 80% amplitude from battery graphite, particularly those recovered in the dispersed fraction, contain a higher amount of graphene.

4.9. DRX Analysis

Figure 4 presents the diffractograms obtained for the different materials. All samples exhibit several common peaks associated with the characteristic crystalline phases of carbonaceous materials. The most prominent peaks correspond to the (002), (100), (101), (004), and (103) planes, located at 2θ angles of 26.1°, 42.4°, 44.5°, 54.7°, and 59.9°, respectively. These reflections indicate the laminar structure and degree of crystalline ordering typical of this type of material [11,64,65].

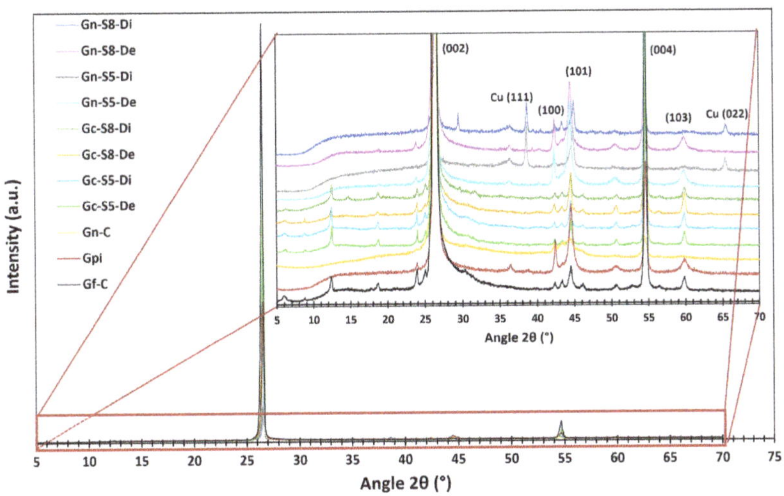

Figure 4. XRD spectra of different materials.

Additionally, in graphene derived from graphite recovered from LIBs, two additional peaks associated with copper oxide (CuO: tenorite) are identified at 2θ angles of 38.7° and 65.5°, corresponding to the (111) and (022) planes, respectively [66,67]. However, these peaks are observed only in the dispersed fraction of these materials (Gn-S5-Di and Gn-S8-Di). This finding is consistent with AAS and XPS results, which identified metal impurities, including Cu. Other metals were not detected by this analysis, likely because their concentrations were below the detection limit of the technique or due to their specific distribution within the graphene structure [68].

The presence of Cu may result from its accumulation in graphite during the electrochemical cycling of charge and discharge in LIBs or from the graphite recovery process from the copper foils on which it is deposited. This hypothesis aligns with AAS analyses, which identified Cu contents ranging from 0.465 to 1.271 wt% in materials recovered from LIBs. The fact that these CuO peaks appear only in the dispersed fraction suggests that CuO nanoparticles may localize during exfoliation. Their weak affinity with exfoliated carbon layers could promote the formation of ordered crystalline domains, increasing the XRD detection limit (~1–5%) despite their low overall concentration. In contrast, in non-dispersed fractions, such as unexfoliated graphite or sedimented residues, Cu likely remains diluted within the graphite matrix or in amorphous forms, preventing its detection [64,66]. Based on this, it can be inferred that the dispersed fraction of LIB-derived graphene exhibits a structure similar to that of commercial graphene, likely to consist of very few stacked layers.

The most prominent peak characteristic of carbonaceous materials appears at 26.1°, corresponding to the (002) plane of graphite, which indicates graphene layer stacking. Figure 4 shows that commercial graphite (Gf-C) exhibits the most intense peak, reflecting a highly ordered structure with minimal defects. In contrast, graphite recovered from LIBs (Gpi) shows a peak intensity reduced to approximately one-third of that of Gf-C, suggesting lower stacking order due to degradation during its use in LIBs. This process induces defects and increases interlayer spacing, leading to a less crystalline and more turbostratic structure [69]. Meanwhile, commercial graphene (Gn-C) presents the lowest (002) peak intensity, indicating a highly defective structure or very few stacked layers with weak interlayer interactions. The progressive decrease in peak intensity reflects a reduction in layer stacking, which aligns with increased turbostratic disorder (order within each plane but with random orientation) [5,70].

Figure 5 presents the second most intense peak, identified at 54.7°, assigned to the higher-order (004) plane. This reflection is particularly useful for estimating the exfoliation level of the evaluated materials, as higher-order planes like (004) require a more periodic stacking of layers to produce constructive interference. This characteristic enhances its sensitivity to layer ordering, especially since reflections at angles greater than 50° are more widely spaced, reducing phase ambiguity. It can be observed that commercial graphite (Gf-C) exhibits the highest intensity for this peak, followed by battery-derived graphite (Gpi), whereas commercial graphene (Gn-C) shows the lowest intensity, consistent with its disordered or few-layered structure.

Additionally, Figure 5 illustrates that sonicated materials display a broad variation in intensity depending on the graphite source and analyzed fraction. For instance, graphene derived from commercial graphite exhibits higher peak intensity than that from LIBs. Moreover, the dispersed fraction of LIB-derived materials presents the lowest intensity, approximately 30% higher than commercial graphene. In contrast, the dispersed fraction of materials originating from commercial graphite exceeds commercial graphene by approximately 500%, suggesting that the recovered dispersed fraction of LIB-derived graphite possesses a layer count approaching that of graphene.

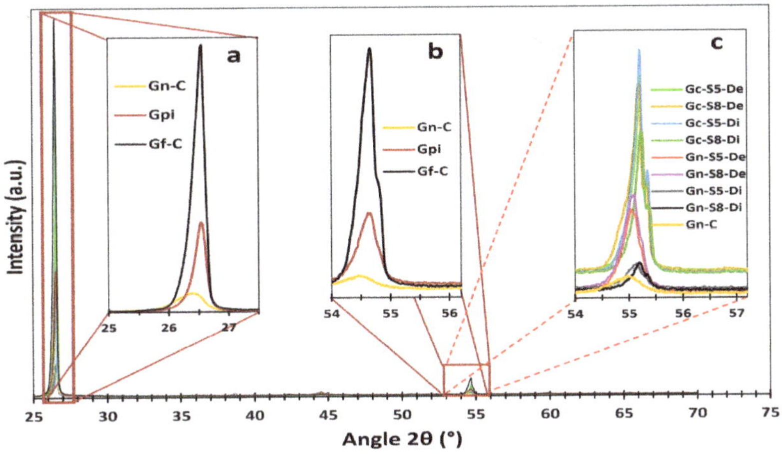

Figure 5. XRD spectra zoom of different peaks. DRX spectra zoom of different materials, (**a**) peak zoom at 26.1 ° of reference materials Gn-C, Gpi and Gf-C, (**b**) peak zoom at 54.7 ° of reference materials Gn-C, Gpi and Gf-C, and (**c**) peak zoom at 54.7 ° of the other sonicated materials.

Furthermore, Figure 4 reveals that the dispersed fraction of LIB-derived graphene does not exhibit the $2\theta \approx 59.9°$ peak corresponding to the (103) plane. This peak serves as a key marker for assessing three-dimensional (3D) order in crystalline graphite, as it corresponds to a non-basal reflection (hkl with h, $l \neq 0$). Its presence requires atomic correlation in three dimensions due to the ordered stacking of layers along the c-axis (ABAB sequence) and lateral alignment in the basal plane (a–b) [71]. Therefore, the presence of this peak confirms a 3D structure, whereas its attenuation or absence indicates disorder or partial to complete exfoliation. Based on this, the complete disappearance of this peak in the dispersed fraction of graphene (Gn-S5-Di and Gn-S8-Di) signifies a significant loss of 3D order, indicating exfoliation approaching few-layer graphene. This result is consistent with the behavior observed in commercial graphene, which lacks the (103) peak due to its inherent 2D or few-layer nature [36,72].

According to the literature [5,70,73,74], XRD analysis is an essential tool for determining key properties such as interlayer spacing (d002) and average crystallite size (DGP). In carbonaceous materials like graphite, these parameters are obtained from the analysis of the (002) peak, which characterizes the stacking of graphene layers, using Bragg's law ($n\lambda = 2d \sin\theta$) and the Scherrer equation ($DGP = K\lambda/\beta\cos\theta$), where K is a shape factor (typically 0.89 for (002) planes), λ is the wavelength of the radiation (0.15406 nm), β is the full width at half maximum (FWHM) of the peak, and θ is the diffraction angle. Based on these values, the approximate number of layers (NL) was determined using the relation $NL = DGP/d_{002}$, where d_{002} was experimentally obtained for each material based on its specific diffraction angle [73].

This approach has been widely applied in the literature for carbonaceous materials derived from commercial sources and even recycled LIBs [10,14,73,75]. However, it is important to consider that the application of the Scherrer equation assumes a spherical crystallite morphology, which introduces uncertainties in layered materials such as graphene. In these materials, anisotropy and the presence of defects may alter the correlation between particle diameter and interlayer spacing. Nonetheless, despite these limitations, this technique remains fundamental for comparative studies and process optimization, provided that its scope and limitations are carefully considered, particularly in the context of highly exfoliated or disordered systems.

Table 5 presents the NL results for each material. Based on these results, the evaluated materials can be categorized into two main groups: those derived from commercial graphite and those obtained from LIBs. The NL values reveal general trends associated with the exfoliation process and the inherent limitations of the XRD technique. First, it is observed that materials derived from commercial graphite exhibit a layer count ranging from 2.5 to 3.7, with a moderate reduction in NL as the sonication amplitude increases from 50% to 80%. For instance, in the sedimented fraction, NL decreases from 3.0 (Gc-S5-De) to 2.5 (Gc-S8-De), while in the dispersed fraction, it decreases from 3.7 (Gc-S5-Di) to 2.7 (Gc-S8-Di). This trend suggests that higher ultrasonic energy (80% amplitude) promotes slightly more efficient exfoliation, although the effect is limited, likely due to a balance between crystallite fragmentation and actual layer separation.

Table 5. Physicochemical Properties of the Materials (Commercial Graphite, LIBs-Derived Graphite, and Graphene).

Material	Graphene Layers Number (XRD)	I_{2D}/I_G	I_D/I_G	Graphene Layers Number (Raman)
Gf-C	~2.0	0.18	0.12	>6.0
GPi	~2.0	0.16	0.19	>6.0
Gn-C	~0.7	0.76	0.46	~2.0
Gc-S5-De	~3.0	0.46	0.08	~3.0
Gc-S5-Di	~3.7	0.52	0.15	~3.0
Gc-S8-De	~2.5	0.51	0.13	~3.0
Gc-S8-Di	~2.7	0.68	0.18	~2.0
Gn-S5-De	~2.0	0.50	0.18	~3.0
Gn-S5-Di	~2.1	0.53	0.48	~3.0
Gn-S8-De	~2.1	0.49	0.19	~3.0
Gn-S8-Di	~2.5	0.74	0.39	~2.0

In contrast, graphene derived from LIB-recovered graphite consistently shows NL values between 2.0 and 2.5, regardless of the applied amplitude. This indicates a slightly higher degree of exfoliation compared to graphene derived from commercial graphite [62], which may be attributed to an enhanced exfoliation capability even at lower sonication amplitudes. This behavior is likely related to the presence of structural defects or impurities in the recycled graphite source, as identified by both AAS and XPS.

A particular aspect of the results is observed in the NL values obtained for Gf-C and GPi graphite (~2.0 layers), which are unexpectedly low for a graphitic material. This can be attributed to specific limitations of the XRD method, as the Scherrer equation assumes the presence of well-defined and separated crystalline domains. However, in graphite, the layered morphology and the possible presence of turbostratic disorder (rotationally misaligned layers) induce an artificial broadening of the (002) peak, which reduces the calculated DGP and, consequently, severely underestimates NL.

Additionally, the technique may not be well-suited for detecting extremely tightly stacked layers, which could explain the observed underestimation in the analyzed graphite. Based on these observations, the XRD-based exfoliation efficiency analysis suggests that the sonication treatments promote layer separation in both the decanted and dispersed fractions of materials derived from LIBs and commercial graphite. However, given the limitations of the technique, complementary analyses using Raman spectroscopy and HRTEM are necessary to achieve a more precise characterization of the exfoliation degree.

4.10. Raman Analysis

The Raman spectra of all materials (Figure 6) exhibited three characteristic bands of crystallized carbonaceous materials with sp^2 hybridization: the D band at ~1360 cm^{-1}, attributed to the A_1g mode activated by structural defects at graphene ring edges; the G band at ~1580 cm^{-1}, associated with the E_2g vibrational mode of C-C bond stretching; and the 2D band at ~2710 cm^{-1}, corresponding to the second harmonic of the D band, generated by double resonant phonon scattering [17,75–77]. These bands provide key structural insights, such as the number of layers and defect density, through the analysis of their relative intensities. The I_{2D}/I_G ratio is a critical parameter for determining the exfoliation degree, where values close to or above 1.6 indicate monolayer graphene, values around 0.8 suggest few-layer structures (~2 layers), values near 0.3 correspond to multilayer graphene (~3 layers), and a further reduction to ~0.07 reflects graphite-like behavior, indicating insufficient exfoliation [12,75,78,79]. Additionally, an increase in the I_D/I_G ratio signifies a higher density of defects, fractures, or topological imperfections in the carbon network [17,76,80].

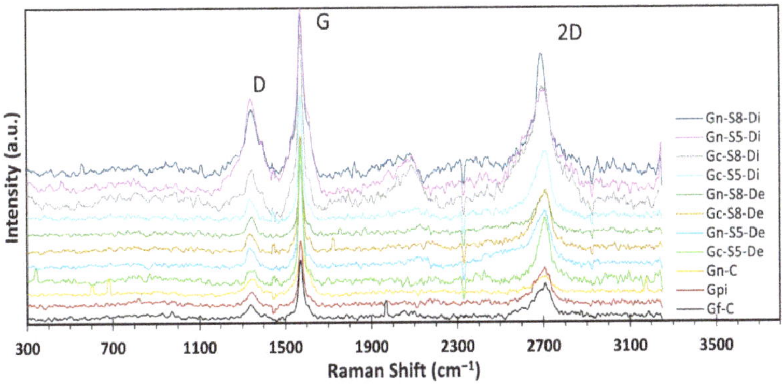

Figure 6. Raman spectra of different materials.

The results presented in Table 5 show the intensity ratio values for all evaluated materials. When comparing the reference materials (Gf-C, Gpi, and Gn-C), it is observed that both commercial graphite and graphite recovered from LIBs exhibit I_{2D}/I_G values significantly lower than 0.3 (0.18 and 0.16, respectively). This indicates that these materials are predominantly composed of highly ordered multilayer structures, whereas commercial graphene presents an I_{2D}/I_G value of 0.76, characteristic of ~2-layer materials [76,81].

Figure 5a,b illustrate the evolution of the I_{2D}/I_G and I_D/I_G intensity ratios, respectively, as a function of the analyzed fraction (decanted or dispersed) for each material. These results provide insight into the number of layers based on these parameters. The I_{2D}/I_G values reveal that all sonicated materials, both from commercial graphite and LIBs, exhibit graphene-like characteristics, even those processed at lower amplitudes (50%). However, this ratio is slightly higher for materials derived from LIB graphite, which may be associated with the presence of impurities that facilitate the exfoliation process during localized sonication.

Additionally, the decanted fractions displayed similar I_{2D}/I_G values (ranging from 0.46 to 0.51), indicating a reduction to a comparable number of layers (~3 layers) [76], compared to the original multilayer graphite, regardless of the sonication amplitude or material. In contrast, in the graphene obtained from the dispersed fraction, the effect of increased amplitude on exfoliation is more pronounced. For instance, in LIB-derived materials, this ratio increased from 0.53 to 0.74 when sonicated at 50% and 80%, respectively. Furthermore, in both graphite sources (commercial and LIBs) sonicated at 80%, the dispersed fraction exhibited I_{2D}/I_G values closer to those of commercial graphene (~2 layers), with Gc-S8-Di showing the highest similarity to commercial graphene.

On the other hand, Figure 7b presents the I_D/I_G ratio as an estimate of the defect density or structural disorder [82]. As observed, commercial graphite exhibits an I_D/I_G value of 0.12, indicative of a highly ordered structure, whereas LIB-recovered graphite shows a higher value (0.19), suggesting greater disorder. In contrast, commercial graphene presents an I_D/I_G of 0.46, reflecting a more fragmented structural network [9,77,83], consistent with its low surface area and non-pristine monolayer nature.

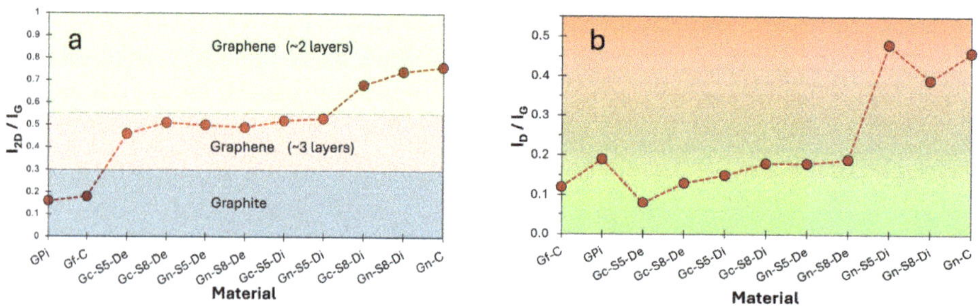

Figure 7. (**a**) I_{2D}/I_G ratio as an estimate of number of layer and (**b**) I_D/I_G ratio as an estimate of defect density or structural disorder.

Figure 7b also reveals that materials derived from commercial graphite retain a superior structural organization in all cases compared to those obtained from LIBs. The higher I_D/I_G values in LIB-derived materials coincide with the presence of impurities and the electrochemical history of the precursor. Additionally, it is observed that the decanted fractions of both materials exhibit similar I_D/I_G values regardless of sonication amplitude but consistently lower than their corresponding dispersed fractions. This suggests that the decanted material retains a more ordered structure, whereas the dispersed fractions, subjected to more intense exfoliation, accumulate greater fragmentation and defects.

These findings, together with the calculated number of layers, confirm the successful exfoliation of graphene in both fractions across all materials. However, the resulting graphene varies in quality. Mechanical exfoliation effectively reduces the number of layers while increasing structural disorder, a trend more pronounced in graphene derived from LIBs. This behavior may be attributed to the presence of residual metallic impurities and the degradation of the carbon network during battery life cycles. Notably, defect density could be a critical factor in applications sensitive to electronic, mechanical, or catalytic properties. Although sonication at higher amplitude (80%) enhances exfoliation in dispersed LIB fractions, bringing them closer to commercial graphene, the significant increase in defect density implies a trade-off between exfoliation efficiency and structural preservation. This balance could be strategically managed depending on specific application requirements.

4.11. HRTEM Analysis

Figure 8 presents the HRTEM micrographs corresponding to commercial graphene and the samples obtained in this study. The images reveal thin sheets with regular edges in the case of commercial graphene, while the graphene obtained through exfoliation exhibits irregular edges, characteristic of the two-dimensional structure of graphene. The presence of a few layers, generally between one and three, is a key indicator of graphene formation, although it is important to note that ideal graphene is monolayer, whereas multilayer graphene exhibits properties more like graphite.

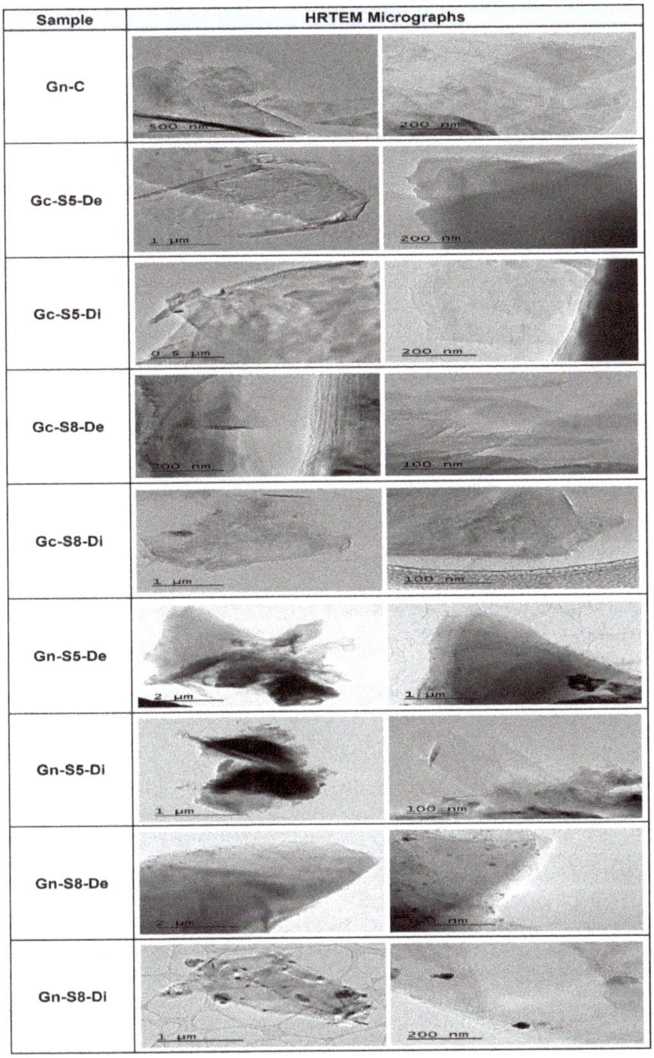

Figure 8. HRTEM micrographs of commercial graphene and the samples obtained.

In all images, the graphene sheets appear partially overlapped or folded, attributed to surface interactions and the exfoliation process. However, commercial graphene shows a greater number of semitransparent regions compared to the samples obtained from both types of graphite, confirmed by the presence of areas with lower contrast, indicating a smaller number of layers in these samples. In contrast, the exfoliated samples display more dark areas corresponding to stacks of multiple layers. These findings are consistent with results obtained through XRD and Raman spectroscopy, indicating several layers approximately between 1 and 3 for all samples.

Additionally, the micrographs reveal significant differences in the purity of the obtained materials. Commercial graphene reflects high purity, evidenced by the absence of impurities, indicating superior quality, likely due to a controlled synthesis process. In contrast, graphene derived from the graphite of LIBs shows the presence of various nanoparticles, in some cases homogeneously distributed, attributable to residual metals such as Cu, Mn, and Co from the batteries. These metals were identified through techniques like AAS and XPS, indicating lower quality in these materials. However, the presence of impurities in graphene is not always a negative factor, as they could have a positive effect if these materials are intended for catalytic applications, given that these nanoscale metals can catalyze various reactions.

It is important to note that HRTEM analysis did not reveal significant differences between the various materials obtained from LIBs, commercial graphene, and those derived from commercial graphite, except for the presence of impurities, which are clearly observed in graphene from LIBs. Additionally, it was not possible to identify differences between the materials obtained at different amplitude percentages. These results are consistent with previous studies that have reported the presence of metallic impurities on graphene obtained from recycled LIBs. For example, in the study by [35,59], the presence of metallic nanoparticles in recycled graphene was observed, which influenced the electrochemical properties of the material.

On the other hand, the HRTEM micrographs provide visual evidence of the structural and purity differences between commercial graphene and those obtained by exfoliation of graphite from spent batteries and commercial graphite. These observations are fundamental to understanding the properties and potential applications of the materials obtained.

4.12. H_2-TPR Analysis

Figures 9 and 10 present the general and deconvoluted H_2-TPR profiles of the materials, respectively, providing insight into their reducibility. Furthermore, Table 6 summarizes the H_2 consumption associated with each reduction event, enabling a quantitative comparison of the materials' reduction behavior. The results reveal notable differences between the samples derived from commercial graphite and those obtained from recycled graphite from spent LIBs. In general, the observed reduction events can be attributed to the presence of oxygen-containing functional groups and the reduction of metal impurities, particularly copper and manganese oxides in the case of materials derived from LIB graphite.

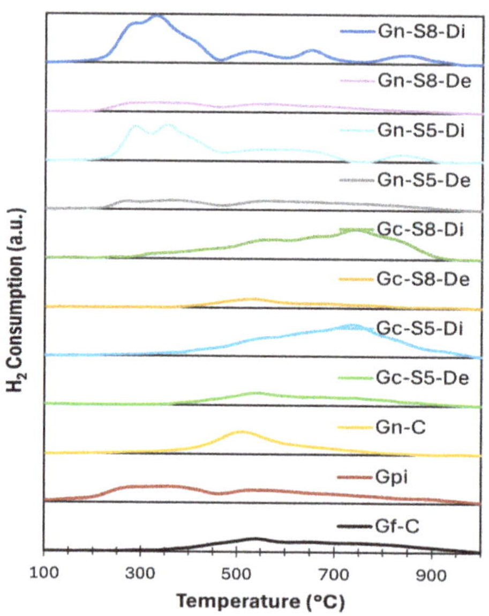

Figure 9. Deconvoluted H_2-TPR profiles of different materials.

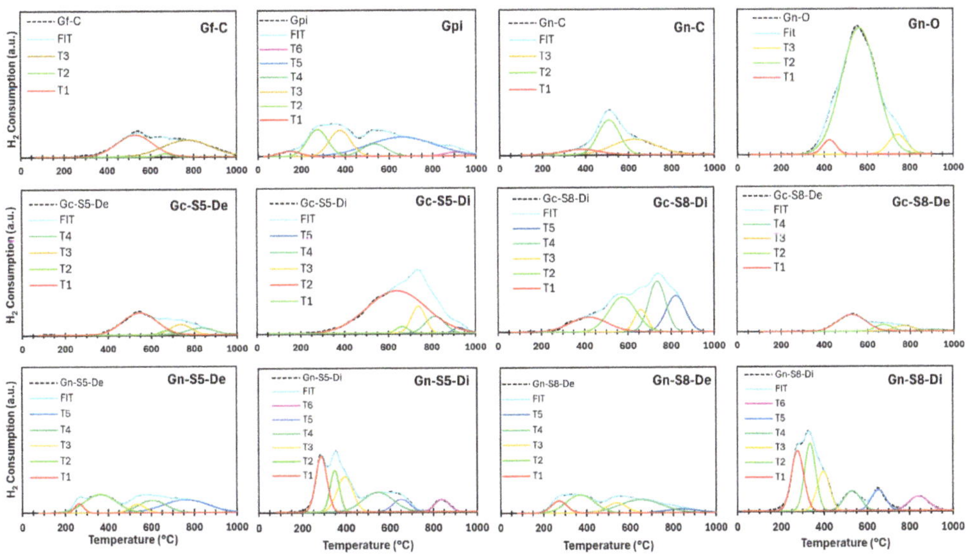

Figure 10. Deconvoluted H_2-TPR profiles of different materials.

Table 6. Associated H_2 consumption with each reduction event for the materials.

Material	Reduction Temperature Event (°C)						Hydrogen Consumption (mmol $H_{2/g}$ Material)						
	T1	T2	T3	T4	T5	T6	T1	T2	T3	T4	T5	T6	Total
Gf-C	527	669	769	-	-	-	0.012	0.005	0.012	-	-	-	0.028
Gpi	147	273	376	533	659	905	0.006	0.009	0.009	0.007	0.016	0.005	0.053
Gn-C	380	508	625	-	-	-	0.006	0.011	0.010	-	-	-	0.028
Gn-O	424	562	742	-	-	-	0.009	0.107	0.013	-	-	-	0.129
Gc-S5-De	546	669	736	832	-	-	0.012	0.005	0.007	0.007	-	-	0.030
Gc-S5-Di	639	663	737	815	924	-	0.026	0.006	0.008	0.008	0.005	-	0.053
Gc-S8-De	526	674	769	912	-	-	0.009	0.006	0.006	0.005	-	-	0.026
Gc-S8-Di	419	574	659	735	822	-	0.010	0.013	0.007	0.012	0.011	-	0.053
Gn-S5-De	263	363	535	604	761	-	0.005	0.010	0.006	0.008	0.010	-	0.038
Gn-S5-Di	286	348	394	545	651	839	0.011	0.009	0.010	0.011	0.007	0.006	0.054
Gn-S8-De	272	370	537	650	844	-	0.006	0.009	0.006	0.009	0.006	-	0.037
Gn-S8-Di	276	335	395	530	652	841	0.012	0.012	0.010	0.008	0.008	0.008	0.058

The samples obtained from commercial graphite exhibited reduction events at higher temperatures, exceeding 500 °C. This behavior suggests a lower concentration of oxygen-containing functional groups and a more ordered structure, resulting in reduced reactivity toward hydrogen. In these samples, the total hydrogen consumption ranged from 0.026 to 0.053 mmol H_2/g (Table 6), indicating a relatively low number of reducible species.

In contrast, the samples obtained from recycled LIBs graphite displayed multiple reduction events at lower temperatures, around 300 °C. This behavior suggests a higher content of oxygen-functionalized groups, such as epoxides and hydroxyls, which typically reduce within this temperature range. Moreover, these samples exhibited higher hydrogen consumption, with values ranging from 0.028 to 0.058 mmol H_2/g (Table 6). This increased reactivity could be related to a higher concentration of metal impurities and a more defective structure compared to the materials derived from commercial graphite. These results can be explained by several key factors related to the nature of the material, the presence of impurities, and the conditions of the exfoliation process.

The fact that graphene materials derived from commercial graphite exhibit reduction peaks at higher temperatures (>500 °C), while those from spent LIBs show reduction events at lower temperatures (~300 °C), suggests significant differences in chemical composition and structural characteristics. One plausible explanation is the presence of metal impurities in the graphene derived from LIBs, which may act as catalytic centers, facilitating the reduction of oxygen-functionalized groups at lower temperatures. Indeed, AAS analysis revealed the presence of Mn, Li, Cu, and Al in variable concentrations, with higher levels in the dispersed fraction. Based on the comparison of the TPR profiles of materials obtained from LIBs and commercial graphite, it is evident that the presence of these metals influences the reduction behavior, either by forming oxidized species that are reduced within specific temperature ranges or by modifying the hydrogen-carbon matrix interactions. These effects are discussed in detail in the TPR results section.

Moreover, in LIBs, the anode graphite can be contaminated with residues from the cathode or electrolyte, which contain metal oxides (e.g., MnO_2 from $LiMn_2O_4$ cathodes, Cu from current collectors). Although the material was not intentionally subjected to oxidation processes or high temperatures, it is possible that the liquid-phase exfoliation using ultrasound may have induced slight surface oxidation of exposed metals due to conditions inherent to the process itself. This phenomenon has been reported in studies on colloidal dispersion and nanoparticle synthesis. Additionally, in aqueous environments, Mn and Cu can exist in different oxidation states depending on pH and redox potential. Cu can oxidize to Cu^{2+} and form species such as $Cu(OH)_2$ or CuO under certain conditions, whereas Mn can exist in soluble forms or as colloidal MnO_2. Furthermore, the detection of metal impurities by AAS and XPS suggests that these metals may be present in various forms, including metallic nanoparticles, oxides, or even species adsorbed on the graphene surface.

According to the literature, oxygen-functionalized groups in carbonaceous materials undergo reduction within distinct temperature ranges. Specifically, epoxide and hydroxyl groups typically reduce between 200 and 400 °C, while carbonyl and carboxyl groups generally reduce within the 400–600 °C range [84,85]. Additionally, the presence of metal impurities can modify the reduction profiles, as certain metals, such as copper (Cu) and manganese (Mn), form oxides that reduce at characteristic temperatures. In this regard, copper oxides (CuO, Cu_2O) commonly reduce between 200 and 300 °C, whereas manganese oxides (MnO_2, Mn_2O_3) typically reduce within the 300–500 °C range [84,86]. Therefore, the observed reduction peaks in the samples can be attributed to both the decomposition of these oxygen-functionalized groups and the reduction of metal oxides present as impurities.

On the other hand, the use of water as a solvent in the liquid-phase exfoliation process may have played a significant role in the surface composition of the materials obtained. In the presence of water, hydrolysis could promote the formation of additional oxygen-containing functional groups, such as hydroxyls and carboxyls, which would explain the higher hydrogen consumption observed during the reduction process. Furthermore, the ultrasonic treatment applied during exfoliation induces structural defects, which can enhance the material reactivity.

An additional factor influencing the reducibility of the materials is the sonication applied during sample preparation. A higher sonication amplitude tends to increase the number of structural defects and the presence of oxygen-containing functional groups, which in turn leads to higher hydrogen consumption during the reduction process. This effect is more pronounced in samples derived from recycled graphite, where the exfoliation induced by sonication enhances the accessibility of reducible species to hydrogen gas.

4.13. Surface Analysis by DRIFT-ATR

The results of the surface analysis by FTIR-ATR are shown in Figure 11. The spectra of all materials reveal the presence of different bands. A broad band between 3200 cm^{-1} and 3500 cm^{-1}, centered at 3450 cm^{-1}, is associated with the stretching of hydroxyl (-OH) groups and adsorbed water. This band appears more intense in materials derived from commercial graphite (Gf-C, Gpi, Gc-S5-De, and Gc-S8-De) compared to those obtained from LIB-recovered graphite (Gn-S5-Di, Gn-S8-Di, and Di), suggesting that the commercial material has a greater tendency to retain these functional groups after the applied treatments. Additionally, the suspended fraction in sonicated materials exhibits greater intensity in this region than the decanted fraction, which may be attributed to a higher specific surface area and a greater dispersion of oxygen-containing groups in this fraction.

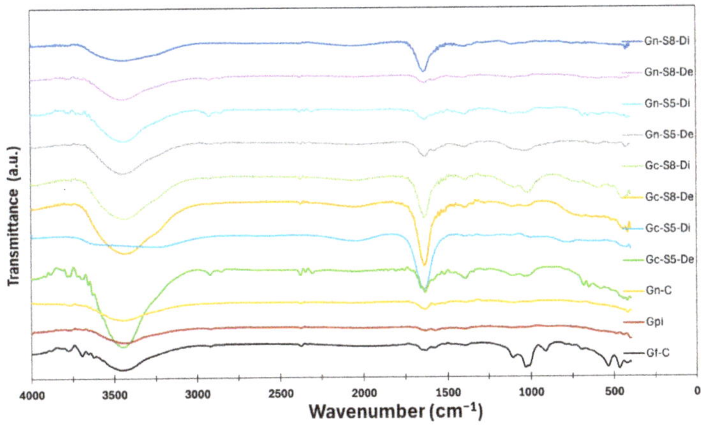

Figure 11. FTIR-ATR spectra of the materials.

A second band, located between 1600 cm^{-1} and 1700 cm^{-1} and centered at 1650 cm^{-1}, corresponds to the stretching of C=O bonds. This band is more intense in materials derived from commercial graphite compared to those from LIBs, indicating a higher degree of functionalization with carbonyl groups. Among these, Gn-S8-Di exhibits the highest intensity, suggesting that higher-power sonication promotes more effective oxidation. The difference in C=O band intensity may be related to the initial structure of the precursor material since LIB-recovered graphite may contain fewer surface defects susceptible to oxidation. In contrast, commercial graphite, having undergone industrial treatment, might have greater accessibility for carbonyl group formation.

Additionally, weak bands are observed between 2925 cm^{-1} and 2850 cm^{-1} in some materials (Gf-C, Gpi, Gn-S5-Di, Gc-S5-De, Gn-S8-De, and Gc-S8-De), attributed to the C-H stretching of methyl and methylene groups. The presence of these bands suggests that, despite the applied treatments, residual aliphatic structures persist, likely originating from the graphite precursor of the graphene. The intensity of these bands is low in all cases but slightly more pronounced in materials derived from commercial graphite, suggesting that these materials retain a higher surface composition of such compounds. Furthermore, the band at 1580 cm^{-1}, related to C=C stretching, is more prominent in graphene derived from commercial graphite in both decanted and suspended fractions, indicating a greater presence of surface functional groups.

At around 1400 cm^{-1}, most materials, except for commercial graphene, exhibit a very weak band, which is slightly more evident in LIB-recovered materials subjected to 80% sonication. This suggests that sonication may induce structural changes that enhance the presence of vibrational modes associated with OH or CH bond deformation. Finally, in the 1100–1050 cm^{-1} region, two small bands appear, associated with C=O and C-C bonds. These bands are more pronounced in Gf-C and, to a lesser extent, in Gc-S5-De and Gc-S8-De, whereas they appear with even lower intensity in Gn-S5-Di. The presence of these bands suggests that some materials retain residual oxygenated structures after the applied treatments. Their relative intensity may be linked to the oxidation degree induced by the sonication process and the nature of the precursor material.

These results suggest that sonication has a significant effect on surface functionalization and the intensity of the observed bands. In general, sonicated materials exhibit more pronounced bands, indicating that this treatment enhances the exposure and generation of functional groups. Additionally, the difference between the suspended and decanted fractions suggests that sonication may induce greater fragmentation, increasing the accessibility of oxygen-containing groups in the more dispersed fraction.

For materials derived from LIBs, the lower intensity of carbonyl groups compared to those from commercial graphite suggests that the initial structure of the precursor plays a key role in the final surface composition. The fact that Gn-S8-Di exhibits higher carbonyl band intensity compared to other recycled materials indicates that the combination of high-power sonication and the chemical history of the starting material can influence the evolution of functional groups. This differential oxidation could be strategically exploited for specific applications.

5. Conclusions

These findings demonstrate that tip-sonotrode-assisted sonication is highly effective in modifying the textural properties of carbon-based materials. The production of dispersed fraction graphene with characteristics similar to commercial graphene (two layers) highlights the efficiency of the process, particularly for commercial graphite. Meanwhile, LIB-derived graphene offers additional sustainability advantages by contributing to the valorization of hazardous waste.

These results highlight that the combination of high sonication amplitude (80%), dispersed fraction (Di), and recycled battery graphite maximizes the conditions for producing conductive graphene. This not only validates LIB recycling as a sustainable strategy but also suggests that "used" materials can outperform their virgin counterparts in specific applications. This finding highlights a potential advantage in leveraging battery waste for the sustainable synthesis of advanced nanomaterials, offering promising prospects for large-scale graphene production.

In summary, increasing the sonication amplitude from 50% to 80% results in progressive exfoliation, reducing the number of layers. The sonication treatment significantly modifies the structure of both processed graphite, promoting mechanical exfoliation and graphene layer separation. Notably, LIB-derived graphite is more susceptible to this treatment. These findings emphasize the critical influence of sonication amplitude in achieving efficient exfoliation, particularly in the dispersed fraction, to obtain few-layer graphene structures.

Although water is generally considered an optimal dispersion medium due to its ubiquity and cost-effectiveness, most studies on liquid phase exfoliation have been carried out using organic solvents and surfactants as they have shown that dispersion stability of graphene derived from commercial graphite is so low, this is due to the hydrophobic nature of commercial graphite. Here, graphite is exfoliated from LIBs directly in pure water, circumventing the need for chemical agents or surfactants. The presence of impurities in graphite sourced from LIBs, such as metals and organic solvents, has been found to promote both exfoliation and the stable dispersion of graphene in water. This work demonstrates that it is possible to valorize the anodic graphitic material waste from spend LIBs as inputs for multilayer graphene production by exfoliation in the liquid phase; a simple, economical and environmentally friendly method. In contrast to the productivity exhibited by graphene derived from pristine graphite powders procured from Sigma-Aldrich, the graphene extracted from battery anodes showcased a remarkable fourfold increase in yield.

This underscores the paramount significance of graphene sourced from batteries in terms of productivity enhancement.

These results confirm that combining recycled graphite from LIBs, high sonication amplitudes, and appropriate fraction selection creates optimal conditions for producing conductive graphene. Notably, recycled materials can outperform their virgin counterparts in terms of yield and aqueous dispersion behavior, making them highly relevant for real-world applications such as printed electronics, conductive inks, sensors, and energy storage devices.

Despite the promising results, this study has certain limitations. A more detailed mechanistic analysis of the role of impurities in the exfoliation behavior is still needed. Furthermore, while the process shows potential for scale-up, additional work is required to optimize energy consumption and assess the process from a life cycle perspective. Future research should focus on a deeper physicochemical characterization of impurity–graphite interactions, process evaluation at pilot scale, and validation for industrial applications.

Overall, this work presents a practical, economical, and environmentally friendly strategy for producing multilayer graphene from battery waste, aligning with circular economy principles and advancing the sustainable development of advanced nanomaterials. Moreover, it expands existing knowledge by validating the use of pure water as a dispersion medium and demonstrating the potential of waste-derived precursors for functional graphene production.

Author Contributions: Conceptualization, A.N.A.A., E.A.-V., S.B.B. and L.M.O.-C.; methodology, A.N.A.A., S.B.B. and E.A.-V.; validation, A.N.A.A., S.B.B. and E.A.-V.; formal analysis, A.N.A.A., S.B.B. and E.A.-V.; investigation, A.N.A.A., S.B.B. and E.A.-V.; resources, A.N.A.A., L.M.O.-C., T.A.Z.P., S.A.G.T. and G.A.F.Z.; writing—original draft preparation, A.N.A.A., S.B.B. and E.A.-V.; writing—review and editing, A.N.A.A., E.A.-V., S.B.B., L.M.O.-C., T.A.Z.P., S.A.G.T. and G.A.F.Z.; visualization, A.N.A.A. and E.A.-V.; supervision, A.N.A.A.; project administration, A.N.A.A., L.M.O.-C. and T.A.Z.P.; funding acquisition, A.N.A.A., L.M.O.-C. and T.A.Z.P. All authors have read and agreed to the published version of the manuscript.

Funding: This research was made possible through funding from Minciencias-Colombia, Politécnico Colombiano Jaime Isaza Cadavid, and PREI-DEGAPA-UNAM, Mexico. We acknowledge Minciencias for supporting the research project "Comprehensive valorization of post-consumer and industrial waste for the development of materials with catalytic potential under a circular economy approach" (code 82312). We also extend our gratitude to Universidad Autónoma Metropolitana Iztapalapa for providing the necessary infrastructure and facilitating analyses in various laboratories, particularly those conducted at the Laboratorio Central de Microscopía Electrónica UAM-I de la CB.

Data Availability Statement: The original contributions presented in this study are included in the article. Further inquiries can be directed to the corresponding author.

Conflicts of Interest: The authors declare no conflict of interest.

Appendix A

Figure A1 shows the behavior of the different materials upon contact with a 200 μL water droplet. The images were captured exactly one second after the droplet was deposited, allowing the observation of their hydrophilic or hydrophobic characteristics based on the resulting contact angle. These differences in wettability were analyzed using the LB-ADSA method, providing a visual comparison of surface interactions across the various samples.

Figure A1. Photographs of the contact angle measurements obtained using the LB-ADSA method for the different materials: (**a**) Gf-C, (**b**) GPi, (**c**) Gn-C, (**d**) Gc-S5-De, (**e**) Gc-S5-Di, (**f**) Gc-S8-De, (**g**) Gc-S8-Di, (**h**) Gn-S5-De, (**i**) Gn-S5-Di, (**j**) Gn-S8-De, and (**k**) Gn-S8-Di. The green line shown in each image is generated by the LB-ADSA software and is used to define the droplet profile for accurate contact angle measurement.

References

1. Ramezani, M.J.; Rahmani, O. A Review of Recent Progress in the Graphene Syntheses and Its Applications. *Mech. Adv. Mater. Struct.* **2024**, 1–33. [CrossRef]
2. Shahzad, T.T.; Nawaz, S.; Jamal, H.; Shahzad, T.T.; Akhtar, F.; Kamran, U. A Review on Cutting-Edge Three-Dimensional Graphene-Based Composite Materials: Redefining Wastewater Remediation for a Cleaner and Sustainable World. *J. Compos. Sci.* **2025**, *9*, 18. [CrossRef]
3. Ramachandran, T.; Roy, N.; Hegazy, H.H.; Yahia, I.S.; Kumar, Y.A.; Moniruzzaman, M.; Joo, S.W. From Graphene Aerogels to Efficient Energy Storage: Current Developments and Future Prospects. *J. Alloys Compd.* **2025**, *1010*, 177248. [CrossRef]
4. Divya, M.L.; Natarajan, S.; Aravindan, V. Graphene from Spent Lithium-Ion Batteries. *Batter. Supercaps* **2022**, *5*, e202200046. [CrossRef]
5. Ma, H.; Shen, Z.; Yi, M.; Ben, S.; Liang, S.; Liu, L.; Zhang, Y.; Zhang, X.; Ma, S. Direct Exfoliation of Graphite in Water with Addition of Ammonia Solution. *J. Colloid Interface Sci.* **2017**, *503*, 68–75. [CrossRef]
6. Tyurnina, A.V.; Tzanakis, I.; Morton, J.; Mi, J.; Porfyrakis, K.; Maciejewska, B.M.; Grobert, N.; Eskin, D.G. Ultrasonic Exfoliation of Graphene in Water: A Key Parameter Study. *Carbon N. Y.* **2020**, *168*, 737–747. [CrossRef]
7. Turner, P.; Hodnett, M.; Dorey, R.; Carey, J.D. Controlled Sonication as a Route to In-Situ Graphene Flake Size Control. *Sci. Rep.* **2019**, *9*, 8710. [CrossRef]
8. Han, J.T.; Jang, J.I.; Kim, H.; Hwang, J.Y.; Yoo, H.K.; Woo, J.S.; Choi, S.; Kim, H.Y.; Jeong, H.J.; Jeong, S.Y.; et al. Extremely Efficient Liquid Exfoliation and Dispersion of Layered Materials by Unusual Acoustic Cavitation. *Sci. Rep.* **2014**, *4*, 5133. [CrossRef]
9. Güler, Ö.; Tekeli, M.; Taşkın, M.; Güler, S.H.; Yahia, I.S. The Production of Graphene by Direct Liquid Phase Exfoliation of Graphite at Moderate Sonication Power by Using Low Boiling Liquid Media: The Effect of Liquid Media on Yield and Optimization. *Ceram. Int.* **2021**, *47*, 521–533. [CrossRef]
10. Gürünlü, B.; Taşdelen-Yücedağ, Ç.; Bayramoğlu, M. Graphene Synthesis by Ultrasound Energy-Assisted Exfoliation of Graphite in Various Solvents. *Crystals* **2020**, *10*, 1037. [CrossRef]
11. Wang, S.; Wang, C.; Ji, X. Towards Understanding the Salt-Intercalation Exfoliation of Graphite into Graphene. *RSC Adv.* **2017**, *7*, 52252–52260. [CrossRef]
12. Hu, K.; Brambilla, L.; Sartori, P.; Moscheni, C.; Perrotta, C.; Zema, L.; Bertarelli, C.; Castiglioni, C. Development of Tailored Graphene Nanoparticles: Preparation, Sorting and Structure Assessment by Complementary Techniques. *Molecules* **2023**, *28*, 565. [CrossRef] [PubMed]
13. Haar, S.; El Gemayel, M.; Shin, Y.; Melinte, G.; Squillaci, M.A.; Ersen, O.; Casiraghi, C.; Ciesielski, A.; Samorì, P. Enhancing the Liquid-Phase Exfoliation of Graphene in Organic Solvents upon Addition of n-Octylbenzene. *Sci. Rep.* **2015**, *5*, 16684. [CrossRef] [PubMed]

14. Gürünlü, B.; Taşdelen-Yücedağ, Ç.; Bayramoğlu, M. One Pot Synthesis of Graphene through Microwave Assisted Liquid Exfoliation of Graphite in Different Solvents. *Molecules* **2022**, *27*, 5027. [CrossRef]
15. Phiri, J.; Gane, P.; Maloney, T.C. High-Concentration Shear-Exfoliated Colloidal Dispersion of Surfactant–Polymer-Stabilized Few-Layer Graphene Sheets. *J. Mater. Sci.* **2017**, *52*, 8321–8337. [CrossRef]
16. Montana, F.; Cellura, M.; Di Silvestre, M.L.; Longo, S.; Luu, L.Q.; Riva Sanseverino, E.; Sciumè, G. Assessing Critical Raw Materials and Their Supply Risk in Energy Technologies—A Literature Review. *Energies* **2025**, *18*, 86. [CrossRef]
17. He, K.; Zhang, Z.Y.; Zhang, F.S. Synthesis of Graphene and Recovery of Lithium from Lithiated Graphite of Spent Li-Ion Battery. *Waste Manag.* **2021**, *124*, 283–292. [CrossRef]
18. Wang, D.; Gu, Y.; Chen, Q.; Tang, Z. Direct Conversion of Syngas to Alpha Olefins via Fischer–Tropsch Synthesis: Process Development and Comparative Techno-Economic-Environmental Analysis. *Energy* **2023**, *263*, 125991. [CrossRef]
19. Yu, J.; Lin, M.; Tan, Q.; Li, J. High-Value Utilization of Graphite Electrodes in Spent Lithium-Ion Batteries: From 3D Waste Graphite to 2D Graphene Oxide. *J. Hazard. Mater.* **2021**, *401*, 123715. [CrossRef]
20. Liivand, K.; Kazemi, M.; Walke, P.; Mikli, V.; Uibu, M.; Macdonald, D.D.; Kruusenberg, I. Spent Li-Ion Battery Graphite Turned Into Valuable and Active Catalyst for Electrochemical Oxygen Reduction. *ChemSusChem* **2021**, *14*, 1103–1111. [CrossRef]
21. Alcaraz, L.; Díaz-Guerra, C.; Fernández-Martínez, R.; Gómez-Mancebo, M.B.; Sotillo, B.; Llorente, I.; López, F.A. Second Life of Recycled Graphite for Sustainable Production of Multilayer Graphene Related Materials with Multiple Potential Applications. *Mater. Charact.* **2024**, *209*, 113695. [CrossRef]
22. Beletskii, E.V.; Pakalnis, V.V.; Lukyanov, D.A.; Anishchenko, D.V.; Volkov, A.I.; Levin, O.V. Recycling Spent Graphite Anodes into a Graphite/Graphene Oxide Composite via Plasma Solution Treatment for Reuse in Lithium-Ion Batteries. *J. Environ. Chem. Eng.* **2023**, *11*, 109234. [CrossRef]
23. Kumari, P.; Samadder, S.R. Urban Mining for Graphite Recovery from Retired Lithium-Ion Batteries and Its Valorization to Valuable Graphene Materials: A Comprehensive Characterization Study. *J. Energy Storage* **2024**, *102*, 113873. [CrossRef]
24. Li, H.; Peng, J.; Liu, P.; Li, W.; Wu, Z.; Chang, B.; Wang, X. Re-Utilization of Waste Graphite Anode Materials from Spent Lithium-Ion Batteries. *J. Electroanal. Chem.* **2023**, *932*, 117247. [CrossRef]
25. Sandhya, M.; Ramasamy, D.; Sudhakar, K.; Kadirgama, K.; Harun, W.S.W. Ultrasonication an Intensifying Tool for Preparation of Stable Nanofluids and Study the Time Influence on Distinct Properties of Graphene Nanofluids—A Systematic Overview. *Ultrason. Sonochem.* **2021**, *73*, 105479. [CrossRef]
26. An, C.; Wang, T.; Wang, S.; Chen, X.; Han, X.; Wu, S.; Deng, Q.; Zhao, L.; Hu, N. Ultrasonic-Assisted Preparation of Two-Dimensional Materials for Electrocatalysts. *Ultrason. Sonochem.* **2023**, *98*, 106503. [CrossRef]
27. Godoy, A.P.; Ecorchard, P.; Beneš, H.; Tolasz, J.; Smržová, D.; Seixas, L.; Pedrotti, J.J.; de Souza, E.A.T.; El Seoud, O.A.; Donato, R.K. Ultrasound Exfoliation of Graphite in Biphasic Liquid Systems Containing Ionic Liquids: A Study on the Conditions for Obtaining Large Few-Layers Graphene. *Ultrason. Sonochem.* **2019**, *55*, 279–288. [CrossRef]
28. Anaklı, D.; Erşan, M. Optimization of Reduced Graphene Oxide Yield Using Response Surface Methodology. *Diam. Relat. Mater.* **2024**, *148*, 111524. [CrossRef]
29. Hu, L.H.; Hsiao, K.S.; Wang, C.T. Dual-Functional Electrochromic Green Glass Incorporated with PEDOT:PSS/Electrochemically Exfoliated Graphene and Cesium Tungsten Oxide for Transmittance Modulation and near Infrared Shielding. *Electrochim. Acta* **2025**, *524*, 146055. [CrossRef]
30. Ghanbari, H.; Shafikhani, M.A.; Daryalaal, M. Graphene Nanosheets Production Using Liquid-Phase Exfoliation of Pre-Milled Graphite in Dimethylformamide and Structural Defects Evaluation. *Ceram. Int.* **2019**, *45*, 20051–20057. [CrossRef]
31. Bandara, T.M.W.J.; Thennakoon, T.M.A.A.B.; Gamachchi, G.G.D.M.G.; Bandara, L.R.A.K.; Pemasiri, B.M.K.; Dahanayake, U. An Electrochemical Route to Exfoliate Vein Graphite into Graphene with Black Tea. *Mater. Chem. Phys.* **2022**, *289*, 126450. [CrossRef]
32. Zhang, Y.; Xu, Y.; Zhu, J.; Li, L.; Du, X.; Sun, X. Electrochemically Exfoliated High-Yield Graphene in Ambient Temperature Molten Salts and Its Application for Flexible Solid-State Supercapacitors. *Carbon N. Y.* **2018**, *127*, 392–403. [CrossRef]
33. Amin, Y.; Nugroho, N.; Bahtiar, E.T.; Dwianto, W.; Lubis, M.A.R.; Adzkia, U.; Karlinasari, L. Surface Roughness, Dynamic Wettability, and Interphase of Modified Melamine Formaldehyde-Based Adhesives on Jabon Wood. *Polymers* **2024**, *16*, 1084. [CrossRef]
34. Ribeiro, J.S.; Freitas, M.B.J.G.; Freitas, J.C.C. Recycling of Graphite and Metals from Spent Li-Ion Batteries Aiming the Production of Graphene/CoO-Based Electrochemical Sensors. *J. Environ. Chem. Eng.* **2021**, *9*, 104689. [CrossRef]
35. Zhang, Y.; Song, N.; He, J.; Chen, R.; Li, X. Lithiation-Aided Conversion of End-of-Life Lithium-Ion Battery Anodes to High-Quality Graphene and Graphene Oxide. *Nano Lett.* **2019**, *19*, 512–519. [CrossRef]
36. Alcaraz, L.; Díaz-Guerra, C.; Calbet, J.; López, M.L.; López, F.A. Obtaining and Characterization of Highly Crystalline Recycled Graphites from Different Types of Spent Batteries. *Materials* **2022**, *15*, 3246. [CrossRef]
37. He, H.; Zhang, R.; Zhang, P.; Wang, P.; Chen, N.; Qian, B.; Zhang, L.; Yu, J.; Dai, B. Functional Carbon from Nature: Biomass-Derived Carbon Materials and the Recent Progress of Their Applications. *Adv. Sci.* **2023**, *10*, 2205557. [CrossRef]

38. Chen, W.; Ball, A.S.; Cole, I.; Yin, H. Metal-Doped Carbon Dots as Fenton-like Catalysts and Their Applications in Pollutant Degradation and Sensing. *Sustainability* **2025**, *17*, 3642. [CrossRef]
39. Pan, J.; Li, C.; Peng, Y.; Wang, L.; Li, B.; Zheng, G.; Song, M. Application of Transition Metal (Ni, Co and Zn) Oxides Based Electrode Materials for Ion-Batteries and Supercapacitors. *Int. J. Electrochem. Sci.* **2023**, *18*, 100233. [CrossRef]
40. Spirescu, V.A.; Chircov, C.; Grumezescu, A.M.; Andronescu, E. Polymeric Nanoparticles for Antimicrobial Therapies: An up-to-Date Overview. *Polymers* **2021**, *13*, 724. [CrossRef]
41. Konkena, B.; Vasudevan, S. Understanding Aqueous Dispersibility of Graphene Oxide and Reduced Graphene Oxide through p*K*a Measurements. *J. Phys. Chem. Lett.* **2012**, *3*, 867–872. [CrossRef] [PubMed]
42. Foti, A.; Calì, L.; Petralia, S.; Satriano, C. Green Nanoformulations of Polyvinylpyrrolidone-Capped Metal Nanoparticles: A Study at the Hybrid Interface with Biomimetic Cell Membranes and In Vitro Cell Models. *Nanomaterials* **2023**, *13*, 1624. [CrossRef] [PubMed]
43. Parviz, D.; Das, S.; Ahmed, H.S.T.; Irin, F.; Bhattacharia, S.; Green, M.J. Dispersions of Non-Covalently Functionalized Graphene with Minimal Stabilizer. *ACS Nano* **2012**, *6*, 8857–8867. [CrossRef] [PubMed]
44. Baskoro, F.; Wong, C.B.; Kumar, S.R.; Chang, C.W.; Chen, C.H.; Chen, D.W.; Lue, S.J.; Jessie, S. Graphene Oxide-Cation Interaction: Inter-Layer Spacing and Zeta Potential Changes in Response to Various Salt Solutions. *J. Memb. Sci.* **2018**, *554*, 253–263. [CrossRef]
45. Gaur, D.K.; Agrahari, K.; Singh, B.P.; Alam, M.B.; Parmar, A.S.; Manohar, R.; Singh, S. Optical Properties and Zeta Potential of Polyvinyl Pyrrolidone Capped Gold Nanoparticles Dispersed Nematic Liquid Crystal Mixture E7. *Opt. Mater.* **2023**, *145*, 114317. [CrossRef]
46. Liu, H.; Li, L. Graphitic Materials: Intrinsic Hydrophilicity and Its Implications. *Extrem. Mech. Lett.* **2017**, *14*, 44–50. [CrossRef]
47. Romadiansyah, T.Q.; Lakuy, F.R.A.; Iqbal, R.M.; Zulfiani, U.; Pratama, A.W.; Purnomo, A.S.; Subaer, S.; Gunawan, T.; Rahmawati, Z.; Asranudin; et al. Functionalized Graphite into Hydrophilic Activated Carbon: Synthesis, Characterization, and Adsorption Studies. *Case Stud. Chem. Environ. Eng.* **2025**, *11*, 101035. [CrossRef]
48. Kim, E.; Kim, D.; Kwak, K.; Nagata, Y.; Bonn, M. Perspective Wettability of Graphene, Water Contact Angle, and Interfacial Water Structure. *Chem. Perspect.* **2022**, *8*, 1187–1200. [CrossRef]
49. Nowak, E.; Combes, G.; Stitt, E.H.; Pacek, A.W. A Comparison of Contact Angle Measurement Techniques Applied to Highly Porous Catalyst Supports. *Powder Technol.* **2013**, *233*, 52–64. [CrossRef]
50. Gu, Y.; Zhang, W.; Mou, J.; Zheng, S.; Jiang, L.; Sun, Z.; Wang, E. Research Progress of Biomimetic Superhydrophobic Surface Characteristics, Fabrication, and Application. *Adv. Mech. Eng.* **2017**, *9*, 168781401774685. [CrossRef]
51. Jishnu, A.; Jayan, J.S.; Saritha, A.; Sethulekshmi, A.S.; Venu, G. Superhydrophobic Graphene-Based Materials with Self-Cleaning and Anticorrosion Performance: An Appraisal of Neoteric Advancement and Future Perspectives. *Colloids Surf. A Physicochem. Eng. Asp.* **2020**, *606*, 125395. [CrossRef]
52. Sotoudeh, F.; Mousavi, S.M.; Karimi, N.; Lee, B.J.; Abolfazli-Esfahani, J.; Manshadi, M.K.D. Natural and Synthetic Superhydrophobic Surfaces: A Review of the Fundamentals, Structures, and Applications. *Alexandria Eng. J.* **2023**, *68*, 587–609. [CrossRef]
53. Avrămescu, R.E.; Ghica, M.V.; Dinu-Pîrvu, C.; Udeanu, D.I.; Popa, L. Liquid Marbles: From Industrial to Medical Applications. *Molecules* **2018**, *23*, 1120. [CrossRef]
54. Emani, P.S.; Maddah, H.A.; Rangoonwala, A.; Che, S.; Prajapati, A.; Singh, M.R.; Gruen, D.M.; Berry, V.; Behura, S.K. Organophilicity of Graphene Oxide for Enhanced Wettability of ZnO Nanorods. *ACS Appl. Mater. Interfaces* **2020**, *12*, 39772–39780. [CrossRef]
55. Avramescu, R.E.; Ghica, M.V.; Dinu-Pîrvu, C.; Prisada, R.; Popa, L. Superhydrophobic Natural and Artificial Surfaces-A Structural Approach. *Materials* **2018**, *11*, 866. [CrossRef]
56. Sun, Y.; Zheng, Y.; Liu, C.; Zhang, Y.; Wen, S.; Song, L.; Zhao, M. Liquid Marbles, Floating Droplets: Preparations, Properties, Properties, Operations and Applications. *RSC Adv.* **2022**, *12*, 15296–15315. [CrossRef]
57. Ren, S.; Chen, J.; Jiang, M.; Wang, S.; Wan, Z.; Xie, Y.; Li, L. The Effect of Drop Volume on the Apparent Contact Angle of Hierarchical Structured Superhydrophobic Surfaces. *Colloids Surfaces A Physicochem. Eng. Asp.* **2021**, *611*, 125849. [CrossRef]
58. Ruan, D.; Wang, F.; Wu, L.; Du, K.; Zhang, Z.; Zou, K.; Wu, X.; Hu, G. A High-Performance Regenerated Graphite Extracted from Discarded Lithium-Ion Batteries. *New J. Chem.* **2021**, *45*, 1535–1540. [CrossRef]
59. Lee, X.J.; Hiew, B.Y.Z.; Lai, K.C.; Lee, L.Y.; Gan, S.; Thangalazhy-Gopakumar, S.; Rigby, S. Review on Graphene and Its Derivatives: Synthesis Methods and Potential Industrial Implementation. *J. Taiwan Inst. Chem. Eng.* **2019**, *98*, 163–180. [CrossRef]
60. Hernandez, Y.; Nicolosi, V.; Lotya, M.; Blighe, F.M.; Sun, Z.; De, S.; Mcgovern, I.; Holland, B.; Byrne, M.; Gun'ko, Y.; et al. High-Yield Production of Graphene by Liquid-Phase Exfoliation of Graphite. *Nat. Nanotechnol.* **2008**, *3*, 563–568. [CrossRef]
61. Franco, C.; Lozano-pérez, A.S.; Mendieta-reyes, N.E.; Guerrero-fajardo, C.A. Graphene Oxide Obtention via Liquid Phase Exfoliation from High-Rank Coal: A Comparison of Mineral Matter Removal by Alkaline Bath. *MethodsX* **2023**, *10*, 102147. [CrossRef] [PubMed]

62. Chen, X.; Zhu, Y.; Peng, W.; Li, Y.; Zhang, G.L.; Zhang, F.; Fan, X. Direct Exfoliation of the Anode Graphite of Used Li-Ion Batteries into Few-Layer Graphene Sheets: A Green and High Yield Route to High-Quality Graphene Preparation. *J. Mater. Chem. A* **2017**, *5*, 5880–5885. [CrossRef]
63. You, A.; Be, M.; In, I. Ultraviolet-Visible Spectroscopy of Graphene Oxides. *AIP Adv.* **2012**, *2*, 032146. [CrossRef]
64. Morsy, S.W.; Refai, M.K.; El-Mahdy, A.N.; Kandil, M.M.; El-Mahdy, A.N. Fabrication of a thin film gamma radiation sensor using nano metal oxides. *J. Al-Azhar Univ. Eng. Sect.* **2024**, *19*, 1–14. [CrossRef]
65. Stobinski, L.; Lesiak, B.; Malolepszy, A.; Mazurkiewicz, M.; Mierzwa, B.; Zemek, J.; Jiricek, P.; Bieloshapka, I. Graphene Oxide and Reduced Graphene Oxide Studied by the XRD, TEM and Electron Spectroscopy Methods Journal of Electron Spectroscopy and Graphene Oxide and Reduced Graphene Oxide Studied by the XRD, TEM and Electron Spectroscopy Methods. *J. Electron Spectros. Relat. Phenom.* **2014**, *195*, 145–154. [CrossRef]
66. Pitiphattharabun, S.; Auewattanapun, K.; Sato, N.; Janbooranapinij, K.; Ratchatee, T.; Gasidit, P.; Ohta, J.; Jongprateep, O. Fe-Doped CuO/MWCNT as a Sensing Material for Electrochemical Detection of Nitrite. *Crystals* **2022**, *12*, 1536. [CrossRef]
67. Siddiqui, H.; Parra, M.R.; Qureshi, M.; Malik, M.; Haque, F. Studies of Structural, Optical, and Electrical Properties Associated with Defects in Sodium-Doped Copper Oxide (CuO/Na) Nanostructures. *J. Mater. Sci.* **2018**, *53*, 8826–8843. [CrossRef]
68. Ali, A.; Chiang, Y.W.; Santos, R.M. X-Ray Diffraction Techniques for Mineral Characterization: A Review for Engineers of the Fundamentals, Applications, and Research Directions. *Minerals* **2022**, *12*, 205. [CrossRef]
69. Narayan, R.; Lim, J.; Jeon, T.; Li, D.J.; Kim, S.O. Perylene Tetracarboxylate Surfactant Assisted Liquid Phase Exfoliation of Graphite into Graphene Nanosheets with Facile Re-Dispersibility in Aqueous/Organic Polar Solvents. *Carbon N. Y.* **2017**, *119*, 555–568. [CrossRef]
70. Cherakkara, A.; Zafar, S.; Izwan, I.; Yang, C. Graphite from Biomass: A Review on Synthetic Feasibility. *J. Ind. Eng. Chem.* **2025**, *145*, 75–98. [CrossRef]
71. Stefanescu, D.M.; Alonso, G.; Suarez, R. Recent Developments in Understanding Nucleation and Crystallization of Spheroidal Graphite in Iron-Carbon-Silicon Alloys. *Metals* **2020**, *10*, 221. [CrossRef]
72. Howe, J.Y.; Rawn, C.J.; Jones, L.E.; Ow, H. Improved Crystallographic Data for Graphite. *Powder Diffr.* **2003**, *18*, 150–154. [CrossRef]
73. Huh, S.H. X-Ray Diffraction of Multi-Layer Graphenes: Instant Measurement and Determination of the Number of Layers. *Carbon N. Y.* **2014**, *78*, 617–621. [CrossRef]
74. Ratih, D.; Siburian, R.; Andriayani. The Performance of Graphite/n-Graphene and Graphene/n-Graphene as Electrode in Primary Cell Batteries. *Rasayan J. Chem.* **2018**, *11*, 1649–1656. [CrossRef]
75. Kumar, V.; Kumar, A.; Lee, D.; Park, S. Estimation of Number of Graphene Layers Using Different Methods: A Focused Review. *Mater. Rev.* **2021**, *14*, 4590. [CrossRef]
76. Martínez-González, J.; Delfino, R.C.; Vigueras Santiago, E.; Garcia Orozco, I. Chapter 3 Determining Crystallite Size and Density of Recovered Graphite Defects from New and Used Zn-C Batteries. In *Handbook T-II Ingeniería y Ciencias Aplicadas*; Chapter 2; ECORFAN-México: Mexico City, Mexico, 2021; Volume 2, pp. 18–27.
77. Cascales, J. Estudio de Capas Nanométricas de H-BN y Grafeno Para Su Apilamiento En Multicapas. Ph.D. Thesis, Universidad Autónoma de Madrid, Madrid, Spain, 2016. Volume 1.
78. Bleu, Y.; Bourquard, F.; Loir, A.S.; Barnier, V.; Garrelie, F.; Donnet, C. Raman Study of the Substrate Influence on Graphene Synthesis Using a Solid Carbon Source via Rapid Thermal Annealing. *J. Raman Spectrosc.* **2019**, *50*, 1630–1641. [CrossRef]
79. Papageorgiou, D.G.; Kinloch, I.A.; Young, R.J. Mechanical Properties of Graphene and Graphene-Based Nanocomposites. *Prog. Mater. Sci.* **2017**, *90*, 75–127. [CrossRef]
80. Saravanan, A.; Kumar, P.S.; Srinivasan, S.; Jeevanantham, S.; Vishnu, M.; Amith, K.V.; Sruthi, R.; Saravanan, R.; Vo, D.V.N. Insights on Synthesis and Applications of Graphene-Based Materials in Wastewater Treatment: A Review. *Chemosphere* **2022**, *298*, 134284. [CrossRef]
81. Chabot, V.; Kim, B.; Sloper, B.; Tzoganakis, C.; Yu, A. High Yield Production and Purification of Few Layer Graphene by Gum Arabic Assisted Physical Sonication. *Sci. Rep.* **2013**, *3*, 1378. [CrossRef]
82. Chaliyawala, H.A.; Narasimman, R.; Pati, R.K.; Mukhopadhyay, I.; Ray, A. Effect of Copper Pretreatment on Optical and Electrical Properties of Camphor-Based Graphene by Chemical Vapour Deposition. *J. Mater. Sci. Mater. Electron.* **2022**, *33*, 8397–8408. [CrossRef]
83. Natarajan, S.; Bajaj, H.C.; Aravindan, V. Template-Free Synthesis of Carbon Hollow Spheres and Reduced Graphene Oxide from Spent Lithium-Ion Batteries towards Efficient Gas Storage. *J. Mater. Chem. A* **2019**, *7*, 3244–3252. [CrossRef]
84. Ren, Z.; Qiu, X.; Wang, B.; Wang, R. Cu Induced Reconstruction of Microporous Structure of Tungstophosphoric Acid Decorated Popcarbon and the Synergistic Effects on NH3-SCR Performance. *J. Environ. Chem. Eng.* **2024**, *12*, 112428. [CrossRef]

85. Aboul-Enein, A.A.; Haggar, A.M.; Awadallah, A.E.; Azab, M.A. Synthesis of Polypropylene Waste-Derived Graphene Sheets in the Presence of Metal Oxide Nanoparticles. *Mater. Chem. Phys.* **2024**, *315*, 129021. [CrossRef]
86. Miao, Y.; Zheng, J.; Liu, Y.; Xiang, N.; Li, Y.; Han, X.; Huang, Z. Relationship Between Oxygen-Containing Groups and Acidity of Graphene Oxide Supported Mn-Based SCR Catalysts and the Effects on the Catalytic Activity. *Catal. Lett.* **2020**, *150*, 3243–3255. [CrossRef]

recycling

MDPI

Article

Graphite Separation from Lithium-Ion Battery Black Mass Using Froth Flotation and Quality Evaluation for Reuse as a Secondary Raw Material Including Non-Battery Applications

Johannes Rieger [1,*], Stephan Stuhr [2], Bettina Rutrecht [1], Stefan Morgenbesser [3], Thomas Nigl [3], Astrid Arnberger [4], Hartwig Kunanz [5] and Stefanie Lesiak [1]

1 K1-MET GmbH, 4020 Linz, Austria; bettina.rutrecht@k1-met.com (B.R.); stefanie.lesiak@k1-met.com (S.L.)
2 UVR-FIA GmbH, 09599 Freiberg, Germany; stuhr@uvr-fia.de
3 Chair of Waste Processing Technology and Waste Management, Montanuniversitaet Leoben, 8700 Leoben, Austria; stefan.morgenbesser@unileoben.ac.at (S.M.); thomas.nigl@unileoben.ac.at (T.N.)
4 Saubermacher Dienstleistungs AG, 8073 Feldkirchen bei Graz, Austria; a.arnberger@saubermacher.at
5 RHI Magnesita GmbH, 8700 Leoben, Austria; hartwig.kunanz@rhimagnesita.com
* Correspondence: johannes.rieger@k1-met.com; Tel.: +43-664-8832-2499

Abstract: This study investigates graphite separation from Lithium-Ion Battery (LIB) black mass (which is a mixture of anode and cathode materials) via froth flotation coupled with an open-loop recycling approach for the graphite (froth) product. Black mass samples originating from different LIB types were used to produce a carbon-poor and a carbon-enriched fractions. The optimization of the flotation parameters was carried out depending on the black mass chemistry, i.e., the number of flotation stages and the dosing of flotation agents. The carbon-enriched product (with a carbon content of 92 wt.%, corresponding to a recovery of 89%) was subsequently used as a secondary carbon source for refractory material (magnesia carbon brick). Analyses of brick chemistry, as well as thermo-mechanic properties in terms of density, porosity, cold crushing strength (CCS), hot modulus of rupture (HMOR—the maximum bending stress that can be applied to a material before it breaks), and thermal conductivity showed no negative influence on brick quality. It could be demonstrated that flotation graphite can principally be used as a secondary source for non-battery applications. This is a highly valuable example that contributes to a more complete closure of a battery's life cycle in terms of circular economy.

Keywords: lithium-ion batteries; black mass; froth flotation; graphite; recycling; refractory industry; open-loop recycling; circular economy; critical raw materials

1. Introduction

Currently, graphite is the most widely used anode material for Lithium-Ion Batteries (LIBs). Its low electrochemical potential, low cost, low toxicity, high energy density (high capacity with a low de-/lithiation potential), and very long life cycle make it ideally suited for a variety of applications, such as batteries for devices, transportation, and grid-based storage [1–3]. Within an LIB, the graphite anode is the negative electrode that is responsible for storing and releasing electrons during the charging and discharging process; a typical electrical vehicle battery contains around 50–100 kg of graphite [4].

Graphite is of high economic importance not only for electric vehicles but also for other industries (e.g., refractory materials or bearings) and due to its global availability, graphite is defined as a critical raw material by the European Commission [5]. Therefore,

https://doi.org/10.3390/recycling10020075

the recovery of graphite from spent LIBs is an essential aspect to save primary graphite resources and to close material cycles with regard to LIB recycling.

Reusing recovered graphite from LIBs can generally follow a closed-loop or an open-loop approach. A closed-loop approach means the direct reuse of the graphite for the anode of an LIB. Although closed-loop recycling is obvious and there are enhanced levels of recycling targets set by the European Parliament within the latest regulation amendment concerning batteries and waste batteries (Regulation EU 2023/1542 [6]), it is difficult to meet the strict quality requirements for LIB-grade graphite with the current methods. For this reason, a significant amount of research efforts is aimed at open-loop recycling as a feasible and necessary alternative to closed-loop recycling. An open-loop approach includes the use of secondary graphite for other purposes. An exemplary open-loop approach is the use of graphite for organic wastewater treatment. In this case, graphite (or generally carbon) can support the catalytic degradation of pollutants; however, the mechanism of its influence on preparation and catalysis has not been studied in depth at present [7].

The mechanical recycling processes of graphite from active material (also called black mass, which is a fine-grained fraction originating from mechanical–thermal end-of-life battery treatment processes) represent one possibility to recover graphite. The centrifugal fractionation of an aqueous anode slurry using a decanter centrifuge containing carbon black mass and graphite to separate the graphite from the carbon black mass revealed a graphite recovery level of up to 90% [8]. However, this technology was not applied to an actual black mass slurry from end-of-life LIBs to separate the carbon black mass (graphite). Another mechanically based approach to reuse graphite from spent LIBs is directly recycling LIB electrode materials. Electrodes are comminuted in a cutting mill down to ~20 mm pieces and are put into a stirred vessel for a solvent-based recovery, whereas water is used as a solvent for the anode. A graphite recovery rate of 96% was found in small-scale tests (50 g batch per trial). A sieved fraction (500 μm grain size) was used for electrode production and was tested regarding its electrochemical performance. Regarding the cell performance of the LIBs produced, cells with a recyclate share of 10% achieved similar performances to the respective reference cells without recyclate [9].

The coupling of mechanical and pyrometallurgical treatments of spent LIBs focuses on recovering electrode materials including graphite. An LIB discharged in a 15% sodium chloride solution and disassembled after it was air-dried was treated in a pyrolysis reactor at 650 °C under a nitrogen atmosphere, whereas the anode and cathode were processed separately. The pyrolyzed electrode plates were ground in a ball mill and were subsequently sieved. Analyses revealed an anode graphite recovery rate of 86.9% [10]. It was also determined that the reductive pyrolysis gases reduced the chemical valences of the recovered metals. In contrast, the graphitization degree of the recovered graphite decreased dramatically because of the presence of pyrolysis char [10]. This reduced graphitization degree may also affect the electrochemical performance of the graphite when reusing it in an LIB, which was not in the scope of the study by Zhao et al. [10].

Additionally, the recycling of carbon from other industrial sectors for possible reuse in LIBs has been explored [11]. Exemplarily, dust from a blast furnace cast house (dust generated during hot metal tapping from the blast furnace) was purified by applying a flotation–acid leaching treatment process, where a mixture of hydrochloric acid and hydrofluoric acid (HCl-HF) was used. A maximum graphite content of 95.6% was reached in a separate concentrate. The use of the recovered graphite in an LIB anode was also studied, revealing a high reversible capacity of ~370 $mAh \cdot g^{-1}$ and a coulombic efficiency of 99.6% after 350 cycles [11]. Although promising electrochemical properties were demonstrated, blast furnace cast house dust availability is somewhat limited on the market since it is mainly internally recycled within an integrated steel plant (reuse in the sinter plant or the blast furnace [12]).

A closed-loop approach was explored by treating spent graphite separated in a lab-scale microwave oven to remove impurities and to expand the graphite interlayers, whereas puffed graphite was produced. Accompanied treatment was performed with $K_2S_2O_8$ and H_3PO_4 to support the interlayer expansion. This method of graphite processing aimed to provide a graphite quality that is suitable for anode production. In a further step of the study conducted by Jia et al. [13], $FeCl_3$ 6 H_2O was added to the puffed graphite in an ultrasonic bath to integrate Fe_2O_3 into the carbon layers, forming a sandwich composite with a high Lithium storage capacity. Electrochemical investigations revealed a specific capacity of 1100 mAh g^{-1} at 200 mA g^{-1}, delivering a stable performance for ~500 cycles. The potential of this technology in larger-scale trials has not yet been proven. Another research approach related to direct graphite recycling separated the graphite via washing in water and dimethyl carbonate followed by thermal treatment for graphite purification and re-graphitization. A total of 53% of the initial graphite was recovered [14]. Electrochemical properties were also tested, showing a battery capacity of 80% after 170 continuous charge/discharge cycles [14]. No validation on a larger scale was carried out within this study. Apart from LIBs, Sodium-Ion Batteries (SIBs) are considered as an alternative to traditional LIBs due to their potential advantages, such as abundance, the uniform geographical distribution of sodium, and low cost per kWh [15]. Related research was conducted to quantify the reuse of spent graphite separated via treatment in a dimethyl formamide bath before being ground. A SIB was produced using the recovered graphite as the anode and a carbon-coated $Na_3V_2(PO_4)$ as the cathode. Electrochemical properties were investigated using voltammetry and cyclic charging/discharging, whereas a stable performance was found for a duration of 300 cycles [15].

To recover graphite from spent LIBs, hydrometallurgical treatment (acid leaching using HCl or sulfuric acid—H_2SO_4), heat treatment (pyrolysis), or water leaching, as well as flotation, are known exemplary techniques [7]. Leaching, grinding, and pyrolysis can be used as supporting technologies for flotation processes. Froth flotation using the wettability properties of material surfaces as a basis is also a known flotation technique to separate graphite from the black mass [16]. For leaching-assisted flotation, a postprocessing of the leaching solution is important since soluble Lithium salt is being transferred into the leaching solution; however, a graphite purity of 84% was reported in the literature [7]. In recent years, froth flotation has been applied to the <100 μm black mass fraction to obtain a high-metal-grade product before hydrometallurgical treatment. Some authors have also proposed the use of froth flotation to directly recover lithium metal oxide particles enabling high material recovery from black mass when combining flotation and hydrometallurgy. Recovering graphite together with metal-rich fractions from black mass increases considerably the overall recovery efficiency during LIBs recycling as graphite represents 14–22 wt.% of a LIB [17]. The use of frothing agents, such as Methyl Isobutyl Carbinol (further referred as MIBC), increases the hydrophobicity of graphite particles supporting one advantage of the flotation technique for enhanced graphite separation. Further advantage of froth flotation is lower energy demand compared to heat treatment (pyrolysis) for graphite separation. Studies exist using synthetically produced black mass (mixing of pure NMC (Nickel–Manganese–Cobalt Oxide) cathode and anode materials) together with a mixture of MIBC and kerosene as collector [17]. Similar results were found out in other studies [18]; however, synthetically mixed black mass was used in many other research works, and not black mass from real industrial LIB processes.

Closed-loop recycling approaches reported for the black mass coming from spent LIBs in the literature regarding the reuse of secondary graphite from flotation for battery production are often hindered by the quality of the derived flotation product. The graphite for LIB anodes should meet the highest quality standards.

While previous studies have demonstrated the feasibility of graphite recovery from spent LIBs using various mechanical, hydrometallurgical, and pyrometallurgical approaches, there is limited research on the scalability and optimization of froth flotation for graphite recovery. Whilst closed-loop recycling practices might be most favorable, certain shares of the other elements, e.g., Ni, Co, Mn, Al, and Cu, are transferred into the carbon-rich froth product. This may hinder the direct reuse of the flotation graphite for battery anodes, which gives way to the potential for open-loop recycling applications, such as the use of recovered graphite in refractory materials; this remains underexplored, highlighting the need for comprehensive studies that bridge this knowledge gap. This challenge of transferring metals into the graphite-rich fraction was also discovered by studies in which industrially produced black mass was used in a froth flotation together with ESCAID^{TM} (hydrocarbon fluid from Exxon Mobil) as collector and MIBC as frother. The resulting graphite product comprised metals from the black mass feed inducing the need of a chemical post-processing (reaction with caustic soda at ~500 °C) to derive a graphite product being useable for LIB again [19]. There is a lack of research work to better understand the behavior of graphite separated from froth flotation in non-battery applications. Additionally, it was not yet clearly demonstrated how the flotation graphite can be used without any other subsequent treatment, such as hydroleaching, or another physical treatment e.g., with caustic soda, at higher temperatures. Therefore, the focus of this study is to evaluate the open-loop recycling possibilities of the graphite separated from LIB black mass (a mixture of anode and cathode materials) in refractory applications without any other treatment than froth flotation. In the current study, froth flotation is applied to black mass samples from different LIB types to produce a carbon-poor and a carbon-enriched fraction and to vary flotation parameters and flotation reagents to optimize the graphite fraction purity (i.e., the flotation efficiency). Froth flotation is a wet mechanical sorting process in which different solid–fine disperse materials are separated due to their different surface wettability. The most hydrophobic materials attach to rising air bubbles in the aqueous media and are recovered in the froth product, whereas hydrophilic materials remain in the pulp (cell product) [20,21]. The graphite contained in the black mass is naturally hydrophobic; therefore, it is expected to recover it in the froth product [22]. The sorting can be enhanced by the addition of reagents (surfactants). Reagents, which enhance the natural hydrophobic surface of the graphite, such as kerosene, diesel or n-dodecan, are called collectors. With other reagents, the froth is stabilized for process control [23]. Sometimes dispersants are added to prevent undesired aggregates [24]. For the refractory industry, carbon represents an important raw material to produce magnesia carbon (MgO-C) bricks. Such bricks are used for the steel industry in metallurgical aggregates, such as for crude steelmaking using a Basic Oxygen Furnace (BOF) or an Electric Arc Furnace (EAF), as well as in metallurgical ladles that are used for crude steel refining (secondary metallurgical operation). The carbon-enriched product is evaluated in terms of its use as a secondary carbon source for a refractory material (magnesia carbon bricks, which are used in the iron and steel industry).

Finally, an outlook also demonstrates the required use of carbon sources for future steelmaking, even after the ongoing transformation to low-carbon steelmaking processes is complete.

2. Materials and Methods

Five black mass samples, comprising a mixture of anode and cathode materials, from a mechanical recycling process of thermally pre-treated end-of-life LIBs were provided by an industrial project partner and used for the froth flotation tests [25]. The sample nomenclature is as follows:

- NMC (Nickel–Manganese–Cobalt Oxide) cathode material + graphite anode material with Copper (Cu) and Aluminum (Al) as conductive foils, whereas two different black mass batches used (denoted as NMC 1 and NMC 2).
- LFP (Lithium–Iron–Phosphate) cathode material + graphite anode material with Cu and Al as conductive foils (denoted as LFP).
- LIBs from power tools, pedelecs, and e-bikes, as well as a mixture of NMC and LCO (Lithium–Cobalt–Oxide), cathode materials + graphite anode material with Cu and Al as conductive foils (denoted as PSP).
- LIBs from mobile phones and laptops, as well as a mixture of NMC and LCO, cathode materials + graphite anode material with Cu and Al as conductive foils (denoted as HL).

2.1. Material Characterization

The characterization of the black mass mainly comprised particle size distribution, evaluation, chemical analysis, and density measurements. The particle size distribution was determined using a dry ultrasonic-assisted sieve analysis according to DIN ISO 3310-1. X-ray diffraction (XRD, Siemens D5000 system, Siemens AG, Munich/Germany), and X-ray fluorescence (XRF, NITON XL3t 980, Thermo Fishher Inc., Munich/Germany) as well as a carbon analyzer (LECO CS744, combustion method) were used for qualitative chemical analysis and quantitative elemental analysis, respectively. The latter carbon analyzer was also used to quantify the carbon contents in the flotation products. Finally, density was determined by applying helium pycnometry according to DIN 66137 with a Multivolume Pycnometer from Micromeritics.

2.1.1. Particle Size Distribution (PSD)

Figure 1 shows the particle size distributions of the five black mass samples. The particle size distribution shows that approx. 90% of the NMC 1, LFP, and PSP samples are <45 μm. The HL and NMC 2 samples are coarser, with 90% < 90 μm.

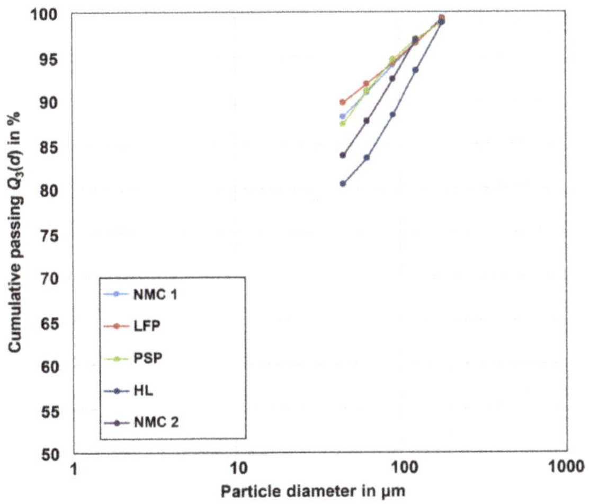

Figure 1. Particle size distributions of the black mass samples derived from sieve analyses.

2.1.2. Chemical Composition and Density

From the investigation of the element distribution in the particle size fractions, it was found that there is an enrichment for Al and Cu in fractions >45 μm. For each of the 5 black mass samples, the share of Al in fractions >45 μm was approx. 32–49% and the share of copper was approx. 36–60%. The carbon share for all the samples >45 μm was approx. 1–4%. A minor recovery of 4–7% nickel, cobalt, and manganese (Ni+Co+Mn) in fractions >45 μm was found for NMC1, NMC2, PSP, and HL samples. For the LFP sample, approx. 4% of Iron (Fe) was recovered in fractions >45 μm. In summary, the majority (>90–95%) of the metal content in the cathode active materials, as well as in the anode active materials, accumulates in the fractions <45 μm. In contrast, the residues of current collector foils, consisting of Al and Cu, are enriched in fractions >45 μm. For this reason, the samples were sieved with a mesh size of 45 μm prior to flotation in order to reduce the aluminum and copper content for the flotation feed.

Table 1 lists the chemical composition and the density of the black mass samples used for the investigations with particle size <45 μm. The main chemical elements underline the cathode materials that were present in the samples (e.g., PSP and HL black mass samples mainly contain NMC, with LCO as the second most abundant cathode active material). From the XRD measurements, it can also be seen that during thermal treatment, active materials NMC and LCO were reduced to NiO, MnO, and CoO, or were even further reduced to metallic Co and Ni.

Table 1. Chemical analysis (extraction showing some of the elements) and density of the black mass samples with particle size <45 μm.

Property	Unit	NMC 1	NMC 2	LFP	PSP	HL
Nickel + Cobalt + Manganese (Ni+Co+Mn)	wt.%	38.0	28.4	0.6	35.5	41.0
Iron + Phosphorus (Fe+P)	wt.%	1.8	1.7	33.3	3.9	1.3
Aluminum + Copper (Al+Cu)	wt.%	8.3	19.3	8.5	9.8	10.4
Carbon	wt.%	37.6	n.a. *	32.6	33.9	36.5
Density	g/cm^3	3.1	n.a. *	2.8	3.1	3.4

* not explicitly analyzed.

2.2. *Froth Flotation for Graphite Separation*

A laboratory Denver D-12 mechanical flotation machine (Metso AG, former Denver Ltd., Sparks, NV, USA) with up to 2.4 L of cell volume (see Figure 2) was used for the froth flotation tests. At the beginning of the test, the dry sample was dispersed in deionized water, with a resulting solid content of 200 g solids per liter of suspension. Then, the conditioning of the reagents was carried out with a rotational speed of 2000 rpm. The conditioning times for the dispersant were 10 min and 5 min for the collector and 1 min for the frother. The same conditioning times and rotational speed were used for subsequent reagent doses. After the first conditioning, deionized water was added to the suspension to achieve a solid content of 100 g L^{-1} for the froth flotation. During froth flotation, a rotational speed of 1000 rpm was applied. Regarding the air flow rate, 1.5 L min^{-1} was used for the 1.2 L cell, and 2 L min^{-1} was used for the 2.4 L cell. All recovered products were dewatered via vacuum filtration and were subsequently dried at approx. 105 °C. Finally, the dried products were deagglomerated and homogenized with a mortar and pestle for the following sample splitting and analysis. Two primary goals should be achieved during the tests, which are the production of a carbon-rich fraction (a high-carbon product, which is denoted as the froth product since the carbon is expected to accumulate in the froth) and a carbon-poor fraction (a low-carbon product, which is denoted as the cell product).

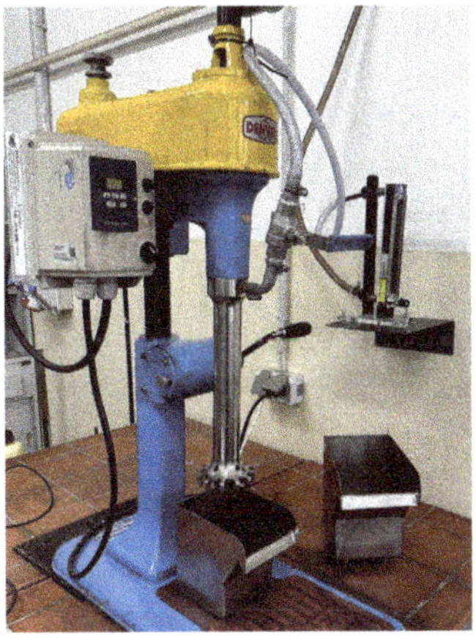

Figure 2. Laboratory Denver D-12 flotation machine (with 1.2 and 2.4 L cells).

The flotation tests are performed according to the general flotation scheme shown in Figure 3 to achieve a low-carbon product and a high-carbon product. The flotation scheme consists of different flotation stages—the rougher, scavenger, and cleaner stages. The test procedure is defined in a way to reach the desired froth and cell products. The upper part of Figure 3 represents the process steps for the low-carbon (Low-C) product, whereas the lower part depicts the process steps for the carbon-rich (High-C) product.

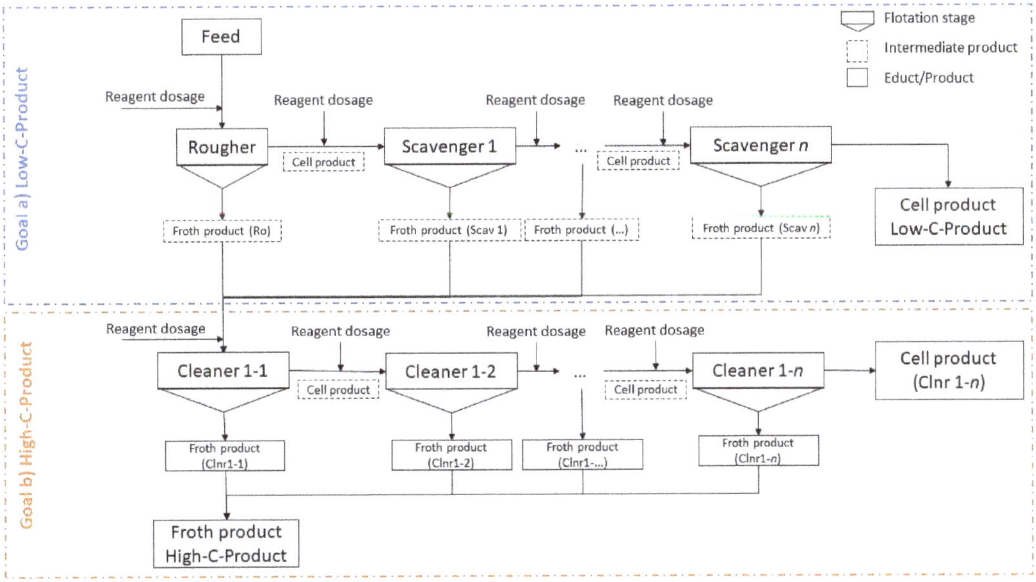

Figure 3. General scheme of the performed flotation tests for graphite separation.

To achieve a low-carbon product (the cell product), the samples are subjected to one rougher and several scavenger stages until there is no further froth formation. After these flotation stages, several intermediate froth products (carbon-enriched) and one cell product (Low-C) are obtained. With the froth products from the rougher and scavenger stages, a cleaner flotation is performed in sub-stages. The obtained froth products subsequently form the final high-carbon (High-C) froth product. The cell products derived from the cleaner stages are considered as intermediate products. As shown in Figure 3, flotation reagents are added at different steps during the flotation process. Based on findings from a prior study, kerosene and diesel were used as collectors, while FLOTANOL™ 7026 from Clariant (Muttenz, Switzerland) (referred to as pine oil) and Methyl–Isobutyl–Carbinol (referred to as MIBC) from Merck (Darmstadt, Germany) were used as frothers [26–30]. A ligninsulfonate from Borregaard (Sarpsborg, Norway) (Pionera™ DP-750) was applied as the dispersant.

To evaluate the use of the flotation graphite as a secondary raw material in refractory products representing an open-loop recycling approach for the flotation graphite, a Magnesia Carbon brick was chosen in this study (MgO-C). Two MgO-C bricks were compared, one standard MgO-C brick mixture (low-carbon MgO-C brick with ~3 wt.% carbon) and a mixture in which 1 wt.% of the carbon was substituted by the flotation graphite. Results of material analyses in terms of thermal and mechanical strength are discussed below (Section 3.3).

3. Results

In the following section, the results, especially in relation to the qualities of the froth product (carbon-rich fraction) and the cell product (carbon-poor fraction) obtained, are analyzed and compared.

3.1. Flotation Results for the NMC, PSP, and LFP Black Mass

Figure 4 shows an exemplary illustration of the flotation test results for the NMC 1 and NMC 2 black mass samples, illustrated in the form of Mayer upgrading curves. This kind of diagram is a visualization of the separation efficiency that enables the evaluation of the flotation tests without the need to execute an explicit carbon analysis. The corresponding test results in the form of the grade of the final froth product (all froth products together) are given for the best sorting results in Table 2.

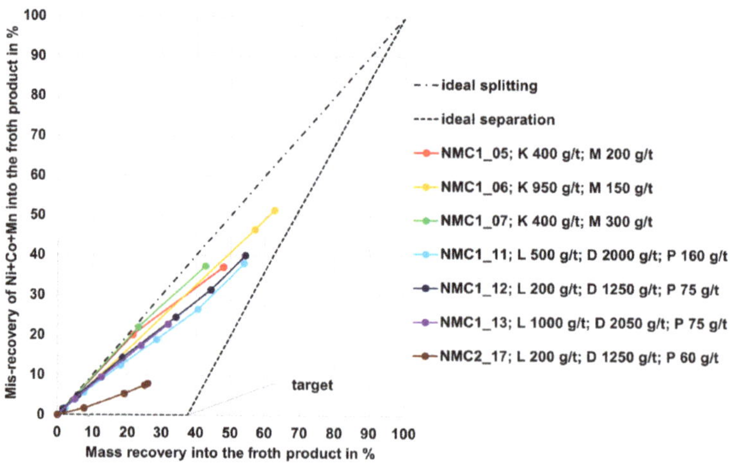

Figure 4. Flotation results for NMC 1 and 2 samples in the form of Mayer upgrading curves. Abbreviations: K—kerosene; D—diesel; M—MIBC; P—Pine oil; L—ligninsulfonate.

In Figure 4, the mis-recovery of Ni+Co+Mn (which is the undesirable recovery of these elements into the froth product) from the NMC samples is represented on the y-axis, while the mass recovery into the froth product is represented on the x-axis. Two conceptual curves are shown in Figure 4. One of them is the ideal splitting (straight dashed-dotted line) representing a separation process without any enrichment of the carbon in the froth product (worst case). The other curve represents the ideal separation (dashed line) and ranges horizontally in the first part up to the optimum mass recovery at a mis-recovery of 0%. The optimum mass recovery corresponds to the carbon grade in the flotation feed (NMC1 37.6 wt.% carbon, see Table 1). This point ("target") represents a total recovery of the carbon in the froth product without any impurities of Ni+Co+Mn (best case). In the second part, the curve runs straight to 100% mis-recovery into the froth product and 100% mass recovery into the froth product. This means that any further mass recovery would lead to a mis-recovery of Ni+Co+Mn. The result is a triangular area between the two conceptual curves. Upgrading curves within this area represent sorting results with an enrichment of carbon in the froth product. The closer the upgrading curve is to the ideal separation and the closer it ends to the target point, the better the sorting result. This means that there is less mis-recovery and therefore the graphite is more selectively enriched and recovered into the froth product. Conversely, this means that upgrading curves close to the ideal splitting mean a poor sorting result. In this case, a higher mis-recovery of Ni+Co+Mn takes place. This means a non-selective enrichment of the graphite in the froth product. Upgrading curves to the left of this triangular area would represent a separation result that corresponds to an enrichment of Ni+Co+Mn in the froth product instead of carbon, whereas upgrading curves to the right are not physically possible. Each data point on the upgrading curves shown in Figure 4 represents a rougher and subsequent scavenger stage (further details stages numbers and reagent dosages can be found in Appendix A, Table A1).

Table 2. Chemical analyses (excerpt of analyzed elements) of the froth and cell products for the NMC and PSP black mass samples (values given in wt.%).

Flotation Test	Froth Product			Cell Product		
	C wt.%	Ni+Co+Mn wt.%	Al+Cu wt.%	C wt.%	Ni+Co+Mn wt.%	Al+Cu wt.%
NMC 1_11 [1]	65.6	27.8	3.5	6.6	53.3	16.9
NMC 1_12 [1]	63.2	29.6	4.0	9.8	52.9	17.4
NMC2_17 [2]	n.a.	8.7	5.2	n.a.	35.4	29.6
PSP_02 [3]	63.0	26.8	3.3	5.3	46.8	17.2

[1] Black mass sample NMC 1; [2] black mass sample NMC 2; [3] black mass sample PSP.

The flotation testing for sample NMC1 can be divided into two different reagent regimes. From Figure 4, the trend can be seen that the upgrading curves for the regime that uses ligninsulfonate as dispersant, diesel as collector, and pine oil as frother (further referred as regime LDP) are underneath the upgrading curves for the regime that uses kerosene as collector and MIBC as frother (further referred as regime KM). From this, it can be concluded that the sorting results for regime LDP are better. However, with a mass recovery between 42 and 62% and a mis-recovery of Ni+Co+Mn between 37 and 51%, all upgrading curves for sample NMC1 exceed the optimum mass recovery of the "target point" (37.6%) and are close to the ideal splitting and therefore represent an inadequate sorting result. Nevertheless, it can be stated that the variation of the reagent dosage in regime KM has nearly no influence on the sorting result (in terms of selective enrichment of carbon). For the LDP regime, a higher dosage of lignosulfonate requires an increased dosage of diesel and pine oil. However, the upgrading curves are almost on top of each other in this case, which means that the sorting result (selective enrichment of carbon) is not affected. The upgrading curve for sample NMC2 with reagent regime LDP is clearly below that of sample NMC1 and thus shows the most selective enrichment of carbon in the froth product.

The exemplary chosen grades of the froth and the cell products reflect the findings from Figure 4 in absolute values (see Table 2). The analyses shown here only include the elements of main interest (C, Co, Ni, Mn, Al, Cu). The high mis-recovery of Ni+Co+Mn into the froth product (approx. 40%) for sample NMC 1 corresponds to a grade of approx. 28 to 30 wt.%. As a result, the carbon grade is only enriched to approx. 63 to 65 wt.%. Sample NMC 2 shows a more selectively enriched froth product since the grade of Ni+Co+Mn is only at approx. 9 wt.%. Al+Cu are enriched in the cell product and the final carbon grade in the cell product account in the range between approx. 7 and 10 wt.%. Tests with the LFP sample are unsuccessful since the sample is non-dispersible in water.

For the PSP black mass sample (mixture of graphite anode material and NMC as well as LCO cathode material), the flotation results are given in Figure 5 as MAYER upgrading curves. For the tests, the reagent regime LDP is applied with different dosages and numbers of flotation stages. The resulting upgrading curves are quite similar to those of sample NMC 1. With a mass recovery between 38 and 51% and a mis-recovery of Ni+Co+Mn between 27 and 41%, all curves exceed the optimum mass recovery of 33.9%. Thus, these sorting results are also inadequate regarding the selective enrichment of carbon in the froth product. The absolute values in the form of the grade of the final froth product are given in Table 2 (last row), which are comparable to those from sample NMC 1.

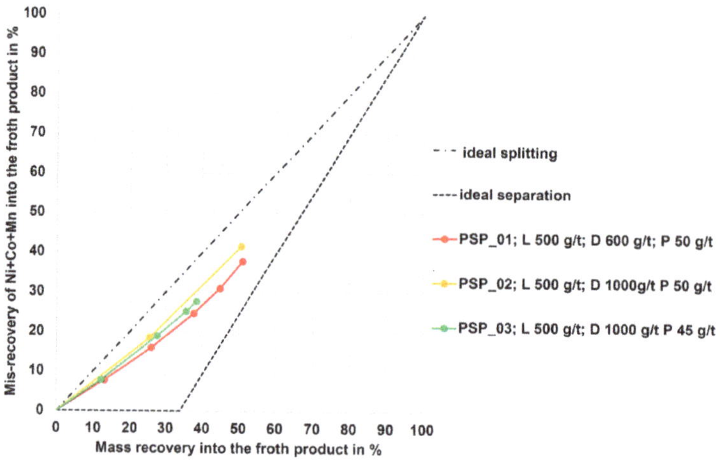

Figure 5. Flotation results for the PSP sample in the form of Mayer upgrading curves. Abbreviations: D—diesel; P—pine oil; L—ligninsulfonate.

3.2. Flotation Results for the HL Black Mass

Table 3 summarizes some exemplary compositions of the froth and cell products for the HL black mass samples (mixture of graphite anode material and NMC as well LCO as cathode material). All flotation test results are shown as MAYER upgrading curves in Figure 6. With a mis-recovery between 15 and 20% for Ni+Co+Mn into the froth product with a corresponding mass recovery into the froth product between 34 and 45% the upgrading curves are close to the ideal separation curve, and they end near the "target point". This shows an acceptable selective enrichment of carbon in the froth product. The use of ligninsulfonate as a dispersant shows that a lower misrecovery and thus a more selective enrichment of carbon is achieved (upgrading curve HL_07 is above the others). Furthermore, the use of pine oil instead of MIBC as a frother shows that the same sorting result is achieved, whereby the dosage of collector and frother can be reduced in total (comparison HL_01). Further optimization of the reagent regime shows no improvement in the sorting result. However, it can also be determined for sample HL that a lower dosage of lignosulfonate leads to a decrease in the required diesel and pine oil dosage.

Table 3. Chemical analyses (excerpt of the analyzed elements) of the froth and cell products for the HL black mass samples (values given in wt.%).

Flotation Test	Froth Product			Cell Product		
	C wt.%	Ni+Co+Mn wt.%	Al+Cu wt.%	C wt.%	Ni+Co+Mn wt.%	Al+Cu wt.%
HL_02	84.8	14.7	3.6	3.6	60.3	22.1
HL_03	86.0	14.4	3.6	4.7.	59.0	21.0

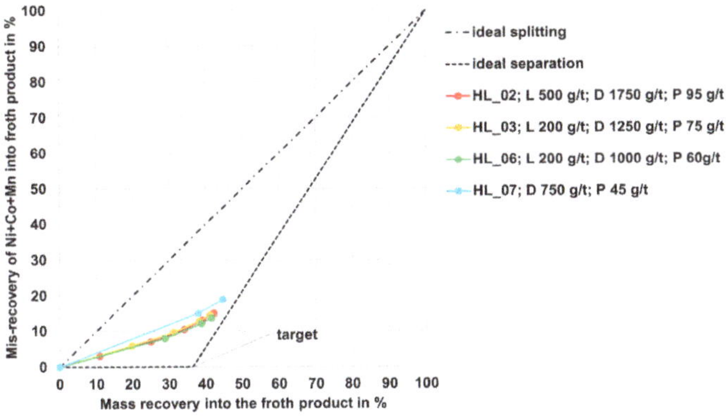

Figure 6. Flotation results for the HL sample in the form of Mayer upgrading curves. Abbreviations: D—diesel; P—pine oil; L—ligninsulfonate.

Froth products with a carbon content of approx. 85 wt.% are thus achieved, which contain impurities of approx. 15 wt.% Ni+Co+Mn.

Based on the findings from the rougher and scavenger stages, a test with a cleaner stage was performed. The test parameters and results can be found in Table 4. With the rougher stage and two scavenger stages, a carbon recovery of 94% with a carbon grade of 86 wt.% could be achieved in the froth product (see columns "Carbon grade" and "Carbon recovery" in the line "Ro-Scav2-Froth" in Table 4). A further improvement in the carbon grade was achievable with the cleaner flotation stage ranging from 94 to 91 wt.%, with recovery from 63 to 89% (see columns "Carbon grade" and "Carbon recovery" in lines "Clnr1-Froth1" to "Cnlr1-Froth4" in Table 4). In the resulting high-carbon product, the main impurities are Ni, Co, Mn, Al, and Cu. The final high-carbon product (denoted as "Clnr1-Froth4" in Table 4) was used as a secondary carbon source for refractory material tests (see Section 3.3).

Table 4. Cleaner flotation results for the HL sample based on the main components. Abbreviations: Ro—rougher; Scav—scavenger; Clnr—cleaner; L—lignisulfonate; D—diesel; P—pine oil. The values of the mass recovery, carbon grade, and carbon recovery for the froth products gathered from the cleaner stages Clnr1-Froth1 to Clnr4-Froth4 are given in cumulative form.

Stage-Product	Reagent Dosage $g\,t^{-1}$			Mass Recovery	Carbon Grade	Carbon Recovery
	L	D	P	%	wt.%	%
Ro-Feed	200			100	37.9	100
Ro-Scav2-Froth		1000	60	41.3	86.3	94.1
Scav2-Cell				58.7	3.8	5.9

Table 4. *Cont.*

Stage-Product	Reagent Dosage g t^{-1}			Mass Recovery	Carbon Grade	Carbon Recovery
	L	D	P	%	wt.%	%
Clnr1-Feed	100			41.3	86.3	94.1
Clnr1-Froth1		125	7.5	15	94.2	37.3
Clnr1-Froth2		100	7.5	25.5	94.0	63.3
Clnr1-Froth3		100	7.5	31.7	93.4	78.3
Clnr1-Froth4		100	7.5	36.7	91.9	89.0
Clnr1-Cell				4.64	42.3	5.2

3.3. Evaluation of the Flotation Graphite as a Secondary Raw Material for Refractory Production

As described and illustrated in Sections 3.1 and 3.2, certain shares of the other elements, Ni, Co, Mn, Al, and Cu, are transferred into the carbon-rich froth product. This may hinder a direct reuse of the flotation graphite for battery anodes. Open-loop recycling practices contribute towards higher recycling rates of LIB fractions. For the refractory industry, carbon represents an important raw material to produce magnesia carbon (MgO-C) bricks. Such bricks are used for the steel industry in metallurgical aggregates, such as for crude steelmaking via a Basic Oxygen Furnace (BOF) or an Electric Arc Furnace (EAF), as well as in metallurgical ladles that are used for crude steel refining (as a secondary metallurgical operation). To evaluate the use of the flotation graphite as a secondary raw material in MgO-C brick products, two MgO-C bricks were compared—one standard MgO-C brick mixture (low-carbon MgO-C brick with ~3 wt.% carbon) and a mixture in which 1 wt.% of the carbon was substituted by flotation graphite. The addition of carbon black, and in this case the flotation graphite, is aiming a porosity reduction in the MgO-C brick. A typical amount for the pore filling materials is 1 wt.%. Adding more of this ultra-fine material (soot fraction as standard source) induces negative production-related issues. This is why 1 wt.% was selected as the ratio for the flotation graphite substitution and not the whole 3 wt.%. Figure 7 shows a cut section of the two studied MgO-C bricks (A: standard refractory brick with 3 wt.% carbon; B: refractory brick with 1 wt.% recycled graphite content). Both samples show a flawless microstructure with an overall homogeneous distribution of the components, which is an important indicator that flotation graphite does not negatively affect the brick structure.

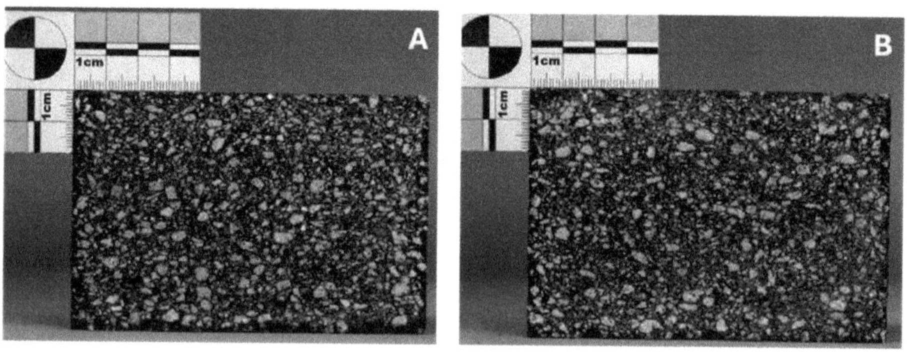

Figure 7. Low-carbon-containing MgO-C brick with standard carbon mixture (**A**); flotation graphite partially substituting the soot fraction (**B**).

Certain relevant properties were determined via chemical analysis and physical parameters, such as density, porosity, cold crushing strength (CCS), hot modulus of rupture

(HMOR—the maximum bending stress that can be applied to that material before it breaks), thermal conductivity, or hot relaxation. Figure 8 shows a comparison of the thermal conductivity for both tested bricks. Thermal conductivity is slightly higher for the brick with flotation graphite (e.g., ~8 W m^{-1} K^{-1} at 600 °C for the brick with the flotation graphite compared to ~7 W m^{-1} K^{-1} for the standard brick) but still within an acceptable range according to the expertise of refractory producers. No further material properties were analyzed since the ones mentioned here represent the main important parameters to compare the principal quality of refractory bricks independent of the raw materials used. From refractory producer side, it can therefore be considered that the brick comprising flotation graphite seems to be mechanically and thermally stable from current state of knowledge.

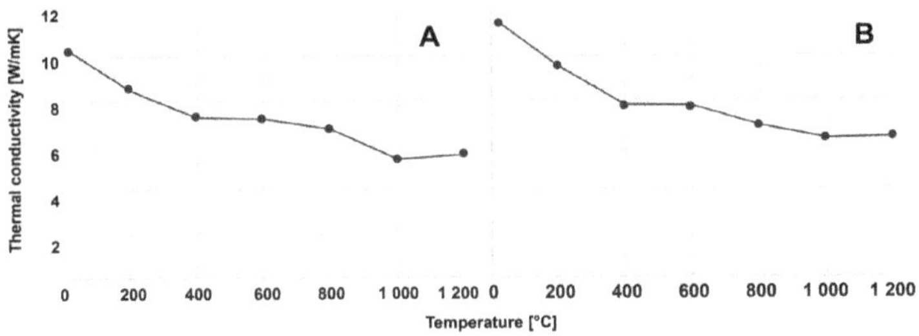

Figure 8. Thermal conductivity of a standard MgO-C brick (**A**) and a brick with flotation graphite (**B**).

Table 5 shows the trends in some of the physical parameters when partially substituting standard carbon (fine-grained pore filling soot fraction) with flotation graphite.

Table 5. Comparison of physical parameters for a MgO-C brick when using flotation graphite instead of the standard carbon fine.

Parameter	**Standard MgO-C Brick**	**MgO-C Brick with 1 wt.% Flotation Graphite**
	Before coking at 1500 °C	
Bulk density [g/cm^3]	3.16	3.17
Apparent porosity [$\%_{vol}$]	5.7	5.9
CCS [MPa]	69	59
HMOR 1400 °C [MPa]	7.8	6.9
HMOR 1500 °C [MPa]	5.1	5.0
	After coking at 1500 °C	
Bulk density [g/cm^3]	3.12	3.13
Apparent porosity [$\%_{vol}$]	10.0%	9.9
CCS [MPa]	38%	44

Although some mechanical parameters that are important for brick strength are lower when using flotation graphite (cf. CCS and HMOR), the values for the coked sample (coking at 1500 °C) lead to the assumption that the recycling graphite from the froth flotation is principally suitable as a carbon source for MgO-C bricks.

4. Discussion

As shown in Section 3, a certain degree of selectivity is reached when applying froth flotation to black mass samples. However, it was also found that not all black mass samples, especially when considering the cathode material fractions, show the same level of selectivity, which leads to a mis-recovery of undesired elements (i.e., Co, Ni, Mn, Al, and Cu) into the carbon-rich flotation product.

4.1. Comparison of Black Mass Types According to the Carbon Content

An extended carbon determination was carried out for the most optimum test results obtained so far. Therefore, specific test results are shown in the form of FUERSTENAU-2 upgrading curves in Figure 9.

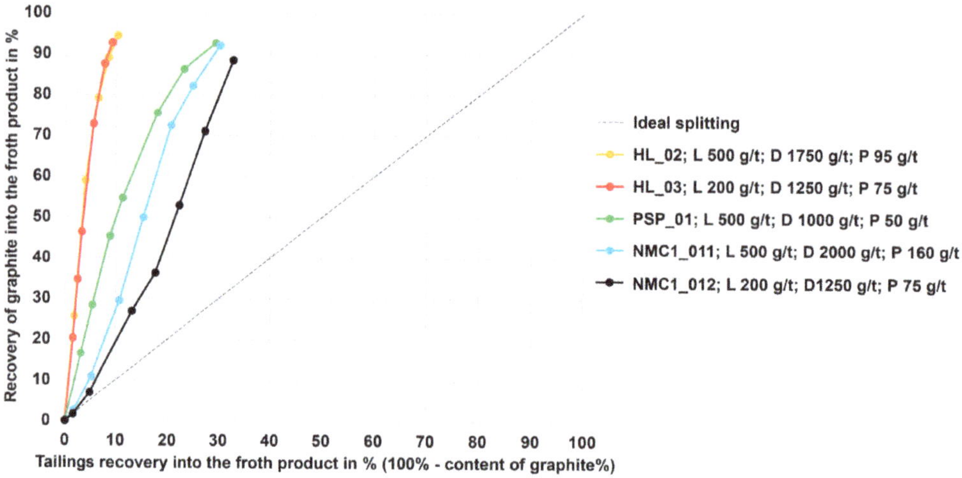

Figure 9. Carbon recovery in the froth product (FUERSTENAU-2 upgrading curves) for the different black mass types; abbreviations of dosing reagents and their quantities used are also mentioned (D—diesel, P—pine oil, L—ligninsulfonate).

In Figure 9, the recovery of carbon into the froth product is shown on the y-axis, while the recovery of tailings (i.e., every other element except carbon) into the froth product is shown on the x-axis. There are two conceptual curves; the first is a diagonal dashed line depicting ideal splitting with the meaning of no selective carbon enrichment in the froth product (worst case). The other ideal curve is represented by the y-axis, depicting an ideal separation (best case). Upgrading curves inside of the triangular area between those two conceptual curves have a selective enrichment of carbon into the froth product. This is because the recovery of carbon is higher than the recovery of the tailings. For selective carbon separation, the aim is that the end of the upgrading curves is near the left upper corner.

The results reveal that each black mass sample needs a different reagent regime and flotation scheme. High carbon recovery rates in the froth product of 90% or even higher are reached for all black mass samples (except the LFP black mass sample as it could not be completely dispersed in water; see Section 3.1). The NMC and PSP samples show a similar tailings recovery rate of ~30%. The HL sample shows the lowest undesired recovery of non-carbon elements in the froth product (~10%), representing the best flotation behavior in terms of a selective carbon enrichment into the froth product. Figure 10 illustrates the carbon recovery in the froth product, depending on the corresponding carbon content (HALBICH upgrading curves).

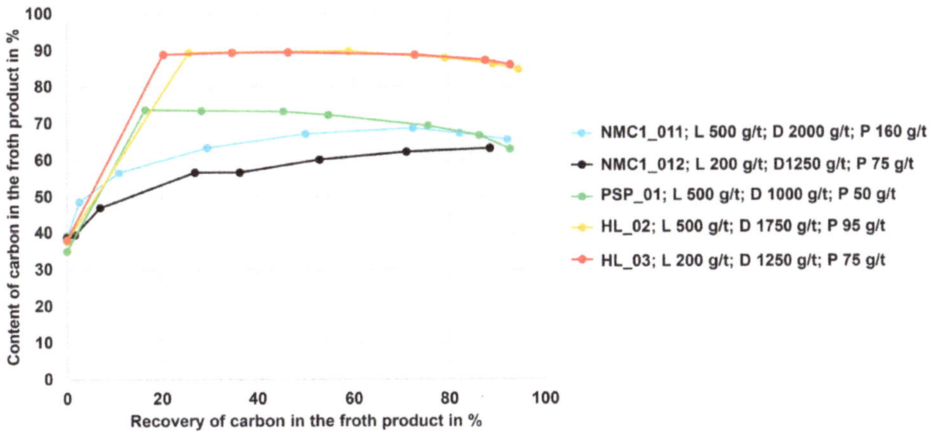

Figure 10. Carbon enrichment (HALBICH upgrading curves; carbon recovery rate versus carbon content) in the froth product of the black mass samples used.

The y-axis shows the carbon content in the froth product, whereas the carbon recovery into the froth product is displayed on the x-axis. The curve begins with a recovery of 0%, representing the feed material (corresponding to an initial carbon content in the black mass samples of ~40 wt.%). The curves must be interpreted from left to right, where each data point on the curve represents an intermediate froth product during the flotation scheme (i.e., after every rougher and scavenger stage). It is evident that the HL black mass sample shows the best separation selectivity for the chosen flotation parameters and reagents. The final carbon contents in the froth product of ~90 wt.% are reached. The PSP black mass shows the second best flotation selectivity, followed by the NMC samples (NMC 1 and NMC 2).

It is also relevant to have a closer look into the quality of the cell product to evaluate flotation efficiency. Therefore, Figure 11 illustrates the flotation results in the form of HALBICH upgrading curves.

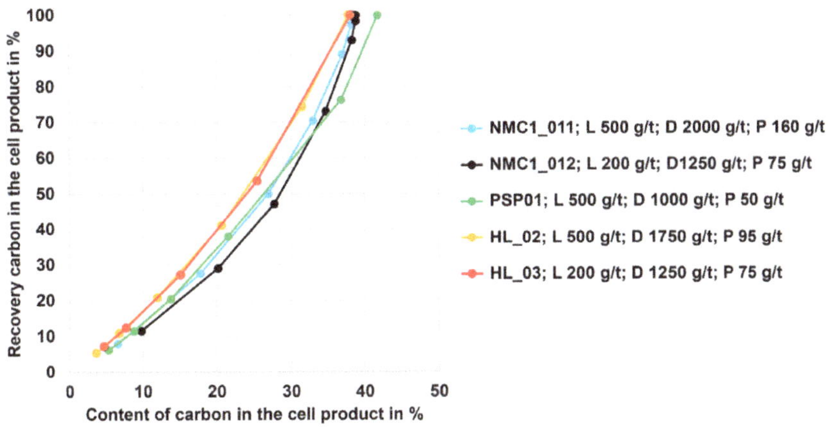

Figure 11. Carbon recovery and content in the cell product (HALBICH upgrading curves) of the different black mass types.

The y-axis displays the recovery of carbon in the cell product, while the carbon content in the cell product is shown on the x-axis. The curve begins with a recovery of 100%, representing the feed material. The curves must be interpreted from right to left, where each data point on the curves represents an intermediate cell product during the flotation scheme (i.e., after every scavenger step). The recovery of carbon was in the range of 5–10% for the different black mass samples with corresponding carbon contents between 4 and 10 wt.%. In case a certain carbon content was valuable, e.g., to reduce the demand of a reducing agent during a subsequent pyrometallurgical black mass treatment (assuming that solid carbon powder is used to pyrometallurgically reduce metals from the black mass), the flotation process can principally be stopped at any position on the curve, i.e., after a certain number of scavenger steps.

4.2. Use of Flotation Graphite as a Secondary Carbon Carrier in Other Energy-Intensive Industries

The principal suitability to use the flotation graphite in the energy-intensive refractory industry as a secondary raw material for MgO-C refractory bricks was demonstrated in Section 3.3. The steel industry is another important energy- and resource-intensive sector that requires certain quantities of carbon. Even in future decarbonized steel production routes, carbon is still required to a certain extent. For example, carbon sources are of great importance in the Electric Arc Furnace (EAF), occupying two main roles. On the one hand, it is required for energetic use as an additional (chemical energy) input alongside electrical energy. On the other hand, carbon sources are used as slag foaming agents. Solid carbon sources are principally used in the EAF in two ways. The charge carbon is used together with scrap or other iron sources (e.g., sponge iron originating from a direct reduction in the briquetted form, i.e., hot briquetted iron—HBI) into the EAF, together with additives, at the beginning of the heating process. This carbon serves to carburize the melt, thereby contributing to slag foaming and, via the direct oxidation of the carbon during meltdown, achieving a chemical energy input. On the other hand, the injection carbon is brought into the EAF via lances or injectors, together with oxygen, to generate CO/CO_2 bubbles within the slag and thereby foaming the slag [31]. According to the Best Available Techniques (BATs) reference document for iron and steel production, ~15 kg of coal is used per ton of steel produced in the EAF (sum of charge and injected carbon) [12]. Slag foaming in the EAF is a well-established and widely used method to significantly increase the efficiency of the energy transfer in the furnace. The shielding of the electric arcs by the foaming slag reduces the energy loss via the water-cooled furnace walls and roof, thus enabling a significantly improved energy transfer from the arc to the melt. In addition, slag foaming has a stabilizing effect on the arcs and reduces the noise emissions of the EAF. Approximately 5–10 kg of injected carbon is used in EAFs per ton of crude steel [32]. In particular, the use of flotation graphite as an injection carbon could be interesting considering the grain size of ~45 µm being principally useful for an injection system. One critical point here is the content of undesired metals coming with the flotation graphite into the EAF process, other than the carbon, which are, depending on the black mass feed material used, Co, Cu, Ni, Mn, or Al. Special attention is necessary since some metals, such as Cu, cannot be removed from the steel melt anymore. Depending on the final steel grade to be produced, different tramp element levels must be kept. Furthermore, tramp elements can negatively influence steel properties during further downstream processing.

5. Conclusions

This study investigated the question of open-loop recycling for graphite separated from LIB black mass (which is a mixture of anode and cathode materials) via froth flotation. Black mass samples originating from different LIB types were used to produce a carbon-poor and a carbon-enriched fraction. The optimization of the flotation parameters was carried out depending on the black mass chemistry, i.e., the number of flotation stages (i.e., carbon separation and cleaning stages) and the dosing of flotation agents. The following main findings were discussed:

- Black mass containing NMC and LCO cathode materials could be treated using froth flotation with graphite recovery rates in the range between 90 and 95%; the low-carbon (cell) products showed final carbon contents between 4 and 10 wt.% (as such, the goal to derive a low-carbon product was reached).
- The currently investigated black mass containing the LFP cathode material did not disperse well in water and therefore showed no selectivity with regard to graphite separation via froth flotation (dispersion may be different with other LFP samples).
- For each of the different black mass samples investigated, a separate flotation scheme including a reagent regime must be developed. There is likely an influence of the conditions during the thermal LIB pre-processing step or another influence of the different cell geometries on the separation efficiency (this hypothesis needs to be investigated more intensively). The black mass sample HL (mixture of NMC and LCO cathode and graphite anode material) and the NMC black mass samples showed the highest selectivity (carbon-enriched product with 8–10 wt.% of undesired metals Co, Cu, Ni, Mn, and Al from the black mass). The PSP black mass (also a mixture of NMC and LCO cathode and graphite anode material) showed less selectivity with ~30 wt.% undesired metal contents in the carbon-enriched product.
- A flotation scheme was elaborated, comprising rougher and scavenger steps, and the separation of carbon-enriched products was achieved. Additionally, the flotation concept allows for a certain flexibility in terms of required product quality, i.e., it is possible to end the flotation process at an earlier stage (i.e., at a lower number of scavenger stages) to end with a higher carbon content in the cell product (if carbon is required as a reducing agent for a subsequent pyrometallurgical treatment to recover metals from the black mass).
- The carbon-enriched product was used as a secondary carbon source for a refractory material (magnesia carbon bricks), whereas an analysis of the brick chemistry, as well as thermo-mechanic properties in terms of density, porosity, cold crushing strength, hot modulus of rupture (the maximum bending stress that can be applied to a material before it breaks), or thermal conductivity, showed no negative influence on brick quality.
- It could be demonstrated that flotation graphite can be used as a secondary source for non-battery applications, representing a valuable positive example that contributes to a more complete closure of the battery life cycle and thus to a more circular economy.

For future research, several following challenges should be considered, whereas more insights into economic and environmental impact analyses would also be required to gain more information about the potential to implement specific flotation steps into existing LIB recycling processes:

- Upscaling of the currently used froth flotation (process sequence and flotation reagents) to obtain more reproducible data about the expected graphite product quality including impurities (tramp elements) present in the flotation graphite.
- Elaboration of a comprehensive connection between operating conditions of the thermal spent LIB pre-processing step and carbon separation efficiencies obtained via froth flotation.
- In case the flotation graphite is considered to be used for open-loop recycling practices in industrial sectors other than battery production, different application-oriented scenarios must be comprehensively investigated; in case the flotation graphite is used

as carbon carrier in an EAF for crude steelmaking, slag foaming efficiency, as well as the transfer of tramp elements into the metal melt, should be comprehensively investigated; in case the flotation graphite is used for refractory production (magnesia-carbon bricks), the high-temperature behavior of the bricks with the flotation graphite should be investigated in long-term trials.

- In-depth discussion on the economic viability of open-loop recycling approaches compared to other graphite recovery methods.
- Comparison of the environmental impact of graphite flotation and subsequent graphite recycling versus primary graphite production to gain added value with regards to circular economy contributions.

Author Contributions: Conceptualization: J.R. and S.S.; methodology: J.R. and S.S.; validation: S.S., B.R., A.A. and H.K.; formal analysis: J.R.; investigation: S.S., S.M. and B.R.; resources: S.S., A.A. and H.K.; data curation: S.S.; writing—original draft preparation: J.R.; writing—review and editing: S.S., B.R., S.M., A.A., H.K, T.N. and S.L.; visualization: S.S. and J.R.; supervision: B.R., A.A. and H.K.; project administration: B.R. and J.R.; funding acquisition: J.R. and S.L. All authors have read and agreed to the published version of the manuscript.

Funding: This research was funded by the Austrian Research Promotion Agency (grant number 888343/FO999888343).

Data Availability Statement: The data are contained in the study.

Acknowledgments: The authors gratefully acknowledge the funding support of K1-MET GmbH, the metallurgical competence center. The Module FuLIBatteR is supported by COMET (Competence Center for Excellent Technologies), the Austrian program for competence centers. COMET is funded by the Federal Ministry for Climate Action, Environment, Energy, Mobility, Innovation and Technology, the Federal Ministry for Labour and Economy, the Federal States of Upper Austria and Styria, as well as the Styrian Business Promotion Agency (SFG). Furthermore, Upper Austrian Research GmbH continuously supports the Module. In addition to the public funding from COMET, this Module is partially financed by the company partners Audi, BRAIN Biotech, Ebner Industrieofenbau, RHI Magnesita, Saubermacher, TÜV SÜD Landesgesellschaft Österreich, VTU Engineering, and voestalpine High-Performance Metals, as well as the scientific partners ACIB, Coventry University, Montanuniversitaet Leoben, BOKU University, and UVR-FIA.

Conflicts of Interest: Astrid Arnberger was employed by the company Saubermacher Dienstleistungs AG. Hartwig Kunanz was employed by the company RHI Magnesita GmbH. The authors declare no conflicts of interest.

Appendix A

The current appendix (Table A1) is dedicated to Section 2.2 and represents an amendment to Figure 3 to explain the different flotation agents used for the black mass samples.

Table A1. Summary of flotation stages, reagents, and dosages (g/t) for the tested black mass samples.

Test No.	Flotation Stage	Reagent Type and Dosage in g/t	Test No.	Flotation Stage	Reagent Type and Dosage in g/t
NMC_1_05	Ro *	K 250; M 150	PSP_02	Ro	L 500; D 500; P 30
	Scav1	K 150; M 50		Scav1	D 500; P 15
	Total	K 400; M 200		Total	L 500; D 1000; P 45

Table A1. *Cont.*

Test No.	Flotation Stage	Reagent Type and Dosage in g/t
NMC_1_06	Ro	K 800; M 150
	Scav1	K 150
	Total	K 950; M 150
NMC_1_07	Ro	K 250; M 250
	Scav1	K 150; M 50
	Total	K 400; M 300
NMC_1_11	Ro	L 500; D 500; P 30
	Scav1	D 250; P 15
	Scav2	D 250; P 15
	Scav3	D 250; P 15
	Scav4	D 250; P 15
	Scav5	D 250; P 20
	Total	L 500; D 1750; P 110
NMC_1_12	Ro	L 200; D 500; P 30
	Scav1	D 250; P 15
	Scav2	D 250; P 15
	Scav3	D 250; P 15
	Scav4	D 250; P 15
	Scav5	D 300; P 20
	Total	L 200; D 1800; P 110
NMC_1_13	Ro	L 1000; D 1300; P 30
	Scav1	D 250; P 15
	Scav2	D 250; P 15
	Scav3	D 250; P 15
	Total	L 1000; D 2050; P 75
NMC_2_17	Ro	L 200; D 500; P 30
	Scav1	D 500; P 15
	Scav2	D 250; P 15
	Scav3	D 500; P 15
	Total	L 200; D 1750; P 75
PSP_01	Ro	L 500; D 100; P 10
	Scav1	D 100; P 10
	Scav2	D 100; P 10
	Scav3	D 100; P 10
	Scav4	D 100; P 10
	Total	L 500; D 600; P 50
PSP_03	Ro	D 500; P 10
	Scav1	D 250; P 10
	Scav2	D 250; P 10
	Scav3	D 250; P 10
	Total	D 1250; P 40
HL_01	Ro	L 500; D 800; M 100
	Scav1	D 500; M 100
	Scav2	D 500; M 100
	Scav3	D 500; M 100
	Scav4	D 500; M 100
	Total	L 500; D 2800; M 500
HL_02	Ro	L 500; D 500; P 30
	Scav1	D 250; P 15
	Scav2	D 250; P 15
	Scav3	D 250; P 15
	Scav4	D 500; P 20
	Total	L 500; D 1750; P 95
HL_03	Ro	L 200; D 500; P 30
	Scav1	D 250; P 15
	Scav2	D 250; P 15
	Scav3	D 250; P 15
	Total	L 200; D 1250; P 75
HL_06	Ro	L 200; D 500; P 30
	Scav1	D 250; P 15
	Scav2	D 250; P 15
	Total	L 200; D 1000; P 60
HL_07	Ro	D 500; P 30
	Scav1	D 250; P 15
	Total	D 750; P 45

* Ro—Roughger; Scav—Scavenger.

References

1. Goodenough, J.B.; Park, K.-S. The Li-ion rechargeable battery: A perspective. *J. Am. Chem. Soc.* **2013**, *135*, 1167–1176. [CrossRef] [PubMed]
2. Asenbauer, J.; Eisenmann, T.; Kuenzel, M.; Kazzazi, A.; Chen, Z.; Bresser, D. The success Story of graphite as a Lithium-Ion anode material—Fundamentals, remaining challenges, and recent developments including silicon (oxide) composites. *Sustain. Energy Fuels* **2020**, *4*, 5387–5416. [CrossRef]
3. Larcher, D.; Beattie, S.; Morcrette, M.; Jumas, K.E.; Tarascon, J. Recent findings and prospects in the field of pure metals as negative electrodes for Li-ion batteries. *J. Mater. Chem.* **2007**, *17*, 3759–3772. [CrossRef]
4. European Carbon and Graphite Association. Graphite in Batteries. Available online: https://ecga.net/wp-content/uploads/2023/02/Graphite-in-batteries_Infosheet_final.pdf (accessed on 16 September 2024).
5. European Commission. *Framework for Ensuring a Secure and Sustainable Supply of Critical Raw Materials (COM(2023) 160 Final)*; European Commission: Brussels, Belgium, 2023.

6. European Union. Regulation (EU) 2023/1542 Concerning Batteries and Waste Batteries. 2023. Available online: https://eur-lex.europa.eu/legal-content/EN/TXT/PDF/?uri=CELEX:32023R1542 (accessed on 5 December 2024).
7. Han, S.-J.; Xu, L.; Chen, C.; Wang, Z.-Y.; Fu, M.-L.; Yuan, B. Recovery of graphite from spent lithium-ion batteries and its wastewater treatment application: A review. *J. Sep. Purif. Technol.* **2024**, *330*, 125289. [CrossRef]
8. Yildiz, T.; Wiechers, P.; Nirschl, H.; Gleiß, M. Direct recycling of carbon black and graphite from an aqueous anode slurry of lithium-ion batteries by centrifugal fractionation. *Next Energy* **2023**, *2*, 100082. [CrossRef]
9. Ahuis, M.; Aluzoun, A.; Keppeler, M.; Melzig, S.; Kwade, A. Direct recycling of lithium-ion battery production scrap—Solvent-based recovery and reuse of anode and cathode coating materials. *J. Power Sources* **2024**, *593*, 233995. [CrossRef]
10. Zhao, L.; Xin-yu, Z.; Yi-ye, L.; Hao, J.; Yan-qin, H.; Ming-xin, X.; Qiang, L. Recovery of electrode materials from a spent Lithium-Ion Battery through a pyrolysis-coupled mechanical milling method. *Energy Fuels* **2024**, *38*, 9280–9319. [CrossRef]
11. Rong, T.; Yuan, Y.; Yang, H.; Yu, H.; Zuo, H.; Wang, J.; Xue, Q. Investigation of the enrichment-purification process and electrochemical performance of kish graphite in dust from blast furnace tapping yard. *J. Waste Manag.* **2024**, *175*, 121–132. [CrossRef]
12. Remus, R.; Aguado Monsonet, M.A.; Roudier, S.; Delgado Sancho, L. *Best Available Techniques (BAT) Reference Document for Iron and Steel Production*; European Commission, Joint Research Centre: Brussels, Belgium, 2013.
13. Jia, P.; Sun, J.; Li, S.; Wang, W.; Song, Z.; Zhao, X.; Mao, Y. A feasible recycling route for spent graphite: Microwave-assisted puffing with Fe_2O_3 loading to construct high-performance anode for lithium-ion batteries. *Mater. Today Sustain.* **2024**, *25*, 100620. [CrossRef]
14. Mancini, M.; Hoffmann, M.F.; Martin, J.; Weirather-Kostner, D.; Axmann, P.; Wohlfahrt-Mehrens, M. A proof-of-concept of direct recycling of anode and cathode active materials: From spent batteries to performance in new Li-ion cells. *J. Power Sources* **2024**, *595*, 233997. [CrossRef]
15. Subramanyan, K.; Jyothilakshmi, S.; Ulaganathan, M.; Lee, Y.-S.; Aravindan, V. An efficient upcycling of graphite anode and separator for Na-ion Batteries via solvent-co-intercalation process. *Carbon* **2024**, *216*, 118525. [CrossRef]
16. Milian, Y.e.; Jamett, N.; Cruz, C.; Herrera-Leon, S.; Chacana-Olivares, J. A comprehensive review of emerging technologies for recycling spent lithium-ion batteries. *Sci. Total Environ.* **2024**, *910*, 165843. [CrossRef]
17. Vanderbruggen, A.; Sygusch, J.; Rudolph, M.; Serna-Guerrero, R. A contribution to understanding the flotation behavior of lithium metal oxides and spheroidized graphite for lithium-ion battery recycling. *Colloids Surf. A Physicochem. Eng. Asp.* **2021**, *626*, 127111. [CrossRef]
18. Savihirta, H.; Wilson, B.P.; Lundström, M.; Serna-Guerrero, R. A study on recovery strategies of graphite from mixed lithium-ion battery chemistries using froth flotation. *Waste Manag.* **2024**, *180*, 96–105. [CrossRef] [PubMed]
19. Olutogun, M.; Vanderbruggen, A.; Frey, C.; Rudolph, M.; Bresser, D.; Passerini, S. Recycled graphite for more sustainable lithium-ion batteries. *Carbon Energy* **2024**, *6*, e483. [CrossRef]
20. Drzymała, J.; Swatek, A. *Mineral Processing: Foundations of Theory and Practice of Mineralogy*, 1st ed.; Wroclaw University of Technology: Wroclaw, Poland, 2008.
21. Wills, B.A.; Finch, J.A. Chapter 12—Froth Flotation. In *Wills' Mineral Processing Technology*, 8th ed.; Butterworth-Heinemann: Oxford, UK, 2016; pp. 265–380.
22. Chehreh Chelgani, S.; Rudolph, M.; Kratzsch, R.; Sandmann, D.; Gutzmer, J. A review of graphite beneficiation techniques. *Miner. Process. Extr. Metall. Rev.* **2016**, *37*, 58–68. [CrossRef]
23. Vasumathi, N.; Sarjekar, A.; Chandrayan, H.; Chennakesavulu, K.; Reddy, G.; Ramanjaneya, V.K. A Mini Review on Flotation Techniques and Reagents Used in Graphite Beneficiation. *Int. J. Chem. Eng.* **2023**, *2023*, 1007689. [CrossRef]
24. Andreola, F.; Castellini, E.; Ferreira, J.M.F.; Olhero, S.; Romagnoli, M. Effect of sodium hexametaphosphate and ageing on the rheological behaviour of kaolin dispersions. *Appl. Clay Sci.* **2006**, *31*, 56–64. [CrossRef]
25. Arnberger, A.; Coskun, E.; Rutrecht, B. Recycling von Lithium-Ionen Batterien. In *Recycling und Rohstoffe*, 11th ed.; Thiel, S., Thomé-Kozmiensky, E., Goldmann, D., Eds.; Thomé-Kozmiensky Verlag GmbH: Neuruppin, Germany, 2018; pp. 583–599.
26. Salces, A.M.; Bremerstein, I.; Rudolph, M.; Vanderbruggen, A. Joint recovery of graphite and lithium metal oxides from spent lithium-ion batteries using froth flotation and investigation on process water re-use. *Miner. Eng.* **2022**, *184*, 107670. [CrossRef]
27. Vanderbruggen, A.; Hayagan, N.; Bachmann, K.; Ferreira, A.; Werner, D.; Horn, D.; Peuker, U. Lithium-Ion Battery Recycling—Influence of Recycling Processes on Component Liberation and Flotation Separation Efficiency. *ACS EST Eng.* **2022**, *2*, 2130–2141. [CrossRef]
28. Zhang, Y.; Zhu, H.; Zhu, J.; Min, F.; Chen, J.; Shi, O. Effect of inorganic cations on enhancing graphite/kerosene adsorption and reducing carbon emission in graphite flotation. *Fuel* **2022**, *314*, 122740. [CrossRef]
29. Verdugo, L.; Zhang, L.; Saito, K.; Bruckard, W.; Menacho, J.; Hoadley, A. Flotation behavior of the most common electrode materials in lithium-ion batteries. *Sep. Purif. Technol.* **2022**, *301*, 121885. [CrossRef]
30. Nazari, S.; Zhou, S.; Hassanzadeh, A.; Li, J.; He, Y.; Bu, X.; Kowalczuk, P.B. Influence of operating parameters on nanobubble-assisted flotation of graphite. *J. Mat. Res. Technol.* **2022**, *20*, 3891–3904. [CrossRef]

31. Echterhof, T. Review on the use of alternative carbon sources in EAF steelmaking. *Metals* **2022**, *11*, 222. [CrossRef]
32. Zulhan, Z. Der Einfluss unterschiedlicher Kohlenstoffträger auf die Schaumschlackenbildung im Elektrolichtbogenofen. In *Berichte aus dem Institut fuer Eisenhuettenkunde (German Document)*; Bleck, W., Krupp, U., Muenstermann, S., Senk, D., Eds.; Shaker Verlag: Aachen, Germany, 2006; Volume 8.

recycling

MDPI

Article

Sustainable Upcycling of Spent Battery Graphite into High-Performance PEG Anodes via Flash Joule Heating

Yihan Luo, Jing Sun *, Wenxin Chen, Shuo Lu and Ziliang Wang *

National Engineering Laboratory for Reducing Emissions from Coal Combustion, Engineering Research Center of Environmental Thermal Technology of Ministry of Education, Shandong Key Laboratory of Green Thermal Power and Carbon Reduction, School of Nuclear Science, Energy and Power Engineering, Shandong University, Jinan 250061, China; luoyihan@mail.sdu.edu.cn (Y.L.); chenwx314415@163.com (W.C.); ls13336286329@163.com (S.L.)

* Correspondence: sunjing0108@163.com (J.S.); zwang2022@sdu.edu.cn (Z.W.)

Abstract

The upcycling of spent lithium-ion battery graphite constitutes an essential pathway for mitigating manufacturing expenditures and alleviating ecological burdens. This study proposes an integrated strategy to upcycle spent graphite into high-performance porous expanded graphite (PEG) anodes, leveraging flash Joule heating (FJH) as a core technique for efficient decontamination, interlayer expansion, and active etching. Results show that the binders and impurities are efficiently removed by FJH treatment, and the graphite interlayer spacing is expanded. The iron oxide, which acts as an etching reagent, can then be easily intercalated and laid into the decontaminated graphite for subsequent etching. A subsequent FJH treatment simultaneously releases oxidized intercalants and triggers in-situ metal oxide etching, yielding PEG with a rich porous architecture and enhanced specific surface area. This method successfully prepared high-performance porous expanded graphite anode material with a mesoporous structure. The resulting anode delivers a remarkable capacity retention of 419 mAh·g^{-1} after 600 cycles at 2C, outperforming the performance of commercial graphite anodes. This innovative approach offers a promising route for sustainable graphite reclamation.

Keywords: flash Joule heating; spent graphite; porous expanded graphite; anode; high-value reutilization

1. Introduction

In response to environmental pollution and the energy crisis, electric vehicles (EVs) are rapidly becoming widespread, and lithium-ion batteries (LIBs) are becoming increasingly prevalent [1]. Given their limited service life (typically 3–10 years [2–4]), a surge in spent LIBs is projected over the next 15 years. Improper disposal of spent LIBs likely leads to environmental and safety hazards since mechanical damage may trigger fires or explosions via thermal runaway [5,6] with toxic gases, i.e., phosphorous oxyfluoride (POF_3), hydrofluoric acid (HF) gases, and particulate matter being released during thermal processes [7,8]. Additionally, the leakage of heavy metals (Ni, Co, Mn) and organic electrolytes from discarded LIBs may cause soil and water contamination [9]. As a special waste with abundant strategic metals (i.e., Co and Ni) and valuable graphite, the improper disposal means a waste of valuable resources. Consequently, advanced recycling is imperative not only to mitigate hazards but also to establish closed-loop sustainability for LIBs.

Current recycling paradigms prioritize cathode metal recovery (e.g., Co, Ni) while undervaluing graphite—a critical anode material requiring energy-intensive production (cost USD 8000–13,000/ton [10]). Graphite constitutes 12–21 wt.% of LIB, representing 10% to 15% of total battery production costs. It is projected that global volumes will reach 7.5 million tons by 2037, and the global graphite market is projected to attain a valuation of tens of billions of USD [11,12]. However, most spent graphite is currently either landfilled or downcycled as reducing agents [13,14]. This significant economic value, compounded by potential environmental hazards, underscores the urgency for efficient graphite reclamation technologies.

The recycling of spent graphite (SG) primarily proceeds via three technical pathways: decontamination-regeneration, high-value conversion, and structural modification. The decontamination-regeneration approach focuses on direct anode reuse by eliminating solid electrolyte interphase (SEI) residues and metallic impurities (e.g., Cu, Al) through acid leaching or high-temperature calcination treatment. Yu et al. [15] implemented ultra-high-temperature calcination (3000 °C, 6 h, N_2 atmosphere) for lattice reconstruction; the obtained regenerated graphite has an initial charge capacity of 352.5 mAh·g^{-1} at a current density of 0.1 A·g^{-1} and a capacity retention rate of 93.52% after 500 cycles. Zhang et al. [16] proposed an environmentally friendly and economical modification method based on sulfate roasting, using sodium fluoride as an auxiliary additive. By low-temperature roasting at 250 °C to recycle waste graphite, graphite with a purity as high as 99.55% can be obtained, with an initial charge capacity of 333.9 mAh·g^{-1}, and a capacity retention rate of 91.2% after 400 cycles. Gao et al. [17] used low-temperature sulfation roasting followed by acid leaching to produce recovered graphite (RG) with a purity of 99.13%. This RG exhibited an initial irreversible capacity of 353.4 mAh·g^{-1} at 0.1C. While effective, these methods incur prohibitive energy/reagent consumption, require complex infrastructure, and their recycled graphite underperforms commercial graphite, failing to match its electrochemical properties.

High-value conversion routes transform SG into advanced materials like graphene, typically using modified Hummers' methods. For instance, Xie et al. [18] used an improved Hummers method to prepare graphene from SG. They optimized the ratio of SG to oxidizing agent ($KMnO_4$) and found the best mass ratio to be 1:3. They oxidized SG with the most appropriate proportion and thermally reduced the resulting graphene oxide, successfully producing reduced graphene oxide with a disordered few-layer structure. Zhang et al. [19] treated the lithiated graphite of spent LIBs under the conditions of a controlled flow mixture gas of H_2O/Ar and then extracted lithium by the dissolution–vapor method. After that, holey graphene was prepared using the common Hummer's method. Although value-added graphene was ultimately produced, these processes remain fundamentally hampered by inefficient exfoliation. Furthermore, the exfoliated graphene tends to restack due to strong π-π interactions and van der Waals forces between adjacent nanosheets [20–22], significantly limiting its practical applications and industrial scalability.

Structural modification strategies represent compromise solutions that enhance anode performance through microstructural regulation via surface coating [23] or elemental doping (e.g., N-doping) [24] or interlayer expansion [25] and pore engineering. Xiao et al. [23] used an effective graphite modification approach involving liquid-phase impregnation of asphalt onto spent graphite surfaces, followed by carbonization treatment to form a uniform amorphous carbon coating layer. The modified graphite anode achieved a high initial specific capacity of 403 mAh g^{-1} at 0.1C, demonstrating a cycling stability with 97.8% capacity retention after 110 cycles. Xu et al. [24] successfully induced g-C_3N_4 between the layers of acid-treated graphite through physically mixing SG and urea thoroughly, followed by a two-stage thermal treatment (550 °C, 3 h, and 800 °C, 1 h) in an argon atmosphere, in which urea served as a nitrogen source and was completely pyrolyzed to form g-C_3N_4. The novel regenerated graphite (N-RG) with nitrogen doping and enlarged interlayer spacing anode delivered a capacity of 465.8 mAh g^{-1} at 0.1 A g^{-1} after 200 cycles. Gong et al. [25] regenerated SG into high-performance expanded graphite anode material through oxidation intercalation. This approach enhances ion transport by expanding interlayer spacings to 0.392 nm (vs. 0.338 nm in SG) and generating hierarchical pore networks, effectively improving the performance of the regenerated anode. These structural modification strategies demonstrate viable pathways for enhancing anode kinetics, with capacity achievements approaching 370 mA g^{-1} after 1000 cycles at 1C. Obviously, expanded graphite, due to its larger interlayer spacing, is conducive to ion transmission, effectively improving the performance of the regenerated anode.

Building upon the aforementioned work, a pivotal concept can be established: skipping energy-intensive exfoliation to focus solely on layer expansion and pore engineering sufficiently enhances ion diffusion kinetics while preserving structural integrity. This approach can not only improve the electrochemical performance of recycled graphite materials but also significantly reduce production costs. Different from graphite, the SG contains impurities, which hinder the intercalation. To address these issues and overcome existing challenges—including constrained electrochemical improvements and marginal economic returns from regeneration, and graphene's inherent energy-intensive exfoliation requirements and restacking tendency—we propose an innovative approach to integrate the decontamination and structure modification through Flashed Joule heating (FJH) to achieve simultaneous purification and structural engineering with ultrafast processing and a minimized ecological footprint. Flashed Joule heating (FJH), which is characterized by instant thermal shocks (>2000 K, <100 ms pulse), can eliminate impurities and engineer structures through thermal shock-induced volatilization, layer expansion, and defect-mediated pore formation. Capitalizing on these dual functionalities, we propose a three-stage upcycling protocol. Firstly, primary FJH decontamination yields purified graphite; secondly, the purified graphite is subjected to oxidation intercalation with in situ metal oxide deposition to achieve chemical functionalization; lastly, controlled expansion/etching was triggered by a secondary FJH to fabricate porous expanded graphite (PEG). The efficacy of this SG upcycling method for fabricating high-value PEG and its potential as a battery material in enhancing electrochemical performance was comprehensively evaluated by galvanostatic charge–discharge cycling, electrochemical impedance spectroscopy (EIS), and cyclic voltammetry (CV) with a target to establish a pioneering SG recovery paradigm.

2. Results and Discussion

2.1. Characterization of Materials

Figure 1 shows the SEM images of the key step products. Figure 1a delineates the pristine morphology of SG, revealing a pronounced agglomeration of particulate constituents interwoven with lamellar frameworks, where layered architectures are substantially obscured by pervasive adherent flocculent contaminants. These surface impurities, primarily comprising organic polymeric binders, fragmented graphite debris, electrolyte decomposition byproducts, and residual solid electrolyte interphase (SEI) films, collectively establish tortuous diffusion barriers that disrupt lithium-ion (Li^+) transport pathways. Such microstructural heterogeneity directly contributes to irreversible capacity loss and Columbic efficiency deterioration, as extensively documented in electrochemical degradation studies [26].

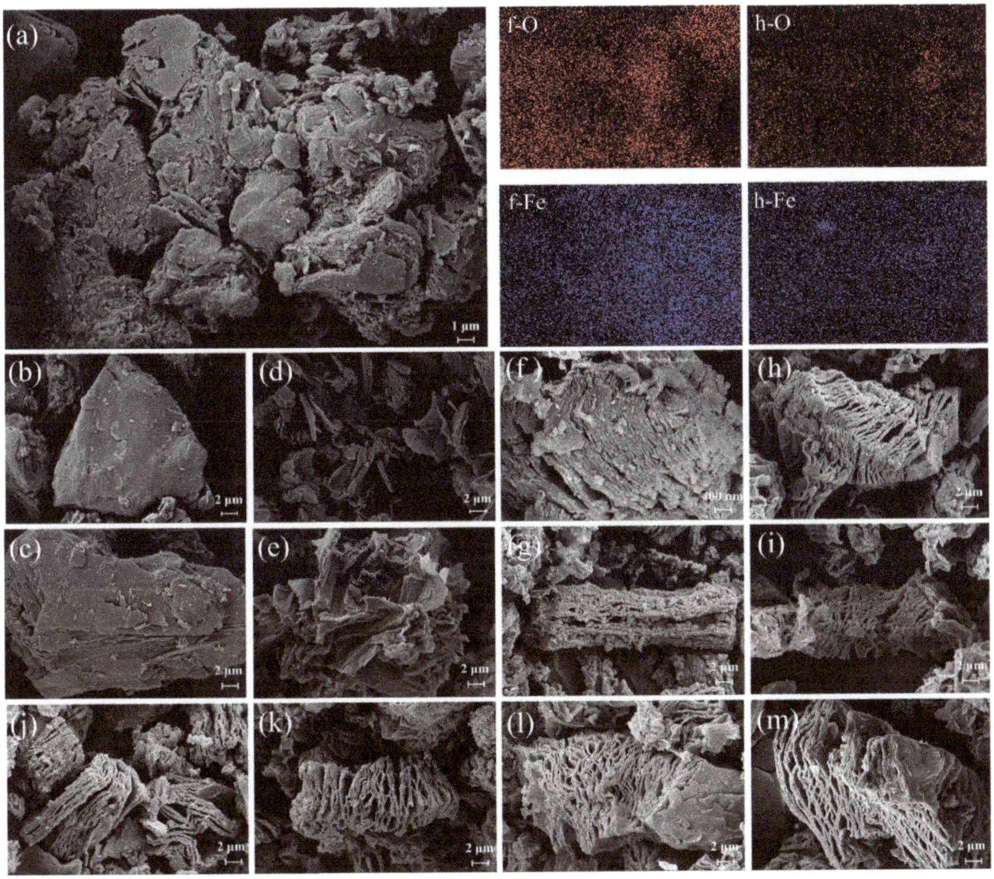

Figure 1. SEM of (**a**) SG, (**b**) FJH-80, (**c**) HTT-1200, (**d**) GO-FJH, (**e**) GO-HTT, (**f**) EG-Fe^{3+}-FJH, (**g**) EG-Fe^{3+}-HTT, (**h**) EG-Fe^{2+}-FJH, (**i**) EG-Fe^{2+}-HTT, (**j**) FF70-Fe^{3+}, (**k**) FF90-Fe^{3+}, (**l**) HF70-Fe^{3+}, (**m**) HF90-Fe^{3+}.

In order to decontaminate SG, HTT and FJH treatments were, respectively, applied to SG. At the processing temperatures of HTT and FJH, the high temperature induces rapid decomposition of the binder, while the metal elements (mainly lithium compounds) will be evaporated. Furthermore, Dong et al. found the conductive agent (i.e., super P) will react with the decomposition products of the binder to form graphene [27]. Crucially, distinct morphological differentiation emerges between treatment modalities: FJH induces conspicuous edge curling and localized delamination (Figure 1b), while high-temperature treatment maintains predominantly planar layer configurations with preserved basal plane integrity (Figure 1c). The FJH treatment induces obvious edge curling and local peeling, forming a microstructure distinct from SG and HTT samples; this morphology is more conducive to the ingress of intercalants and oxidants into graphite layers during subsequent GO preparation. The XRD test results also confirm this point, as shown in Figure S1 [28], the (002) reflection of FJH-80 undergoes progressive displacement toward lower diffraction angles relative to SG, while HTT-1200 exhibits systematic migration toward higher angular positions. These directional shifts indicate significantly different interlayer reconfigurations: expansive dilation in FJH-80 compared with compressive contraction in HTT-1200, as rigorously supported by Bragg's law [29] computations detailed in Table S1.

In order to achieve the expansion and layering of the SG and etching in the second FJH, the oxidation intercalation and the hydrothermal deposition of active metal oxides on the SG were carried out after decontamination. Figures 1d,e and S2 show graphite oxide (GO) synthesized by the modified Hummers' method. This morphology demonstrates a high degree of graphite exfoliation. GO-FJH displays a prominent (001) reflection with substantially greater integrated intensity than GO-HTT in XRD images (Figure S2). Comparative assessment of iron oxide nanocomposites (Figure 1f–i) reveals significant nanoscale particle size differentiation governed by precursor chemistry: hydrothermal treatment by iron (III) acetate precursors yields coarser, irregular iron oxide nanoparticles, contrasting markedly with the finer, more homogeneous crystallites generated from hydrothermal treatment by iron (II) acetate. This valence-dependent nucleation behavior is further corroborated by intensified Fe and O elemental signatures and spatial distribution gradients in corresponding EDS mappings, reflecting distinct growth kinetics during composite formation.

Subsequently, the second FJH was used to achieve the release of intercalated substances and the etching of graphite layers. Structural transformations following flash etching (Figure 1j–m) exhibit substantial layer exfoliation and incipient pore network development, mechanistically attributed to rapid decomposition of intercalated reagents and CO_2 and CO venting during redox reactions between iron oxides and carbon matrices under ultrafast thermal activation. The violent gas evolution generates internal stresses that forcibly separate graphite layers, creating interconnected void spaces essential for enhanced ionic accessibility [24].

Through the structural analysis of the four PEG materials, FF70-3, FF90-3, HF70-3, and HF90-3 (Figure 2a–d), voltage-dependent cavity formation dynamics is further elucidated: 90V treatment generates larger porous structures (0.1–5 μm) with branched channel structures, while 70V processing establishes smaller porous structures (4–500 nm) exhibiting higher areal density. Figure 2e–h show the pore size distribution diagrams of FF70-3, FF90-3, HF70-3, and HF90-3, respectively. The size distribution data of the aperture was collected and analyzed by Nano Measurer 1.2 software, while the porosity was statistically calculated by Fiji ImageJ 1.54p software. It is not difficult to observe that the GO composites treated by FJH formed a hierarchical pore structure. Pores produced by 90 V flash evaporation etching are exclusively macropores (>50 nm). In contrast, pores formed at 70 V show a broader distribution, with nearly half concentrated below 50 nm. Thanks to the existence of the dual lithium storage mechanism, this synergistic structural optimization established abundant lithium-ion adsorption sites and shortened the length of the diffusion path [23,30].

According to the area proportion of pore structure on the graphite layer in Figure 2i, FF70-3 and FF90-3 are much larger than HF70-3 and HF90-3, respectively, indicating that the proportion of pore area of the materials prepared from FJH decontamination is generally greater than that of HTT decontamination. This is attributed to the higher oxidation degree of GO-FJH, which has more oxidation functional groups to facilitate the deposition of metal oxides [31,32]. The increase in the amount of deposited metal oxides resulted in more holes formed during the subsequent etching process and a larger proportion of combined hole area. Moreover, within the same decontamination method, the etching degree of 90 V is larger than that of 70 V. This is because a higher etching voltage brings a higher etching temperature, which leads to a larger etching range, thereby resulting in a larger proportion of pore area. Therefore, the higher oxidation degree, combined with higher flash etching temperature, results in more etching pores and a larger proportion of pore area in the later stage.

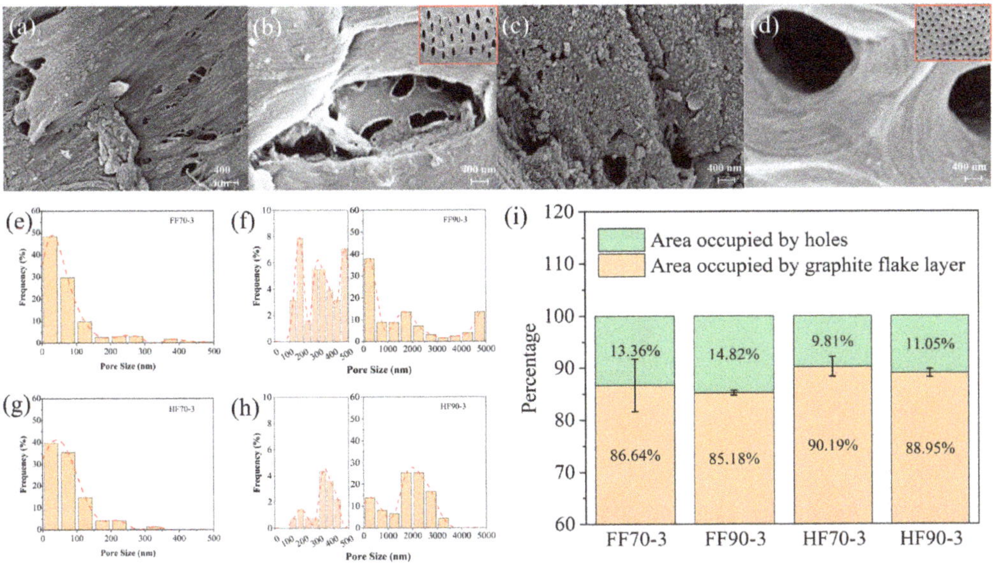

Figure 2. SEM of (**a**) FF70-3, (**b**) FF90-3's local morphology (and overall morphology in the red box), (**c**) HF70-3, (**d**) HF90-3's local morphology (and overall morphology in the red box). Pore size distribution plots for the following four samples: (**e**) FF70-3, (**f**) FF90-3, (**g**) HF70-3, (**h**) HF90-3, (**i**) the ratio of the holes in the four samples of FF70-3, FF90-3, HF70-3, and HF90-3 to the surface area of the graphite flake layer, and the number of samples N = 5.

Figure 3 presents some TEM images of more microscopic structures to complement the deficiencies of the SEM images. Comparative examination of Figure 3a–d reveals distinct iron oxide nucleation patterns dictated by precursor chemistry. Iron (III) acetate-assisted hydrothermal treatment (Figure 3a,b) promotes the conspicuous formation of nano-scaled iron oxide particulates uniformly dispersed across graphene lamellae, exhibiting well-defined crystalline interfaces with carbon matrices. In contrast, iron (II) acetate systems (Figure 3c,d) generate significantly refined nanoparticles, with EG-Fe^{2+}-FJH and EG-Fe^{2+}-HTT measuring approximately 1.3 nm and 1.8 nm, respectively—dimensions approaching quantum-confined regimes. As shown in Figure S3, iron (II) acetate composites (EG-Fe^{2+}-FJH/HTT) present featureless diffraction patterns despite confirmed nanoparticle existence via TEM characterization. This apparent paradox arises from nanoscale confinement effects: critically diminished crystalline domains induce extensive peak broadening that obscures discrete Bragg reflections, causing diffraction signatures to dissipate into the amorphous background continuum [33]. It is found that, whether iron (II) acetate or iron (III) acetate is used as the iron source, the attachment density of metal oxides on the graphite layers is greater when using GO-FJH as the hydrothermal raw material in comparison to GO-HTT. Combined with the previous analysis in the SEM section, this further verifies from another side that the oxidation degree of GO-FJH is higher.

Post-etching microstructures (Figure 3e,f) exhibit two observable characteristics: well-defined nanopores generated through metal oxide-catalyzed gasification and preserved graphitic domain integrity. This structure conclusively validates the efficacy of FJH for precision etching, which targets decomposition of metal oxide–graphene interfaces while maintaining structural coherence in adjacent carbon frameworks. The simultaneous presence of edge-located pores and in-plane defects establishes complementary ion-diffusion pathways, synergistically enhancing electrochemical kinetics as quantified in subsequent performance evaluations.

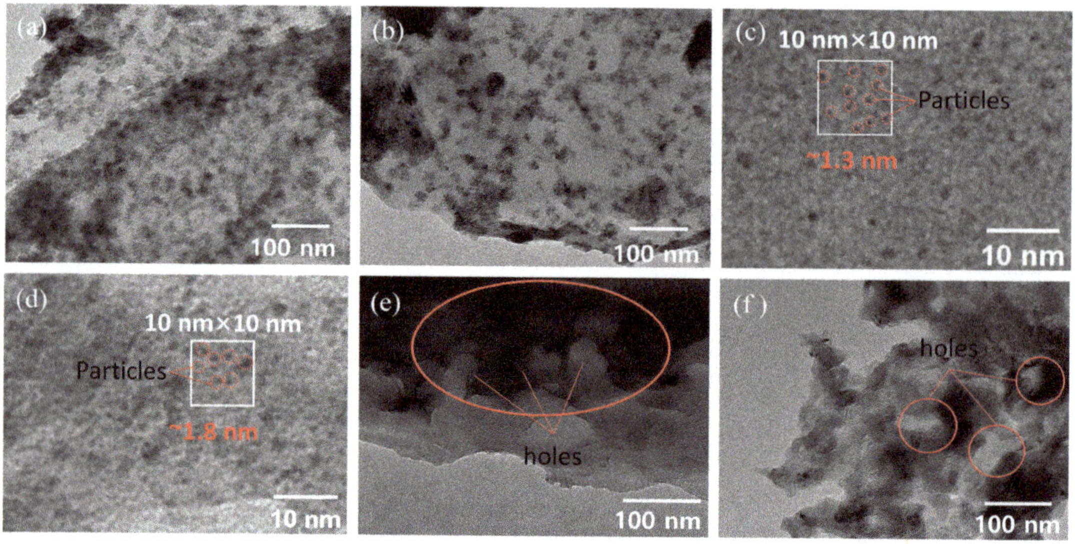

Figure 3. TEM of (**a**) EG-Fe^{3+}-FJH, (**b**) EG-Fe^{3+}-HTT, (**c**) EG-Fe^{2+}-FJH, (**d**) EG-Fe^{2+}-HTT, (**e**) FF70-Fe^{3+}, and (**f**) HF70-Fe^{3+}.

Nitrogen adsorption–desorption isotherms of the PEG materials exhibit two different characteristics. FF70-3, FF90-3, HF70-3, and HF90-3 (Figure 4a,b) consistently exhibit characteristic Type IV profiles with well-defined hysteresis loops, confirming mesoporous dominance across these materials. Comparative assessment reveals significantly intensified hysteresis behavior in FF70-3 and HF70-3 relative to FF90-3 and HF90-3, indicative of greater mesopore volume fractions in the former materials. This structural difference has profound electrochemical implications: mesoporous architectures effectively mitigate ionic diffusion resistance while enhancing electrolyte wettability, which are critical factors governing electrode kinetics [34]. In marked contrast, FF70-2 and HF70-2 display prototypical Type II isotherms diagnostic of microporous frameworks. In addition, all the PEG samples have macropores, and accelerated adsorption at elevated relative pressures ($P/P_0 > 0.9$) occurs for all samples, providing compelling evidence for coexisting macropores within these micro-mesoporous matrices [35]. Pore size distribution analysis further demonstrates better microstructural uniformity in FF70-2 compared with HF70-2, which is manifested through sharper micropore distribution peaks. This enhanced homogeneity aligns precisely with prior TEM observations of more orderly iron oxide nanoparticle distributions on EG-Fe^{2+}-FJH precursors, establishing methodological consistency across complementary characterization techniques.

Specific surface area quantification (Table S2) documents 166.831 $m^2\ g^{-1}$ for FF70-2 and 140.473 $m^2\ g^{-1}$ for HF70-2, stemming from micropore structure, indicating more active sites for lithium storage. In contrast, FF70-3/HF70-3 demonstrates a more mesoporous structure (Figure 4c,d), expanding ion-accessible interfacial domains that can facilitate rapid charge transfer as well as reduce ionic diffusion path lengths through interconnected pore networks. Crucially, FF70-3 and HF70-3 exhibit substantially greater specific areas and mesopore volumes than those of FF90-3/HF90-3, indicating larger electrochemically active surface areas and enhanced site availability for lithium storage. Owing to excessive etching of the graphite layers to generate macropores, the specific surface area and active sites of the FF90-3/HF90-3was remarkably reduced, thereby limiting the lithium storage performance.

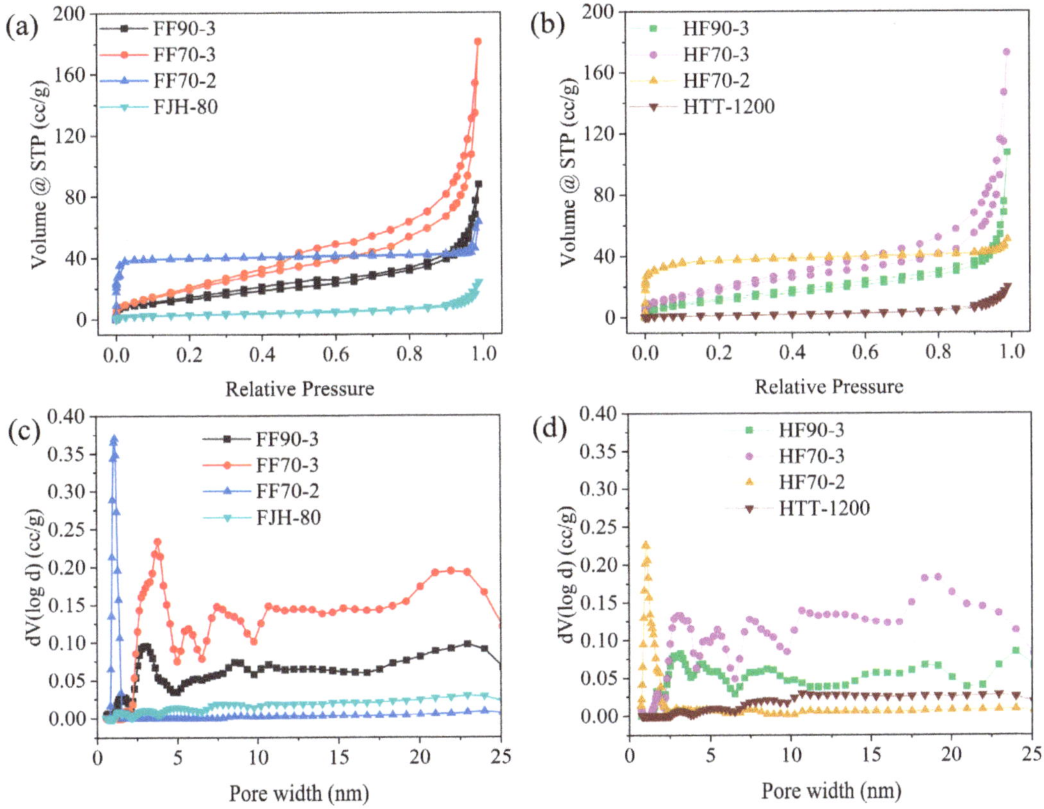

Figure 4. Isothermal adsorption and desorption curves of (**a**) flash Joule heating series: FF70-3, FF90-3, FF70-2, and FJH-80, (**b**) High-temperature treatment: HF70-3, HF90-3, HF70-2, and HTT-1200. DFT Pore size distribution of (**c**) flash Joule heating series: FF70-3, FF90-3, FF70-2, and FJH-80, (**d**) High-temperature treatment: HF70-3, HF90-3, HF70-2, and HTT-1200.

2.2. Electrochemical Characterization

Long-cycle tests under two C-rate regimes (0.1C and 2C, 1C = 600 mAh g^{-1}) were conducted to evaluate the electrochemical performance of the prepared PEG. As demonstrated in Figure 5a, in accordance with the porous architectures derived from FJH-assisted expansion and etching, the PEG exhibited distinct outstanding specific capacities of 550.8 mAh g^{-1} (FF90-3), 633.9 mAh g^{-1} (FF70-3) and 751.5 mAh g^{-1} (FF70-2), in comparison to the 243.7 mAh g^{-1} (FJH-80), indicating the porous structure can greatly enhance the electrochemical performance. This excellent performance of FF70-2 reveals a significant enhancement in lithium storage capability for composites synthesized via hydrothermal processing with iron (II) acetate precursors compared with those incorporating iron (III) acetate under low-rate discharge conditions (0.1C). This is primarily attributed to the large specific surface area of FF70-2, which provides more active sites for lithium storage. It is worth noting that in Figure S4, the apparent "activation" behavior is observed for FF70-2 at 0.1C in long cycling. This is attributed to gradual electrolyte infiltration and wetting in the predominantly microporous FF70-2. This effect can increase the accessible active surface over the first tens of cycles. The sharp decline in capacity around the 90th cycle might be due to a collapse in the material structure, resulting in a sudden reduction in specific surface area and ion transport channels [36].

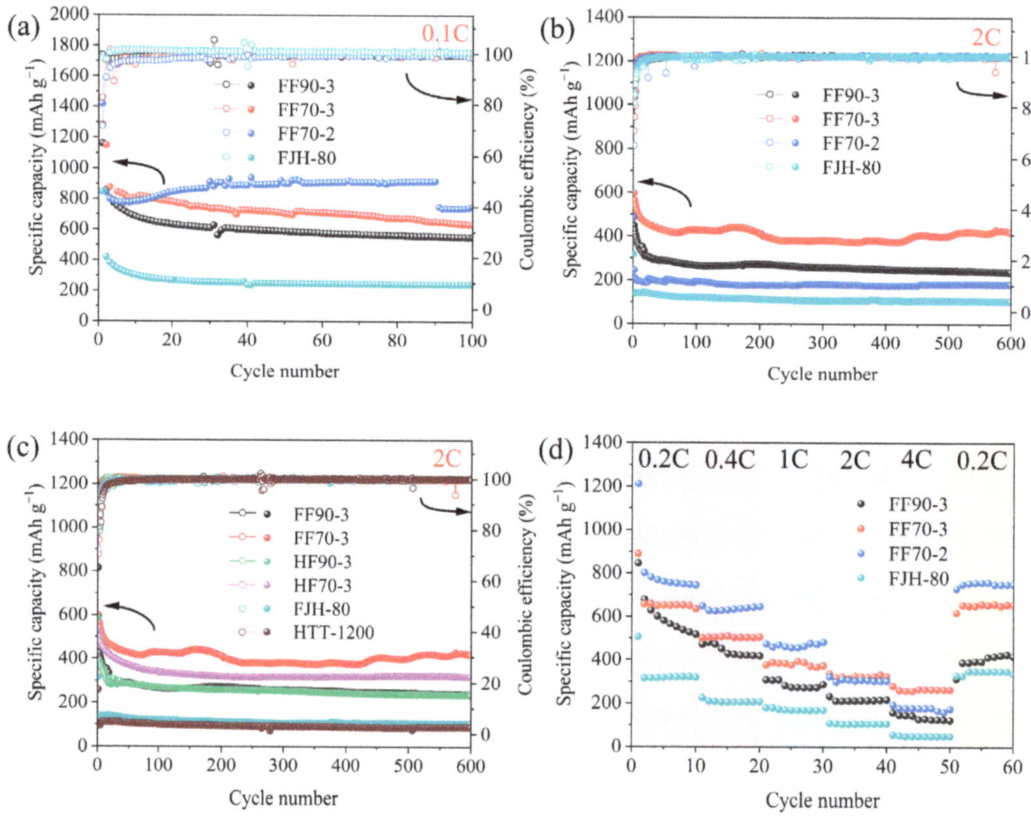

Figure 5. Electrochemical performance of different expanded porous graphite anode materials (The black arrow indicates the corresponding Y-coordinate of the curve.): (**a**) Specific capacity of flash Joule heating series at 0.1C. (**b**) Specific capacity of flash Joule heating series at 2C. (**c**) Specific capacity of 4 groups mixed with iron (III) acetate by hydrothermal method, FJH-80, and HTT-1200 at 2C. (**d**) Specific capacity of 4 groups treated with 70 V Flash Joule Heating etching at various rates.

Under high-rate discharge conditions (2C), as shown in Figure 5b, the specific capacities exhibit significant variation, exhibiting 237 mAh g^{-1} for FF90-3, 419 mAh g^{-1} for FF70-3, 181.9 mAh g^{-1} for FF70-2, and 98.1 mAh g^{-1} for FJH-80. Compared with FF70-2 treated with iron (II) acetate, FF70-3 treated with iron (III) acetate exhibits better performance, which is primarily attributed to the rich mesoporous structure of FF70-3. While sub-2 nm pores enhance low-rate capacity through increased surface adsorption, the promotion of ion transmission is limited. In contrast, the mesoporous structure is more conducive to the transmission of lithium ions, especially under high-rate conditions. As shown in Figure S5, during the testing process, the performance of FF70-3 exhibited a fluctuating state. The fluctuations were modest and consistent with stochastic wetting in porous carbons.

It can be observed from Figure 5c that the initial decontamination method plays an important role in the electrochemical performance of the final PEG. Overall, the PEG prepared from FJH-decontaminated graphite exhibits higher specific capacity than that of HTT-decontaminated graphite, regardless of whether the subsequent FJH for layer expansion and etching occurs at 70 V or 90 V. This increased capacity is mainly attributed to the greater layer spacing produced by FJH to facilitate the introduction of oxygen-containing functional groups in the graphite oxidation process, which will further induce a higher attachment of iron oxide particles and finally result in an increased number of pores within the porous material during the subsequent FJH treatment. Furthermore, for both types of PEG prepared from FJH and HTT decontamination, the subsequent FJH at 70 V exhibits higher specific capacity than that at 90 V. This phenomenon can be explained by the fact that higher voltage leads to higher FJH temperatures, and as a result, the pore structure becomes large in size and more interconnected, leading to a reduction in the number of active sites.

Comprehensive rate capability assessment (Figure 5d) reveals that FF70-3 delivers exceptional electrochemical performance. FF70-3 achieves specific capacities of 653.8 mAh g^{-1} at 0.2C, 503.3 mAh g^{-1} at 0.4C, 388.5 mAh g^{-1} at 1C, 324 mAh g^{-1} at 2C, and 266.4 mAh g^{-1} at 4C. By comparing the performance of FF70-3 with that of commercial graphite measured in our previous work, it can be concluded that the performance of FF70-3 is far superior to that of commercial graphite [37]. Significantly, FF70-3 exhibits an exceptional capacity recovery fidelity of 99.54% during rate cycling transitions, maintaining near-complete capacity restoration upon reverting from high-rate (4C) to low-rate (0.2C) operation. The distinct rate-dependent performance profiles elucidate a critical pore architecture-function interdependence: iron (II) acetate-derived composites (e.g., FF70-2) achieve enhanced low-rate capacities due to optimized nanoscale porosity maximizing interfacial contact area, whereas iron (III) acetate-derived FF70-3 exhibits better electrochemical performance at high rates owing to its hierarchical mesoporous network facilitating rapid ion diffusion. This performance bifurcation conclusively validates the previously postulated mechanism wherein micropore dimensions induce transport limitations during high-current operation through geometric confinement phenomena. In contrast, the mesoporous structure formed in FF70-3 can not only remarkably promote ion diffusion but also accommodate repetitive lithium intercalation/extraction without causing irreversible deformation. This synergistic integration of rapid ion transport capability and mechanical stability establishes a foundational paradigm for advanced energy storage materials.

Electrode reaction kinetics were rigorously interrogated through electrochemical impedance spectroscopy (EIS), revealing fundamental correlations between pore architecture and charge transfer dynamics. As shown in Figure 6a, the composite designated FF70-3 exhibits minimal interfacial charge transfer resistance (R_{ct}). It signifies optimized reaction kinetics at the electrode-electrolyte interface. These enhancements originate in the material's hierarchical mesoporous architecture, which establishes efficient ionic percolation pathways that dramatically reduce lithium-ion diffusion lengths while preserving structural continuity of the conductive network. Through the fitting and analysis of ZView3 software, comparative analysis identifies FF70-2 exhibits intermediate impedance characteristics, with an R_{ct} of 141.6 Ω. While its ultra-high specific surface area derived from nanoscale porosity provides abundant lithium adsorption sites, the diffusion pathways are constrained to impede lithium-ion penetration into graphitic interlayers [38]. The most pronounced impedance occurs in FF90-3, with an R_{ct} of 177.4 Ω, where 90V FJH triggers destructive structural evolution. Excessive thermal energy input during high-voltage etching probably induces carbon lattice fragmentation, generating macropores that disrupt the intrinsic conductive network through graphene sheet disconnection. This irreversible damage creates electron transport bottlenecks, particularly evident in the middle-frequency semicircle expansion. The compromised electrical percolation pathway fundamentally limits charge transfer efficiency.

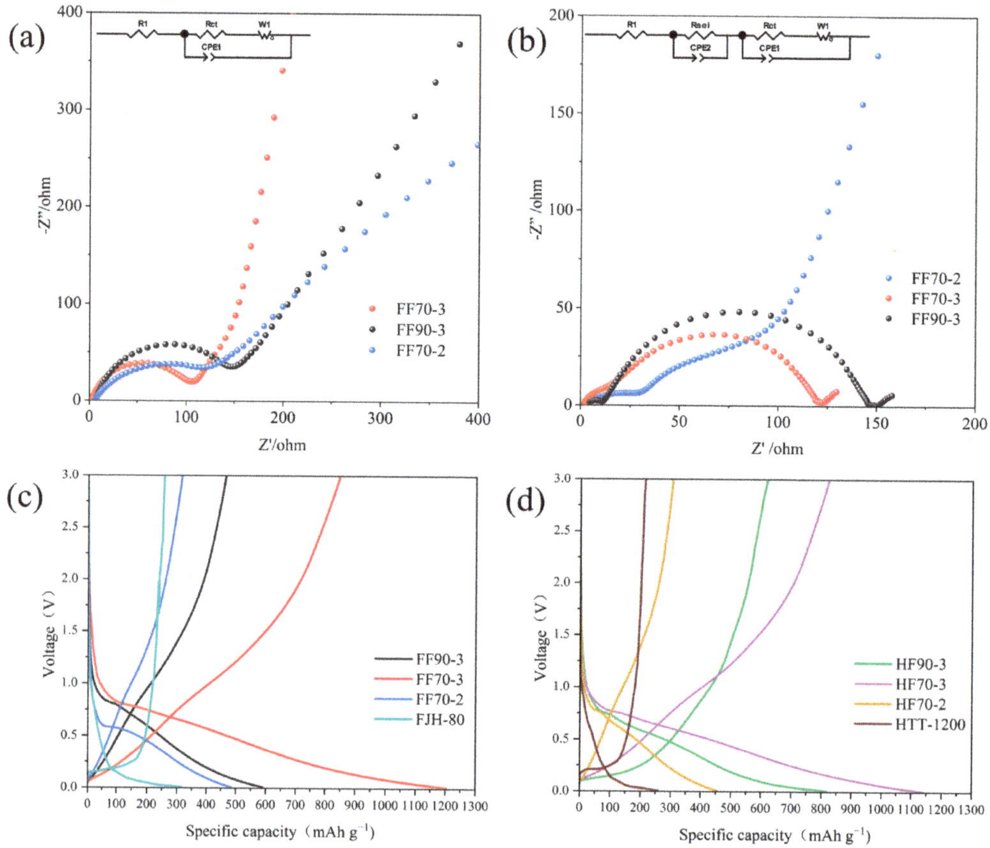

Figure 6. Nyquist plots of flash Joule heating series (**a**) before cycling and (**b**) after 600 cycles at 2C. Voltage-specific capacity diagram in the first complete cycle of (**c**) FJH series and (**d**) HTT series.

Figure 6b shows the EIS plots of the three materials after 600 cycles at 2C. The R_{sei} and R_{ct} after cycling are presented in Table S3. It is obvious that a distinct semi-circle corresponding to R_{sei} can be observed in the high-frequency region for all three samples, which indicates the formation of the SEI layer during the cycling process. Among them, FF70-2 has the largest R_{sei}, which is attributed to its larger specific surface area that generates a thicker SEI layer, resulting in a high resistance in this part. Meanwhile, a relatively large slope in the low-frequency region can be observed for FF70-2, which may be due to the formation of the SEI film blocking the microporous structure and hindering the transport of lithium ions. In contrast, the slopes in the low-frequency region of FF70-3 and FF90-3 are close to the ideal 45°. FF70-3 also shows an almost ideal diffusion state and the smallest R_{ct}, which demonstrates the better electrochemical performance of FF70-3 under high-rate conditions.

Galvanostatic charge–discharge profiles (Figure 6c) corroborate these observations, revealing FF70-2's distinctive voltage plateau depression and premature capacity fade. This performance deterioration stems from compromised structural integrity: the ultrahigh porosity fraction undermines mechanical coherence, inducing localized framework collapse during initial lithium insertion cycles. The resultant loss of percolation pathways and active material isolation creates irreversible capacity loss mechanisms that manifest as abrupt voltage drops during subsequent cycling (Figure 5a). The voltage-specific capacity diagrams of the first cycle for the FJH series and HTT series are shown in Figure 6c,d. It can be observed from the figures that the larger the specific surface area, the smaller the initial coulombic efficiency (ICE). This is because a larger specific surface area consumes more lithium ions when forming the SEI layer in the first cycle, resulting in a lower ICE. The ICE, initial specific capacity, and specific capacity after 600 cycles under 2C conditions for the two groups of eight samples are shown in Table S4.

Cyclic voltametric profiling at 0.1 mV s^{-1} provides complementary kinetic insights (Figure 7a). All samples exhibit characteristic electrochemical signatures: an irreversible reduction peak between 0.5 and 0.75 V during initial cycling corresponds to electrolyte reduction and concomitant SEI formation, while reversible redox pairs below 0.5 V signify lithium intercalation/deintercalation within graphitic domains. The near-perfect superimposition of second and third cycle voltammograms confirms good electrochemical reversibility in FF90-3, FF70-3, and FJH-80.

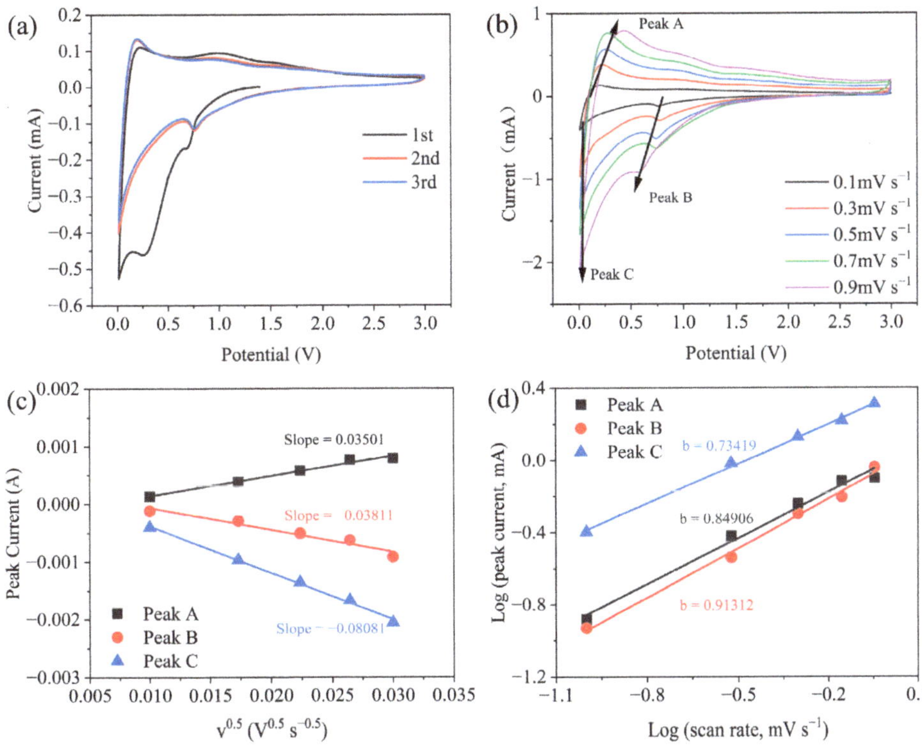

Figure 7. CV curves of (**a**) the first three cycles at 0.1 mV s^{-1} scan rate for FF70-3 and (**b**) FF70-3 at 0.1, 0.3, 0.5, 0.7, 0.9 mV s^{-1} scan rates. (**c**) The fitting curves based on $v^{1/2}$ and peak current from CV results at different sweep rates. (**d**) Log(i) versus log(v) at peak A, peak B, and peak C.

Cyclic voltammetry (CV) analysis spanning scan rates from 0.1 to 0.9 mV s^{-1} (Figure 7b) probes the lithiation–delithiation kinetics. Well-defined redox peaks emerge at characteristic potentials of 0.1 V (corresponding to LiC_6 formation during discharge) and 0.2 V (associated with LiC_6 decomposition during charge), demonstrating highly reversible phase transformation behavior. Progressive CV curve broadening with increasing scan rates signifies good electrochemical stability; at the same time, ΔE_p is also gradually increasing (Table S5). Crucially, Figure S6 quantitatively validates that FF70-3's charge storage capacity and cycling durability consistently surpass alternative regenerated graphite materials documented in prior studies [15,16,19,24,39–41], establishing its performance superiority.

By fitting the peak currents of the three redox peaks and the square roots of their corresponding voltages at different scan rates, three straight lines can be obtained. The slopes of these lines are related to the diffusion coefficient of lithium ions, and the diffusion coefficient can be calculated through the Randles–Ševčík equation.

$$I_P = 269,000n^{1.5}ACD^{0.5}\nu^{0.5}$$

here, ν is the scan rate set in CV, with units of V/s; D is the diffusion coefficient, with units of $cm^2\ s^{-1}$; C is the ion concentration, with units of mol/cm^3; A is the electrode area, with units of cm^2, which can generally be considered as the geometric area of the electrode sheet; and n is the number of electrons transferred in the redox reaction, with n = 1 for a single-electron reaction. $I_P/\nu^{0.5}$ is the slope in Figure 7c. Through calculation, it can be obtained that the diffusion coefficient of FF70-3 is in the range of 10^{-7} to 10^{-8} $cm^2\ s^{-1}$, which is much higher than that of commercial graphite at 10^{-10} $cm^2\ s^{-1}$, proving that its mesoporous structure can effectively enhance the ion transport capacity. As shown in the Figure 7d, by calculating the b values corresponding to the three redox peaks, it can be found that all the b values are between 0.5 and 1.0 and are closer to 1.0. This indicates that the reaction is less limited by ion diffusion kinetics, further demonstrating the impact of the porous structure on improving performance.

3. Materials and Methods

3.1. Materials

Spent LIBs were bought from Shandong Jiuli Electronic Technology Co., Ltd., Jinan, China. Iron (II) acetate tetrahydrate ($C_4H_6O_4Fe$, 95%, Fe 21% min), iron (III) acetate polyhydrate ($C_4H_7FeO_5{\cdot}nH_2O$, AR), and ethylene glycol ($C_2H_6O_2$, AR, 98%) were purchased from Shanghai McLean Biochemical Co., Ltd., Shanghai, China. Potassium permanganate ($KMnO_4$, AR, 99%), sulfuric acid (H_2SO_4, AR), nitric acid (HNO_3, AR), and hydrogen peroxide (H_2O_2, AR, 30%) were purchased from Sinopharm Chemical Reagent Co., Ltd., Shanghai, China.

3.2. Acquisition of Spent Graphite

The spent LIBs were subjected to a deep discharge process, followed by disassembly. Spent LIBs refer to those that have lost their value for continued use. The batteries used in this experiment were originally used in EVs. The copper foil, which had the anode material attached, was immersed in a 1 mol/L KOH solution to facilitate the separation of the anode material from the copper foil. Subsequently, the resulting material was washed with acid and deionized water until its pH reached approximately 7. The powder was then dried to yield SG. Electrochemical tests were conducted on SG, revealing its electrochemical performance is bad, as illustrated in Figure S7.

3.3. Preparation of Materials

The SG was sieved through a 350-mesh sieve and processed via two distinct routes: flash Joule heating at 80 V (denoted as FJH-80) and high-temperature treatment (HTT) $\geq$ 1200 °C (denoted as HTT-1200). These pretreated materials were converted to graphite oxide (GO) using a modified Hummers' method [42]. The GO prepared by using FJH-80 and HTT-1200 was, respectively, denoted as GO-FJH and GO-HTT. Specific methods are detailed in the Supplementary Materials.

For the synthesis of composites, 0.4 g of iron (III) acetate polyhydrate ($Fe(CH_3COO)_3{\cdot}nH_2O$) or iron (II) acetate tetrahydrate ($Fe(CH_3COO)_2{\cdot}4H_2O$) was dissolved in 60 mL of ethylene glycol using ultrasonication for 1 h at 100 W in an ice bath. Simultaneously, 0.5 g of GO-FJH or GO-HTT was dispersed in 100 mL of EG under the same conditions to create 5 mg/mL suspensions. The iron (III) or iron (II) acetate solutions were gradually added to the GO dispersions, followed by 2 h of magnetic stirring. The resulting mixtures were transferred to autoclaves for hydrothermal treatment at 180 °C for 8 h. After cooling, the products were filtered, washed with water, and vacuum-dried at 70 °C for 8 h to yield four composites: EG-Fe^{3+}-FJH, EG-Fe^{3+}-HTT, EG-Fe^{2+}-FJH, and EG-Fe^{2+}-HTT.

Each composite (0.18 g) underwent FJH in an inert atmosphere. The EG-Fe^{3+}-FJH composite also received 15 pulses at 70 V and 90 V, yielding FF70-Fe^{3+} and FF90-Fe^{3+}. The EG-Fe^{3+}-HTT composite received 15 pulses at 70 V and 90 V, resulting in the formation of HF70-Fe^{3+} and HF90-Fe^{3+}. Both the EG-Fe^{2+}-FJH and EG-Fe^{2+}-HTT composites were subjected to 15 pulses at 70 V, producing FF70-Fe^{2+} and HF70-Fe^{2+}, respectively. Following this, the materials were acid-washed with 5 mol/L HNO_3 using magnetic stirring for 2 h. They were then rinsed to neutrality and freeze-dried to obtain porous expanded graphite, designated as FF70-3, FF90-3, HF70-3, HF90-3, FF70-2, and HF70-2. Here, the first "H" or "F" refers to the graphite treated with HTT or FJH for impurity removal, the second "F70" or "F90" indicates the etching voltage of FJH being 70 V or 90 V, and the last "2" or "3" represents the valence state of iron in iron acetate during the loading process. The complete preparation process is shown in Figure 8.

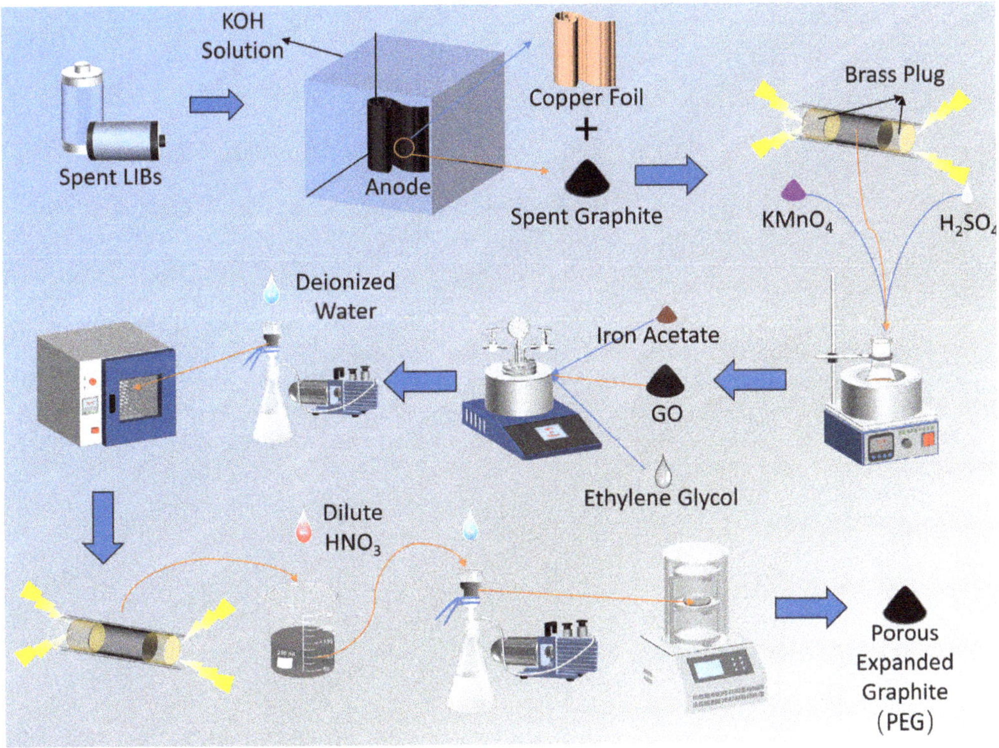

Figure 8. Experimental procedure diagram.

3.4. Characterization and Electrochemical Performance

Surface morphology and microstructure of the graphite specimens were imaged using scanning electron microscopy (SEM; Zeiss SUPRATM55, Oberkochen, Germany) and transmission electron microscopy (TEM; HITACHI HT7700, Tokyo, Japan). Elemental composition across sample surfaces was routinely determined through X-ray energy-dispersive spectroscopy (EDS) integrated with the SEM platform. Crystalline structure properties were characterized by X-ray diffraction (XRD; Rigaku Miniflex600, Tokyo, Japan) employing Cu Kα radiation at a scanning rate of 5°/min over the 2θ range of 10° to 80°. Surface area and pore size distribution were modeled from N_2; adsorption–desorption isotherms (Quanta Chrome Instruments Co., Ltd., Boynton Beach, FL, USA).

Electrochemical assessments employed CR2025 coin cells fabricated in an argon-filled glove box (H_2O and O_2 < 0.1 ppm). Electrode slurries were prepared by blending sodium carboxymethyl cellulose (CMC), Super P carbon, and active material at a 1:2:7 mass ratio. After thorough mixing, homogeneous slurry coatings were applied onto copper foils, followed by vacuum drying (70 °C, 10 h). The resulting electrodes displayed ~20 μm coating thickness with active material loadings of 0.56–0.84 mg. Lithium foil counter electrodes and Celgard 2400 polypropylene separators were utilized. The electrolyte contained 1.0 M $LiPF_6$ in ethylene carbonate/dimethyl carbonate (EC:DMC = 3:7, v/v). Assembled cells underwent 10-h stabilization before testing. Galvanostatic charge–discharge cycling was performed at ambient temperature using a LAND tester (CT3001A, Wuhan Lanhe, Wuhan, China) within 0.01–3.0 V. Cyclic voltammetry (CV) measurements employed CHI 660e electrochemical workstation (Shanghai Chenhua, Shanghai, China) with 0.01–3.0 V (vs. Li^+/Li) sweeps at 0.1–0.9 mV s^{-1} scan rates. Electrochemical impedance spectroscopy (EIS) was subsequently conducted on the same instrument with 10 mV AC amplitude over 100 kHz to 0.01 Hz frequencies.

4. Conclusions

In summary, this research successfully achieves high-value recovery and resource utilization of graphite from spent lithium-ion battery anodes through the innovative integration of flash Joule heating technology with precision structural engineering. The experimental framework systematically examines the influences of three critical processing variables: decontamination methodologies (comparing flash Joule heating with high-temperature treatment), transition metal valence states in acetate precursors (iron (II) versus iron (III)), and etching voltages (70 V versus 90 V). The characterization results show that FJH has a better effect than HTT for SG decontamination and induces a larger interlayer spacing. The GO-FJH prepared in this way has a higher degree of oxidation and can load more iron oxides during the loading process. The iron oxide nanoparticles formed on the graphite layer by the combination of hydrothermal treatment and iron (II) acetate are significantly smaller than those formed by iron (III) acetate. Under the flash etching condition of 70 V, the composites synthesized via hydrothermal processing with iron (III) mainly form meso-macropores. Thanks to the hierarchical structure of mesopores and macropores formed during the process, this structure provides more lithium-ion transmission channels, greatly promoting the transmission of ions and enhancing the long-cycle performance of FF70-3. In the electrochemical test, the capacity of FF70-3 remains as high as 419 mAh g^{-1} after 600 cycles at a high rate of 2C, with a capacity retention rate of 70.18%. This investigation establishes a scientifically grounded and technologically viable pathway for transforming waste graphite into high-performance anodes, providing substantial guidance for sustainable energy material development with promising application prospects.

Supplementary Materials: The following supporting information can be downloaded at https://www.mdpi.com/article/10.3390/recycling10050171/s1, Figure S1: XRD of SG, FJH-80, and HTT-1200; Figure S2: XRD of GO-FJH and GO-HTT; Figure S3: XRD of EG-Fe2+-FJH, EG-Fe2+-HTT, EG-Fe3+-FJH, and EG-Fe3+-HTT; Figure S4: Voltage-specific capacity diagram in the 1st, 10th, 30th, 50th, and 100th cycle of FF70-2 at 0.1C; Figure S5: Voltage-specific capacity diagram in the 1st, 50th,

150th, 300th, and 600th cycle of FF70-3 at 2C; Figure S6: Comparison with previous work performance; Figure S7: Electrochemical performance of SG at 2C; Table S1: SG, FJH-80, HTT-1200 corresponding to the glancing angle and (002) crystal plane spacing; Table S2: Specific surface area of (a) micropore, (b) meso- and macropore, (c) micropore volume, and (d) total pore volume; Table S3: Rsei and Rct values after 600 cycles under 2C conditions for FF70-3, FF90-3, and FF70-2; Table S4: The initial coulombic efficiency, initial specific capacity, and specific capacity after 600 cycles of all products mentioned in this article are based on 2C; Table S5: The relationship between scan rate and ΔEp.

Author Contributions: Conceptualization, Y.L. and J.S.; methodology, Y.L.; validation, Y.L., W.C. and S.L.; formal analysis, Y.L.; investigation, Y.L.; data curation, Y.L.; writing—original draft preparation, Y.L.; writing—review and editing, J.S. and Z.W.; supervision, J.S.; project administration, J.S.; funding acquisition, J.S. and Z.W. All authors have read and agreed to the published version of the manuscript.

Funding: This research was supported by National Natural Science Foundation of China (No. 52370140), and Shandong Province Excellent Youth Science Fund Project (2023HWYQ-022).

Data Availability Statement: The original contributions presented in this study are included in the article. Further inquiries can be directed to the first/corresponding author.

Conflicts of Interest: The authors declare no conflict of interest.

References

1. Wang, W. Green energy and resources: Advancing green and low-carbon development. *Green Energy Resour.* **2023**, *1*, 100009. [CrossRef]
2. Nguyen-Tien, V.; Zhang, C.; Strobl, E.; Elliott, R.J.R. The closing longevity gap between battery electric vehicles and internal combustion vehicles in Great Britain. *Nat. Energy* **2025**, *10*, 354–364. [CrossRef]
3. Li, X.; Deng, C.; Liu, M.; Xiong, J.; Zhang, X.; Yan, Q.; Lin, J.; Chen, C.; Wu, F.; Zhao, Y.; et al. Reutilization and upcycling of spent graphite for sustainable lithium-ion batteries: Progress and perspectives. *eScience* **2025**, *5*, 100394. [CrossRef]
4. Xiao, J.; Li, J.; Xu, Z. Challenges to Future Development of Spent Lithium Ion Batteries Recovery from Environmental and Technological Perspectives. *Environ. Sci. Technol.* **2020**, *54*, 9–25. [CrossRef] [PubMed]
5. Lecocq, A.; Eshetu, G.G.; Grugeon, S.; Martin, N.; Laruelle, S.; Marlair, G. Scenario-based prediction of Li-ion batteries fire-induced toxicity. *J. Power Sources* **2016**, *316*, 197–206. [CrossRef]
6. Wang, Q.; Mao, B.; Stoliarov, S.I.; Sun, J. A review of lithium ion battery failure mechanisms and fire prevention strategies. *Progress. Energy Combust. Sci.* **2019**, *73*, 95–131. [CrossRef]
7. Iturrondobeitia, M.; Vallejo, C.; Berroci, M.; Akizu-Gardoki, O.; Minguez, R.; Lizundia, E. Environmental Impact Assessment of LiNi1/3Mn1/3Co1/3O2 Hydrometallurgical Cathode Recycling from Spent Lithium-Ion Batteries. *ACS Sustain. Chem. Eng.* **2022**, *10*, 9798–9810. [CrossRef]
8. Natarajan, S.; Aravindan, V. An Urgent Call to Spent LIB Recycling: Whys and Wherefores for Graphite Recovery. *Adv. Energy Mater.* **2020**, *10*, 2002238. [CrossRef]
9. Yi, C.; Zhou, L.; Wu, X.; Sun, W.; Yi, L.; Yang, Y. Technology for recycling and regenerating graphite from spent lithium-ion batteries. *Chin. J. Chem. Eng.* **2021**, *39*, 37–50. [CrossRef]
10. Li, Z.; Li, J.; Wang, L. Renewed graphite for high-performance lithium-ion batteries: Catalytic graphitization approach. *J. Mater. Sci. Mater. Electron.* **2024**, *35*, 599. [CrossRef]
11. Shang, Z.; Yu, W.; Zhou, J.; Zhou, X.; Zeng, Z.; Tursun, R.; Liu, X.; Xu, S. Recycling of spent lithium-ion batteries in view of graphite recovery: A review. *eTransportation* **2024**, *20*, 100320. [CrossRef]
12. Yu, W.; Guo, Y.; Xu, S.; Yang, Y.; Zhao, Y.; Zhang, J. Comprehensive recycling of lithium-ion batteries: Fundamentals, pretreatment, and perspectives. *Energy Storage Mater.* **2023**, *54*, 172–220. [CrossRef]
13. Zhang, Y.; Wang, W.; Fang, Q.; Xu, S. Improved recovery of valuable metals from spent lithium-ion batteries by efficient reduction roasting and facile acid leaching. *Waste Manag.* **2020**, *102*, 847–855. [CrossRef]
14. Zhang, Y.; Wang, W.; Hu, J.; Zhang, T.; Xu, S. Stepwise Recovery of Valuable Metals from Spent Lithium Ion Batteries by Controllable Reduction and Selective Leaching and Precipitation. *ACS Sustain. Chem. Eng.* **2020**, *8*, 15496–15506. [CrossRef]
15. Yu, H.; Dai, H.; Zhu, Y.; Hu, H.; Zhao, R.; Wu, B.; Chen, D. Mechanistic insights into the lattice reconfiguration of the anode graphite recycled from spent high-power lithium-ion batteries. *J. Power Sources* **2021**, *481*, 229159. [CrossRef]
16. Zhang, Z.; Zhu, X.; Hou, H.; Tang, L.; Xiao, J.; Zhong, Q. Regeneration and utilization of graphite from the spent lithium-ion batteries by modified low-temperature sulfuric acid roasting. *Waste Manag.* **2022**, *150*, 30–38. [CrossRef] [PubMed]

17. Gao, Y.; Zhang, J.; Chen, Y.; Wang, L.; Wang, C. Graphite regenerating from retired (LFP) lithium-ion battery: Phase transformation mechanism of impurities in low-temperature sulfation roasting process. *Renew. Energy* **2023**, *204*, 290–299. [CrossRef]
18. Xie, X.; Zhang, J.; Chen, Y.; Wang, C. A method for the preparation of graphene from spent graphite of retired lithium-ion batteries. *J. Power Sources* **2024**, *594*, 234023. [CrossRef]
19. Zhang, J.; Lei, Y.; Lin, Z.; Xie, P.; Lu, H.; Xu, J. A novel approach to recovery of lithium element and production of holey graphene based on the lithiated graphite of spent lithium ion batteries. *Chem. Eng. J.* **2022**, *436*, 135011. [CrossRef]
20. Xiao, F.; Chen, X.; Zhang, J.; Huang, C.; Hu, T.; Hong, B.; Xu, J. Large-scale production of holey graphite as high-rate anode for lithium ion batteries. *J. Energy Chem.* **2020**, *48*, 122–127. [CrossRef]
21. Ma, X.; Li, S.; Tang, W.; Liu, R.; Fu, Z.; Wang, S. Laser-Induced Coal-Based Porous Graphene as Anode Toward Advanced Lithium-Ion Battery. *Adv. Sci.* **2025**, *12*, e2504592. [CrossRef] [PubMed]
22. Ma, C.; Zhao, Y.; Li, Y. A facile solution-free etching preparation of porous graphene nanosheets with high performances for lithium storage. *Chem. Eng. J.* **2017**, *320*, 283–289. [CrossRef]
23. Xiao, Y.; Li, J.; Huang, W.; Wang, L.; Luo, J. Green & efficient regeneration of graphite anode from spent lithium ion batteries enabled by asphalt coating. *J. Mater. Sci. Mater. Electron.* **2022**, *33*, 16740–16752. [CrossRef]
24. Xu, C.; Ma, G.; Yang, W.; Che, S.; Li, Y.; Jia, Y.; Liu, H.; Chen, F.; Zhang, G.; Liu, H.; et al. One-step reconstruction of acid treated spent graphite for high capacity and fast charging lithium-ion batteries. *Electrochim. Acta* **2022**, *415*, 140198. [CrossRef]
25. Gong, H.; Xiao, H.; Ye, L.; Ou, X. High-performance expanded graphite regenerated from spent lithium-ion batteries by integrated oxidation and purification method. *Waste Manag.* **2023**, *171*, 292–302. [CrossRef] [PubMed]
26. Zhu, W.; Liu, G.; Hu, J.; Su, G.; Liu, M.; Li, X.; Xu, B. Low-temperature molten salt ion regeneration strategy towards green and efficient spent graphite recycling. *Green. Chem.* **2025**, *27*, 6145–6155. [CrossRef]
27. Dong, S.; Song, Y.; Ye, K.; Yan, J.; Wang, G.; Zhu, K.; Cao, D. Ultra-fast, low-cost, and green regeneration of graphite anode using flash joule heating method. *EcoMat* **2022**, *4*, e12212. [CrossRef]
28. Yi, C.; Ge, P.; Wu, X.; Sun, W.; Yang, Y. Tailoring carbon chains for repairing graphite from spent lithium-ion battery toward closed-circuit recycling. *J. Energy Chem.* **2022**, *72*, 97–107. [CrossRef]
29. Yang, J.-L.; Zhao, X.-X.; Li, W.-H.; Liang, H.-J.; Gu, Z.-Y.; Liu, Y.; Du, M.; Wu, X.-L. Advanced cathode for dual-ion batteries: Waste-to-wealth reuse of spent graphite from lithium-ion batteries. *eScience* **2022**, *2*, 95–101. [CrossRef]
30. Li, L.; Ruan, M.; Tian, D.; Zhang, X.; Hou, X.; Zhong, K.; Cheng, F.; Tong, Y.; Fang, Z. Dual regulation mechanism of MoS2/MoO3 heterostructure for polysulfide and volume effect enables stable and durable lithium storage. *Appl. Surf. Sci.* **2024**, *646*, 158949. [CrossRef]
31. Fan, J.; Yang, J.; Li, H.; Tian, J.; Wang, M.; Zhao, Y. Cryogenic mechanical properties of graphene oxide/epoxy nanocomposites: Influence of graphene oxide with different oxidation degrees. *Polym. Test.* **2021**, *96*, 107074. [CrossRef]
32. Zhang, W.; Xu, H.; Xie, F.; Ma, X.; Niu, B.; Chen, M.; Zhang, H.; Zhang, Y.; Long, D. General synthesis of ultrafine metal oxide/reduced graphene oxide nanocomposites for ultrahigh-flux nanofiltration membrane. *Nat. Commun.* **2022**, *13*, 471. [CrossRef]
33. Al-Tabbakh, A.A.; Karatepe, N.; Al-Zubaidi, A.B.; Benchaabane, A.; Mahmood, N.B. Crystallite size and lattice strain of lithiated spinel material for rechargeable battery by X-ray diffraction peak-broadening analysis. *Int. J. Energy Res.* **2019**, *43*, 1903–1911. [CrossRef]
34. Sun, N.; Li, Z.; Zhang, X.; Qin, W.; Zhao, C.; Zhang, H.; Ng, D.H.L.; Kang, S.; Zhao, H.; Wang, G. Hierarchical Porous Carbon Materials Derived from Kelp for Superior Capacitive Applications. *ACS Sustain. Chem. Eng.* **2019**, *7*, 8735–8743. [CrossRef]
35. Zhao, Y.; Zhang, X. In situ activation graphitization to fabricate hierarchical porous graphitic carbon for supercapacitor. *Sci. Rep.* **2021**, *11*, 6825. [CrossRef]
36. Zhang, X.; Wang, Q.; Li, Y.; Sun, G.; Yu, X.; Li, H. Crack-induced abrupt capacity degradation in commercial LiNi(0.8)Co(0.1)Mn(0.1)O(2) (NCM811)/SiO (x) -graphite pouch batteries. *RSC Adv.* **2024**, *14*, 19116–19123. [CrossRef] [PubMed]
37. Jia, P.; Sun, J.; Li, S.; Wang, W.; Song, Z.; Zhao, X.; Mao, Y. A feasible recycling route for spent graphite: Microwave-assisted puffing with Fe_2O_3 loading to construct high-performance anode for lithium-ion batteries. *Mater. Today Sustain.* **2024**, *25*, 100620. [CrossRef]
38. Rahman, M.A.; Wong, Y.C.; Song, G.; Wen, C. A review on porous negative electrodes for high performance lithium-ion batteries. *J. Porous Mater.* **2015**, *22*, 1313–1343. [CrossRef]
39. Xu, Y.-J.; Song, X.-H.; Chang, Q.; Hou, X.-L.; Sun, Y.; Feng, X.-Y.; Wang, X.-R.; Zhan, M.; Xiang, H.-F.; Yu, Y. The regeneration of graphite anode from spent lithium-ion batteries by washing with a nitric acid/ethanol solution. *New Carbon. Mater.* **2022**, *37*, 1011–1020. [CrossRef]
40. Fan, W.; Zhang, J.; Ma, R.; Chen, Y.; Wang, C. Regeneration of graphite anode from spent lithium-ion batteries via microwave calcination. *J. Electroanal. Chem.* **2022**, *908*, 116087. [CrossRef]

41. Lai, Y.; Zhu, X.; Li, J.; Gou, Q.; Li, M.; Xia, A.; Huang, Y.; Zhu, X.; Liao, Q. Recovery and regeneration of anode graphite from spent lithium-ion batteries through deep eutectic solvent treatment: Structural characteristics, electrochemical performance and regeneration mechanism. *Chem. Eng. J.* **2023**, *457*, 141196. [CrossRef]
42. Kim, J.; Eum, J.H.; Kang, J.; Kwon, O.; Kim, H.; Kim, D.W. Tuning the hierarchical pore structure of graphene oxide through dual thermal activation for high-performance supercapacitor. *Sci. Rep.* **2021**, *11*, 2063. [CrossRef] [PubMed]

recycling

MDPI

Article

Evaluation of the Removal of PVDF Using ToF-SIMS: Comparing Dihydrolevoglucosenone and Pyrolysis as Pretreatments for Cathode Materials of Lithium-Ion Batteries

Marc Simon Henderson [1,2], Aliza Marie Salces [3,4], William D. A. Rickard [5], Denis Fougerouse [5], Álvaro José Rodríguez Medina [3], Elsayed A. Oraby [1], Chau Chun Beh [1], Martin Rudolph [3], Anna Vanderbruggen [4] and Jacques Eksteen [1,2,*]

1 Western Australian School of Mines: Minerals, Energy and Chemical Engineering, Curtin University, Perth, WA 6102, Australia; jane.beh@curtin.edu.au (C.C.B.)
2 Future Battery Industries Cooperative Research Centre, Perth, WA 6102, Australia
3 Helmholtz–Zentrum Dresden–Rossendorf (HZDR), Helmholtz Institute Freiberg for Resource Technology (HIF), Chemnitzer Straße 40, 09599 Freiberg, Germany; m.rudolph@hzdr.de (M.R.)
4 GéoRessources, Université de Lorraine, 54000 Nancy, France; anna.vanderbruggen@univ-lorraine.fr
5 John de Laeter Centre, Curtin University, Kent Street, Bentley, Perth, WA 6102, Australia
* Correspondence: jacques.eksteen@curtin.edu.au

Abstract: Effective and environmentally benign removal of polyvinylidene fluoride (PVDF) binders from spent battery electrodes remains a critical hurdle in sustainable recycling, primarily due to issues related to the mitigation of fluorinated compound emissions. This work evaluates PVDF binder removal from cathode active material using either a green solvent-based dissolution process or pyrolysis, analyzed by time-of-flight secondary ion mass spectrometry (ToF-SIMS). The solvent pretreatment involved mixing dihydrolevoglucosenone (Cyrene™) with PVDF-coated NMC811 at 100 °C, followed by hot filtration to separate the Cyrene-PVDF solution. Pyrolysis was conducted at 800 °C under an argon atmosphere. Positive ToF-SIMS spectra for Cyrene showed characteristic peaks at ketene (42 *m/z*) and 1,3-dioxole (86 *m/z*), along with intense $C_2H_3O^+$, $C_3H_3O^+$, $C_4H_7^+$, and $C_3H_5O^+$ peaks. The characteristic peaks used to identify PVDF were $C_3H_2F_5^+$ (133 *m/z*), $C_3H_2F_3^+$ (95 *m/z*), and $C_3HF_4^+$ (113 *m/z*). Both processes resulted in PVDF removal, with pyrolysis demonstrating higher effectiveness. Particle agglomeration was observed in both pretreated NMC811 samples, however agglomeration was more pronounced with Cyrene pretreatment due to PVDF redeposition. Following pyrolysis, PVDF was transformed into a defluorinated carbonaceous material.

Keywords: dihydrolevoglucosenone; pyrolysis; Tof-SIMS; PVDF; spent LIBs; battery recycling

1. Introduction

The demand for lithium-ion batteries (LIBs) is rapidly increasing, a trend that is expected to continue for the foreseeable future [1]. One major concern faced by the battery industry is the increasing number of spent LIBs entering the recycling value chain, as they pose a risk to the environment and by extension our society [2]. As a result, much research has been conducted on different processing techniques that can be employed when recycling spent LIBs, these include direct recycling, hydrometallurgy and pyrometallurgy [3,4]. Pinegar and Smith [5] reviewed some of the most well-known commercial battery recycling processes, these include the Umicore, Inmetco, Glencore, Accurec, Retriev, Recupyl and

 https://doi.org/10.3390/recycling10020056

EcoBat processes. Regardless of the processing technique chosen, there are a variety of processing stages that need to be employed to deal with the complex nature of battery material. Informative reviews [1,6] which thoroughly discuss the different processing stages that can be employed when recycling spent LIBs are also available. The choice of process is highly dependent on the chemistry of the spent batteries as some batteries are low value (lithium iron phosphate) while others hold significantly more value (nickel and cobalt based batteries), thus it is necessary to understand the market in which they are operating. A life cycle assessment also provides a good resource for better understanding the battery value chain, as such highlighted in the review article of Arshad and coworkers [3]. Finally, to understand the battery industry in an Australian context, the latest market data for the battery industry by Langdon and coworker [7], provided an overview by mapping of battery flows, including battery sales, batteries in use, and spent batteries collection and reprocessing, by battery chemistry, format or size and application.

Lithium-ion batteries contain several components that can negatively impact the environment, including perfluoroalkyl and polyfluoroalkyl substances (PFAS). Perfluoroalkyl and polyfluoroalkyl substances describe a large group of synthetic fluorinated organic compounds, which are chemically stable due to the presence of strong C-F bonds [8]. They are persistent chemicals meaning they do not degrade easily and thus accumulate in the environment [9].

The release of PFAS into the environment can result in devastating environmental, financial, and health implications for the surrounding communities (e.g., requires expensive environmental remediation, can damage the immune system, thyroid, and liver [9,10], particularly when the material is either in a soluble or in a micro-dispersed solid form, making these material highly mobile in the environment. The potential environmental and health risks linked with PFAS have prompted governing bodies like the European Commission to propose banning their usage. If successful, such legislation could be implemented as soon as 2026 [11], however, significant quantities of spent batteries would require safe recycling techniques.

Polyvinylidene fluoride (PVDF), is one such PFAS and has applications in a host of different industries, due to its good physical and chemical properties which make it resistant to chemical and thermal degradation [12,13]. Polyvinylidene fluoride is ideal to be used as a binder in LIBs as it allows for the strong adhesion of cathode active material (CAM) particles to the conductive Al foil and absorption of the electrolyte, thereby allowing for the transfer of lithium (Li) ions to the CAM material, and offering a high degree of electrochemical stability [1,12,14,15]. Water soluble binders such as carboxymethyl cellulose-styrene-butadiene rubber (CMC-SBR) offer alternatives to PFAS binders [14,16], however they are more likely applied to the anode electrodes.

The focus of this study is to compare the effectiveness of greener technologies (i.e., technologies that have a smaller environmental impact than current industrial technologies) in removing binders from cathode materials as found in nickel-manganese-cobalt (NMC) based lithium-ion batteries (LIBs), specifically PVDF as it is the state-of-the-art binder used in CAMs and some anodes [17].

During battery recycling, it is important to remove binders prior to any processing techniques as they can affect the recovery/separation efficiencies of processes such as flotation and leaching. In the case of froth flotation, binders alter the surface wettability of CAMs and anode active materials (AAMs, i.e., graphite), reducing their separation efficiency [18–20]. On the other hand, during leaching, the presence of PVDF on CAMs particles can impact the dissolution of target metals [21,22]. The PVDF binder adheres the CAMs to the Al foil and binds the individual CAM particles to one another resulting in the formation of agglomerates post crushing/shredding [23]. The decrease in the exposed CAM particle surface area leads to reductions in the leaching efficiencies of CAM.

Currently, two main strategies are employed to remove PVDF during LIBs recycling, namely thermal pretreatment or chemical dissolution [24]. In this context, thermal pretreatment describes a process that operates at elevated temperatures (400–600 °C) for the decomposition of an organic binder [17] and is typically performed under an inert atmosphere to prevent graphite oxidation. Pyrolysis is one of the most common thermal pretreatments employed to remove PVDF from spent LIBs/CAM and the decomposition of PVDF occurs at temperatures between 400 °C and 500 °C [17,25,26]. The main advantages of using pyrolysis include its effectiveness in removing PVDF [27] as well as the reduction in overall process complexity as spent batteries can be pyrolyzed without the need for a separate battery discharge stage, as typically seen when using hydrometallurgical routes [28]. However, there are numerous drawbacks, including the generation of toxic off-gases from the decomposition of PVDF and other organics [17]. Off-gases include halogenated hydrocarbons, hydrofluoric acid (HF), polynuclear organic substances and other toxic hydrocarbons [27,29,30], all of which are harmful to the environment and human and animal health. Scrubbing of the gases provides an "end-of-pipe" solution, but transfers many problems to liquid effluents that requires costly downstream treatment.

An alternative to pyrolysis is the dissolution of PVDF in dipolar aprotic solvents, the most common of which include dimethylacetamide (DMA), dimethylformamide (DMF) and *N*-methyl-2-pyrrolidone (NMP) [14,31]. These solvents are often toxic, adversely affect the human reproductive system and cause numerous health issues [32,33]. As such, the use of solvents such as NMP are strictly regulated [24]. Thus, many researchers are investigating greener solvents that can be used for PVDF dissolution.

Dihydrolevoglucosenone (refer to Figure 1), going by the commercial name of "Cyrene™", shows promise in replacing the aforementioned solvents for PVDF dissolution [34–36]. Cyrene is a bio-derived dipolar aprotic solvent that is produced through the pyrolysis of cellulose, e.g., from agricultural and wood-derived wastes, producing levoglucosenone which is then hydrogenated, creating dihydrolevoglucosenone [33]. Cyrene has a high boiling point (227 °C), is miscible in water and thus offers a system in which the solvent polarity can be adjusted by altering the water to Cyrene ratio [31,33]. The advantages of using Cyrene, as opposed to conventional dipolar aprotic solvents, includes no significant toxicity, non-mutagenic, non-genotoxic, and readily biodegradable within 28 days [37].

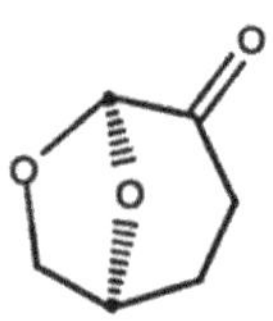

Figure 1. Structure of dihydrolevoglucosenone (adapted from Sigma-Aldrich [37]).

Heating is essential for dissolving PVDF in Cyrene. Previous research [38,39] has demonstrated that temperatures exceeding 80 °C are necessary to achieve dissolution. Once dissolved, PVDF can be recovered through thermally induced phase separation (TIPS) as it precipitates from Cyrene upon cooling [36,38]. For battery recycling applications, it is crucial to consider Cyrene's sensitivity to strongly oxidizing, reducing, acidic, or basic environments [33]. This is particularly relevant because metals like cobalt (Co) and nickel (Ni), commonly found in batteries, are strong reducing agents.

One of the most challenging aspects of the PVDF removal process is in the analysis of PVDF residuals in the CAM or black mass material (fine fraction from LIB comminution).

Common analytical techniques such as inductively coupled plasma mass spectrometry (ICP-MS), energy dispersive X-ray spectroscopy (EDS), X-ray fluorescence spectroscopy (XRF), and X-ray diffraction spectroscopy (XRD) have trouble analysing light elements [40,41] and several of the above techniques (EDS, XRF, and ICP-MS) cannot be applied when analysing organic compounds.

Industrially, there was previously little incentive for companies to analyse residual PVDF, however as the EU tightens its regulations around PFAS materials, companies are becoming more interested in accounting for PVDF in their process streams. Currently, most of the research utilizes combinations of Fourier-transform infrared spectroscopy (FTIR), Raman spectroscopy, and EDS when trying to identify the extent to which PVDF is removed [38,42,43] as well as inductively coupled plasma optical emission spectroscopy (ICP-OES) to identify any deportment of lithium or other metal species during Cyrene pre-treatment [38]. The amount of fluorine in the sample before and after the pre-treatment can be analysed by quantitative EDS [24,38]). FTIR and Raman spectroscopy can be used to identify the presence of PVDF in a sample as well as provide insight into the structure of PVDF [38,43].

Time-of-flight secondary ion mass spectrometry (ToF-SIMS) is a surface-sensitive analytical technique capable of detecting light elements, trace elements, and organic compounds. Sui and co-workers [40] investigated the analysis of battery materials using a combination of ToF-SIMS and focused ion beam —scanning electron microscope (FIB-SEM) to perform nano-scale chemical mapping on LIB CAM and analysis on the mobility or lack thereof of elements such as Li, Co and Mn during the charging and discharging of LIBs.

To the best of authors' knowledge, ToF-SIMS has not been used to qualitatively evaluate the removal of PVDF during LIB recycling through pyrolysis and Cyrene-based pre-treatment techniques. The sensitivity of the technique, coupled with the ability to produce high spatial-resolution chemical/molecular maps makes ToF-SIMS a suitable analysis technique for detecting light elements and organic binders in battery materials. The aim of this research is to investigate the removal of PVDF through either Cyrene or pyrolysis pretreatments. Using ToF-SIMS, this research also aims to identify the unique spectra of pristine Cyrene and PVDF, which can then be applied to identify their presence in battery materials. Techniques such as Raman spectroscopy, FTIR, and SEM are used to support the ToF-SIMS findings.

2. Methods & Materials

2.1. Materials

Various battery materials were studied to determine the presence and removal of PVDF on battery materials namely, pristine CAM without PVDF—NMC 111 ($LiNi_{0.33}Mn_{0.33}Co_{0.33}O_2$, MSE supplies, Product No. PO0126), PVDF-coated CAM—NMC 811 (PVDF—$LiNi_{0.8}Mn_{0.1}Co_{0.1}O_2$, Targray, Lot No. NMC 811-003T) and battery grade PVDF ($\geq$99.6%, MSE supplies, batch No. 22420A2, Tucson, AZ, USA). Cyrene solvent (99.3% dihdyrolevoglucosenone, batch no. DCyD12_220308) with a purity of 99.28% was supplied by the Circa Group, Melbourne, Australia.

2.2. Methods

2.2.1. PVDF Dissolution—Cyrene Pretreatment Experiments

The dissolution of PVDF was initially investigated by dissolving 0.5 g pure PVDF into 20 mL Cyrene at 100 °C. Figure 2 depicts a simple Cyrene-PVDF dissolution experiment which involved mixing PVDF powder with Cyrene at room temperature (A), heating the mixture to 100 °C and achieving complete PVDF dissolution (B) and cooling the solution to

50 °C, resulting in the formation of a gel-like material (C). The room temperature Cyrene-PVDF mixture (D) was centrifuged to recover a single lump of PVDF (E).

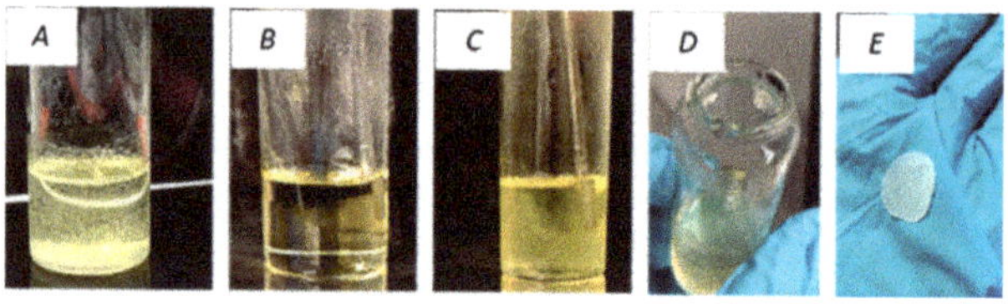

Figure 2. PVDF dissolution with Cyrene: Cyrene-PVDF mixture at room temperature, (**A**) heated solution at 100 °C (**B**), mixture cooled at 53 °C (**C**), mixture at room temperature (**D**), and recovered PVDF after centrifugation of mixture at room temperature (**E**).

To determine the effect of Cyrene pretreatment at elevated temperatures on pristine CAM, 5 g NMC 111 and 30 mL Cyrene were mixed, heated, and held at 100 °C for 1 h. The heated NMC 111-Cyrene mixture was separated by a heated vacuum filtration using a filter paper (Whatman®, GE Healthcare Companies, Chicago, IL, USA) with a pore size of 11 μm. Heated filtration was achieved by placing a typical Büchner funnel filtration set up into a drying oven at 100 °C. To investigate the removal of PVDF from CAM, 5 g of NMC 811 coated with PVDF was contacted with 30 mL of Cyrene following the procedure above. If no heated filtration was used, the Cyrene-PVDF solution was observed to be more viscous than pure Cyrene, leading to a longer filtration time. This causes cooling of the solution to temperatures below 80 °C. After filtration, the Cyrene-PVDF solution was cooled down to room temperature, after which the mixture was vacuum-filtered or centrifuged to recover PVDF.

2.2.2. PVDF Removal—Pyrolysis Pretreatment Experiments

Pyrolysis pretreatment was simulated using thermogravimetric analysis (TGA) equipment. Approximately 30 mg of PVDF-coated NMC 811 was loaded into an alumina crucible and heated to 800 °C. The samples were held at 800 °C for 5 min, ensuring that enough time was given for PVDF decomposition.

The above pre-treatments of pristine materials resulted in the production of: two NMC 111 samples: no pretreatment and pretreated with Cyrene. Along with, four NMC 811 samples, two of which were treated identically to the NMC 111 above, however the third sample was washed post-filtration, and the fourth sample underwent a simulated pyrolysis pretreatment. Figure S1 in the Supplementary Material presents detailed process flow diagrams that describe the preparation of each material examined in this research.

2.3. *Analytical Techniques*

2.3.1. Scanning Electron Microscope (SEM)

The surface morphologies of the pristine, Cyrene pretreated, and pyrolysed CAM were characterized using SEM. The loose powder samples were mounted on aluminium stubs using double-sided carbon tape (PELCO Tabs TM, 6 mm OD from TED PELLA, Inc., Redding, CA, USA). Two SEMs were used, namely the TESCAN LYRA3 field emission SEM (Tescan, Brno, Czech Republic) at the John de Laeter Centre (JdLC), Curtin University, Perth, Australia and the FEI Quanta 600 F SEM (FEI, Hillsboro, OR, USA) at the Helmholtz Institute Freiberg for Resource Technology (HIF), Dresden, Germany. False colour images were produced at HIF by superimposing the secondary and backscattered electron (BSE) images. The BSE image provides information pertaining to material composition (variations in densities), with blue representing elements with high densities (i.e., Ni, Co, Mn) and black colouring the lighter elements such as carbon. The SE image provides topographical information of the analysed surface and has been given an orange colour.

2.3.2. Thermogravimetric Analyser (TGA)

TGA was performed on PVDF and PVDF-coated NMC 811 to determine the decomposition temperature of PVDF. Pyrolysis was simulated using TGA, namely the Mettler Toledo TCA/DSC 1 STAR System, Chicago, IL, USA. 25–30 mg samples were weighed out into platinum crucibles and were heated from 35 °C to 800 °C at a rate of 10 K/min in an argon atmosphere (gas flow rate = 25 mL/min). The samples were held at 800 °C for 5 min, after which they were cooled down to 35 °C at a rate of 10 K/min.

2.3.3. Fourier-Transform Infrared Spectroscopy (FTIR)

FTIR spectroscopy was used to identify the presence of PVDF in both the untreated and Cyrene-pretreated CAM as well as to determine any changes in PVDF and Cyrene (chemical or thermal) during the pretreatment process. A PerkinElmer Springfield the United States Spectrum 100 FT-IR Spectrometer (PerkinElmer, Waltham, MA, USA) with a diamond/ZnSe crystal was used and the analysis was conducted over a wavelength range of 600–4000 cm^{-1} with a resolution of 4 cm^{-1}.

2.3.4. Raman Spectroscopy

Raman spectroscopy was used to determine if Cyrene post-PVDF dissolution contained residual PVDF or if Cyrene or PVDF underwent any chemical or thermal degradation during the pretreatment. The analysis was conducted using a LabRam HR 800 confocal Raman microscope HORIBA France SAS, Palaiseau, France and the spectra were interpreted and analysed using LabSpec5. The excitation wavelength of the laser was 632.817 nm at a laser intensity of 2 mW, a spectrometer grating of 600/mm was used and spectra from Raman shifts of 200 cm^{-1} to 3500 cm^{-1} were acquired. The lens used was at a 10× magnification (NA of 0.25) and the spectra were recorded with 5 s exposures and 5 repetitions.

2.3.5. Time-of-Flight Secondary Ion Mass Spectrometry (ToF-SIMS)

ToF-SIMS was used to investigate the surface of the CAM particles in order to characterise organic materials and map their spatial distribution. Prior to ToF-SIMS analysis, the powdered samples were pressed into pellets. The powder was loaded into a pellet dye and then pressed at 2000 kg for 1 min using the 5T electric press (model YLJ-5TA, KJ MTI Group, St Richmond, CA, USA). The final pellets had a diameter of approximately 5.5 mm and a thickness of 0.5 mm.

The ToF-SIMS analyses were carried out on a M6 instrument (IONTOF GmbH, Münster, Germany) at the JdLC. The instrument was operated in Spectrometry mode with a 30 kV Bi^{3+} primary ion source at a pulsed current of 0.06–0.14 pA. Positive ions were collected using High Mass Resolution mode for the mass analyser at a cycle time of 100 μs. The analytical conditions resulted in a lateral resolution of ~3 μm and a mass resolution of >10,000 M/ΔM. A random primary ion beam raster pattern and an electron flood gun was used to minimise charging effects during the analysis. SIMS maps were collected over a 50 to 200 μm field of view with 128 × 128 pixels in each frame. The mass spectra included all masses up to 250 mass/charge ratio. To remove surface contamination, the field of view was pre-sputtered with a $Ar_{2000}{}^{+}$ cluster ion source (10 kV, 8 nA) over a 500 × 500 μm area for 30 s. Data were analysed with Surface Lab version 7.3.

3. Results and Discussion

3.1. Pure PVDF Dissolution

The dissolution experiment demonstrated the ability of Cyrene to dissolve PVDF at elevated temperatures and the recovery of PVDF from solution upon cooling, which is consistent with the findings of previous research [38,39]. The recovered solution and solid were analysed using Raman spectroscopy to determine whether the Cyrene (post-PVDF dissolution) contained residual PVDF or underwent any changes (i.e., chemical or thermal degradation). The characteristic Raman shifts for different PVDF phases are: α-phases (410, 489, 611/612, 795) cm^{-1}, β-phase (839 cm^{-1}) and γ-phase (874 cm^{-1}) [42,43].

Figure 3A shows a comparison between the Raman shifts obtained from a pure PVDF sample to that of Cyrene post-PVDF dissolution. These results illustrate that no PVDF is detected in the spent Cyrene solvent, suggesting that there is no build-up of PVDF in the Cyrene solvent. To confirm that Cyrene did not undergo any chemical or thermal degradation, a comparison of the Raman shifts for pure Cyrene and Cyrene post-PVDF dissolution are shown in Figure 3B. It was observed that peaks in the fingerprint region (300–1900 cm^{-1}) are retained, and no new peaks appeared, implying that Cyrene did not undergo chemical or thermal degradation on the basis of this analysis. However, a minimal shift to higher energies after PVDF removal is observed in the peaks which can be explained by a change in molecular strains and stresses, as a result of heating Cyrene to 100 °C.

The above results allude to the possibility of recycling Cyrene as there is no build-up of PVDF in solution and no chemical/thermal degradation. Raman analysis on the recovered PVDF (Figure 3C) shows similar Raman peaks, however residual amounts of Cyrene were retained. This is most likely due to the high viscosity of Cyrene, making it difficult to completely remove through centrifugation. The high boiling point of Cyrene (227 °C) also requires elevated temperatures to completely remove Cyrene from PVDF. Thus, any PVDF recovered through this process should be washed thoroughly with water to remove residual Cyrene.

3.2. Cyrene Solvent Treatment

All reported solvents and diluents of PVDF fall under the dipolar aprotic category [36]. Cyrene, a dipolar aprotic solvent, is theoretically capable of dissolving PVDF. The Hansen solubility parameters of Cyrene (dispersion δD: 18.8; polar δp: 10.6; and hydrogen bonding δH: 6.9) are nearly identical to those of the conventional solvent NMP (δD: 18.0; δp: 12.3; and δH: 7.2) [36], suggesting that Cyrene should be able to dissolve PVDF under specific conditions. However, some research [35] has highlighted that the solubility of PVDF can vary significantly with molecular weight. More so, the higher viscosity of the Cyrene-PVDF solution likely played a critical role. It has been reported [44] that a PVDF-Cyrene solution exhibited a high viscosity of 810 mPa·s. As a comparison, PVDF at 10% *w/v* in NMP has viscosities of 120 mPa·s (Mw = 2.7×10^5 g/mol) or 350 mPa·s (Mw = 4.4×10^5 g/mol).

The high viscosity negatively impacted filtration, consequently hindering the separation of the Cyrene-PVDF solution from NMC 811. This led to extended filtration times and cooling of the solution below 80 °C, triggering thermally induced phase separation (TIPS) of PVDF and thus resulted in the non-selective redeposition of PVDF onto the NMC particles. Hot water (100 °C) was added to aid filtration. Figure 4A presents a false-coloured SEM image of pristine NMC 811 particles in comparison with those post-Cyrene pretreatment in Figure 4B,C. In these images, blue particles represent NMC 811, while dark gray material indicates PVDF coating. Figure 4A shows minimal particle agglomeration in the untreated material, with agglomerate sizes not exceeding 50–100 μm. However, Figure 4B–D demonstrate a significant increase in agglomeration post-Cyrene pretreatment, with agglomerates reaching up to 500 μm. This increased agglomeration is attributed to PVDF redeposition during filtration.

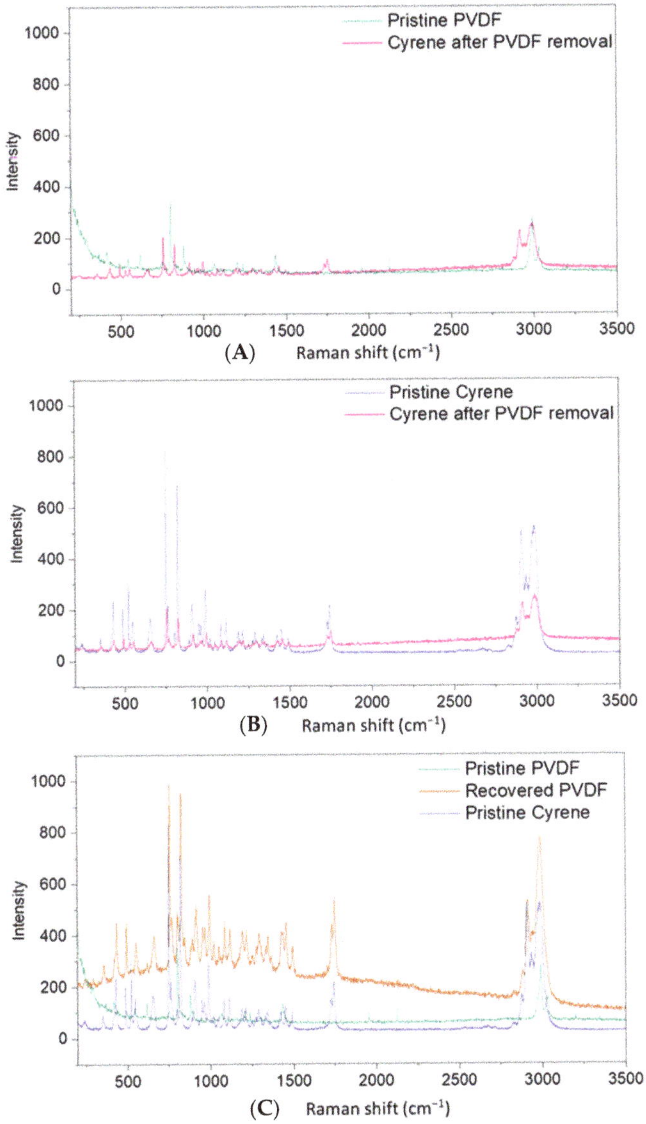

Figure 3. Raman spectra of pristine PVDF and recovered Cyrene after PVDF dissolution (**A**), pristine Cyrene and recovered Cyrene (**B**), and pristine Cyrene and PVDF and recovered PVDF (**C**).

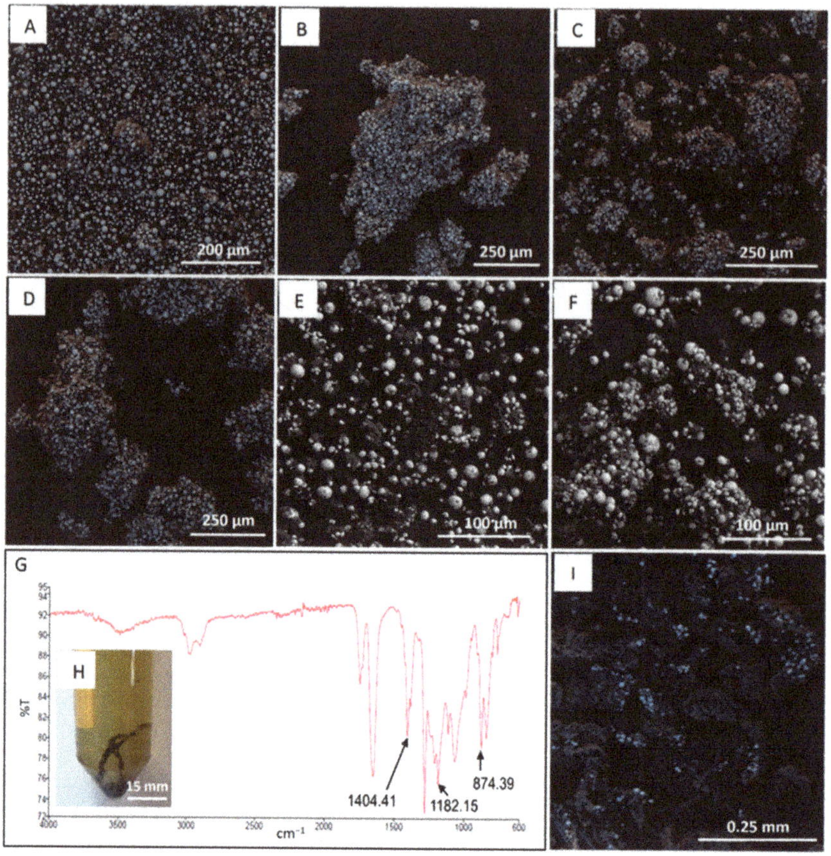

Figure 4. SEM images of NMC and PVDF material. NMC 811 coated with PVDF (**A**), NMC 811 post Cyrene pretreatment (only washing with 100 °C water) (**B–D**), SEM-BSE of NMC 811 post Cyrene pretreatment (**E**,**F**), FTIR spectra of PVDF recovered from Cyrene solution (**G**). Cyrene-water solution and recovered PVDF (**H**). SEM image of recovered PVDF (**I**).

To mitigate PVDF redeposition, hot filtration was performed in an oven. The results are presented in the SEM-BSE images of Figure 4E,F. In these images, lighter (white) particles represent NMC 811, and dark gray regions indicate PVDF. Figure 4E, F, similar to Figure 4A, exhibit particle agglomeration with a maximum agglomerate size of 50–150 μm, demonstrating a significant reduction in agglomeration compared to the non-heated filtration (Figure 4B–D).

Upon cooling the Cyrene-water filtrate to room temperature, a polymer-like material formed (Figure 4H). The SEM image in Figure 4I revealed the presence of fine NMC 811 particles within this material. This contamination can be minimized by using filter paper with a smaller pore size, albeit at the cost of slower filtration rates. The FTIR spectrum in Figure 4G confirmed the material's composition as PVDF, showing three characteristic peaks at 874, 1182, and 1404 cm^{-1}. These peaks, although slightly shifted, confirm partial PVDF removal.

3.3. ToF-SIMS Base Spectra for Cyrene and PVDF

ToF-SIMS was used to characterise both the organic and inorganic components/molecules present in the samples. Molecules fragment during the SIMS ionisation process and as such the resultant mass spectra contains a series of characteristic peaks from the parent and fragment ions. In order to identify the presence of PVDF ($(C_2H_2F_2)_n$) and Cyrene ($C_6H_8O_3$) in CAM, pristine materials were first analysed to determine the characteristic ions. This is illustrated in Figure 5A where the positive ToF-SIMS spectra for PVDF are provided. According to Feng et al. [45], PVDF fragments into carbon-containing molecules with 1, 2 or at most 3 carbon atoms. Feng and co-workers identified $C_3H_2F_5^+$ (133 *m/z*) as a characteristic PVDF peak along with several fluorinated peaks, some of which are present in Figure 5A. In this research, the characteristic $C_3H_2F_5^+$ peak at 133 *m/z* along with two other larger fluorinated molecules, namely $C_3H_2F_3^+$ (95 *m/z*) and $C_3HF_4^+$ (113 *m/z*) were used to identify PVDF.

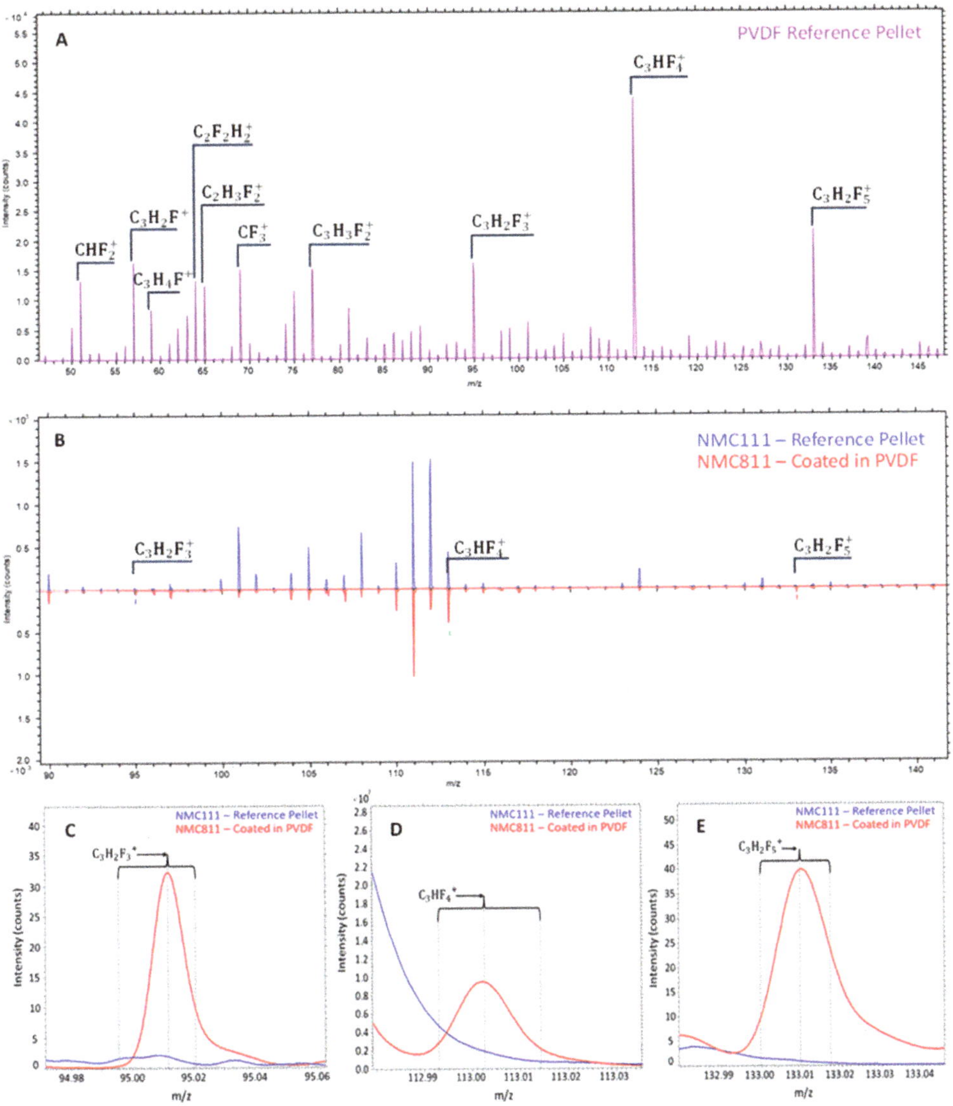

Figure 5. Positive ToF-SIMS spectra of PVDF (**A**), comparison of the spectra for NMC 111 and NMC 811 coated in PVDF (**B**), and characteristic PVDF peaks from Figure 4B (**C**–**E**).

To confirm whether the PVDF peaks identified above could be detected in a battery material, the positive ToF-SIMS spectra from pristine NMC 111 and PVDF-coated NMC 811 powders were compared. As expected, the only difference between NMC 811 and 111 would be the peak intensities for the different metals as well as the PVDF peaks. Figure 5B shows that the base NMC spectra are almost identical, with only varying metal intensities due to different metal concentrations. The spectra for the PVDF-coated NMC 811 contains extra peaks including those of the three heavier fluorinated compounds mentioned above (see Figure 5C–E), confirming the capabilities of ToF-SIMS in detecting PVDF coatings in battery materials.

The positive ToF-SIMS spectra for Cyrene were also determined, which to the best of our knowledge had not yet been reported in literature. Liu et al. [46] proposed that when Cyrene molecule is hit by an electron beam during electron impact mass spectrometry (EI-MS), it fragments into smaller molecules, namely ketene (42 *m/z*), 2-methylene-1,3-dioxole (84 *m/z*), 2-methylene-1,3-dioxolane (85 *m/z*), and 1,3-dioxole (86 *m/z*). When analysing Cyrene with ToF-SIMS, similar compounds to those found by Liu and coworkers [46] were identified. For the analysis, Cyrene in liquid form was applied to silicon wafers and placed into a vacuum chamber to determine its stability under vacuum. After two days, the excess liquid was subsequently removed to obtain a thin layer of Cyrene on the silicon wafer. Figure 6A presents the positive ToF-SIMS spectra for Cyrene, the ketene (42 *m/z*) and 1,3-dioxole (86 *m/z*) molecules were detected along with intense $C_2H_3O^+$, $C_3H_3O^+$, $C_4H_7{}^+$ and $C_3H_5O^+$ peaks which is well comparable to the fragmentation patterns from EI-MS [46].

A molecule with a *m/z* of 146 (higher than that of Cyrene) was also noted and identified as a hydrated Cyrene molecule with the formula $C_6H_{10}O_4$ (Cyrene—$C_6H_8O_3$). Further investigations around the heavier molecules will be conducted as this will shed light on possible compounds/chains that Cyrene may form. Due to Cyrene's low volatility and high viscosity, its presence on the surface of CAMs after exposure/contact is also investigated. Figure 5B compares the spectra for pristine NMC111 to that of NMC111 soaked in Cyrene. This comparison illustrates (see Figure 5C,D) that ketene (42 *m/z*) and 1,3-dioxole (86 *m/z*) were detected on the surface of NMC111, showing that ToF-SIMS can be used when identifying Cyrene in battery materials.

3.4. ToF-SIMS as a Technique for Process Evaluation

Positive ToF-SIMS spectra (Figures 5A and 6A) were used to verify Cyrene or pyrolysis as methods for PVDF removal. Contrary to expectations, the intensity of PVDF peaks was higher after Cyrene pretreatment (Figure 6F–H). Two possible explanations exist for this increase. First, the SIMS matrix effect, where secondary ion intensity depends on both element/molecule concentration and material composition. Second, and more likely, PVDF redeposition onto particle surfaces, as demonstrated in the pure PVDF dissolution experiments showing PVDF segregation from Cyrene below 80 °C. Figure 7A–C show substantial PVDF presence post-Cyrene pretreatment. PVDF redeposition results in particle agglomeration (Figure 4B–D). Uncontrolled PVDF redeposition during cooling leads to thick coatings, unlike those applied by cell manufacturers, resulting in higher PVDF readings. Figure 7C visually demonstrates this, showing a ~4 μm NMC particle coated by ~4 μm of PVDF. To confirm these coatings as PVDF, Figure 7D presents a chemical mapping using ToF-SIMS peak information, with green areas representing Li (NMC 811) and red areas shows $C_3H_4F^+$, representing PVDF.

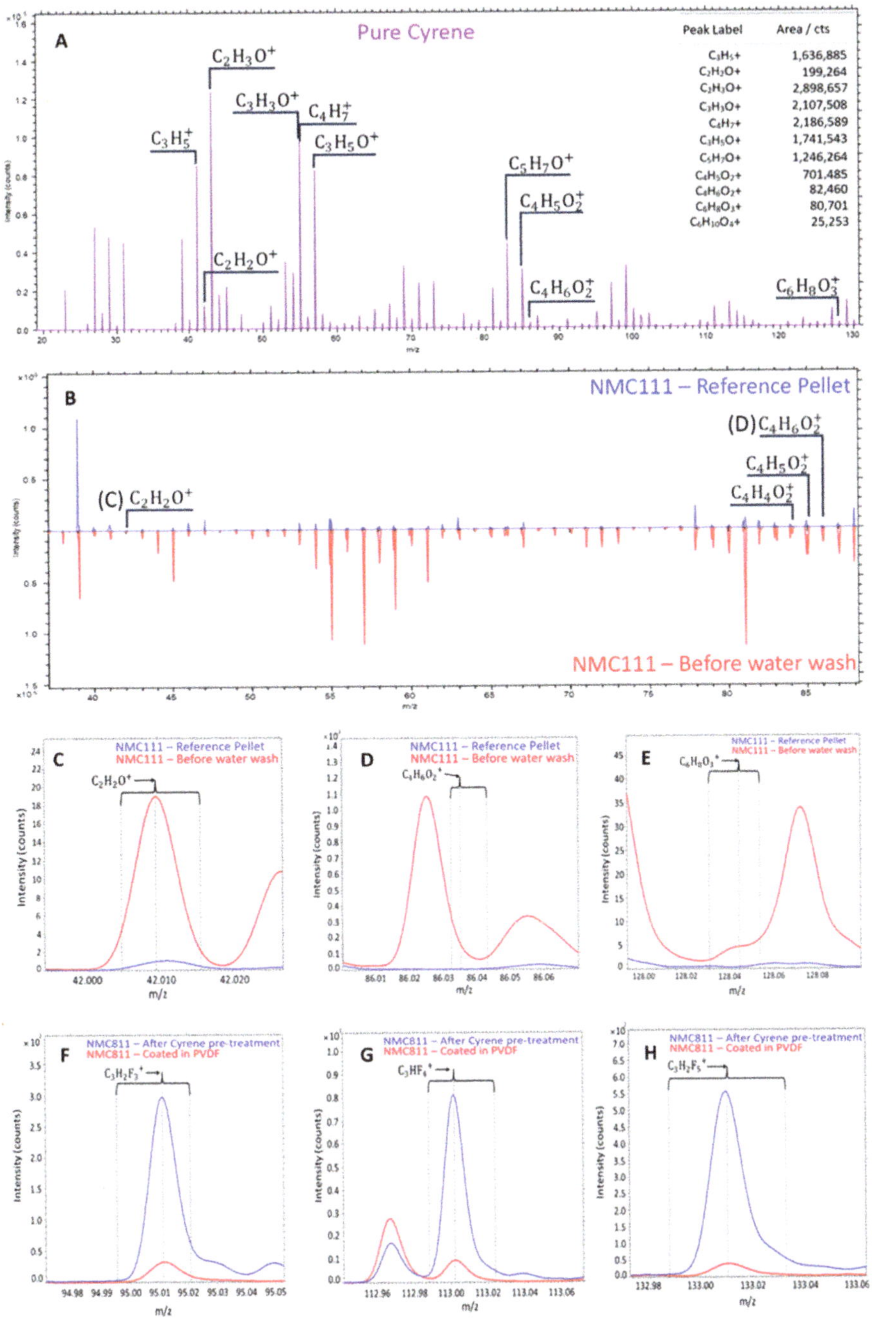

Figure 6. Positive ToF-SIMS spectra of Cyrene (**A**), comparison of the spectra for NMC 111 and Cyrene washed NMC 111 (**B**), characteristic Cyrene peaks from Figure 5B (**C–E**), and characteristic PVDF peaks after Cyrene pretreatment of NMC811 (**F–H**).

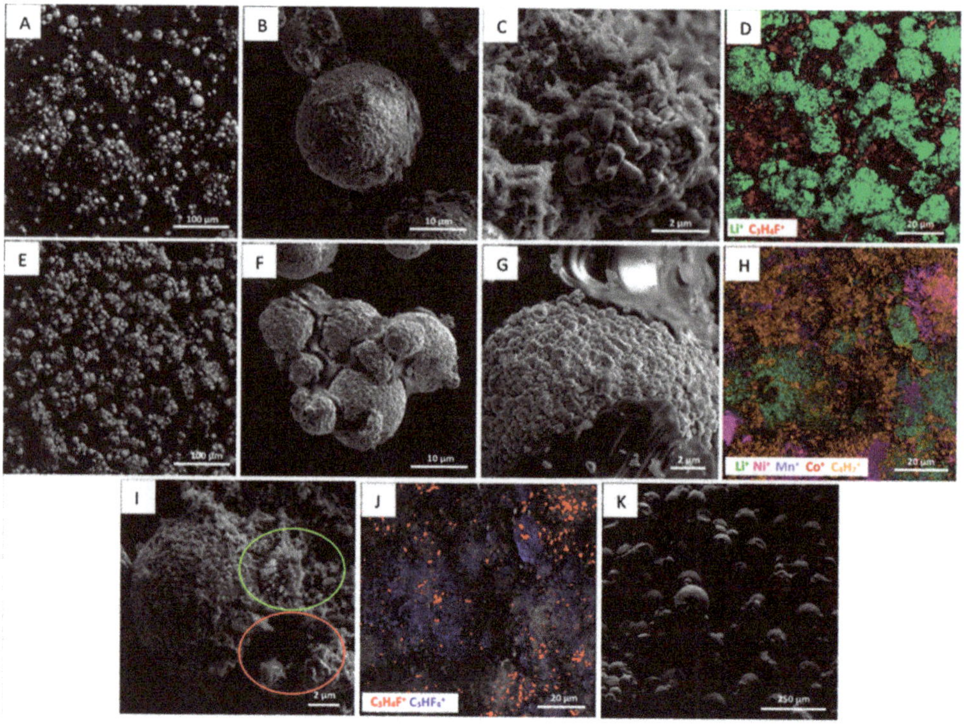

Figure 7. SEM and ToF-SIMS chemical mapping of the treated battery materials. Cyrene pretreated NMC 811 (**A–C**), ToF-SIMS chemical mapping of Cyrene pretreated NMC 811 (**D**), pyrolyzed NMC 811 samples (**E–G**), ToF-SIMS chemical mapping of pyrolyzed NMC 811 (**H**), NMC 811 particle after pyrolysis (**I**), chemical mapping of the characteristic PVDF peaks after pyrolysis (**J**) and residue from PVDF pyrolysis at 800 °C (**K**).

While ToF-SIMS is limited in providing quantitative PVDF or Cyrene concentrations, it can qualitatively indicate pretreatment effectiveness. ToF-SIMS analysis of pyrolyzed samples (Figure S2) showed minimal PVDF peak intensities, suggesting low residual PVDF. ToF-SIMS chemical mapping (Figure 7H) of pyrolyzed material revealed $C_4H_7^+$, consistent with de-fluorination. SEM imaging (Figure 7I) visualized organic matter distribution in pyrolyzed samples, revealing two regions on NMC 811 particles. The red zone, likely organic residue from PVDF decomposition, and the green region, likely residual PVDF (Figure 7J). Chemical mapping (Figure 7J) identified traces of fluorinated hydrocarbons $C_3H_4F^+$ and $C_3HF_4^+$, characteristics of PVDF. Li et al. [47] reported incomplete PVDF pyrolysis, with up to 26% residual mass, consistent with the 29% residue observed in TGA at 800 °C (Figure S3). SEM imaging (Figure 7K) showed that pyrolyzed PVDF particles retained their spherical shape, suggesting minimal change in residual PVDF on NMC 811. Pyrolyzed residue exhibited a melt texture (Figure 7G), binding NMC 811 particles and causing ~100 µm agglomerates. This carbonaceous coating may alter surface properties affecting subsequent recycling processes such as froth flotation.

Comparing Figure 7D,J, pyrolysis demonstrates substantially less PVDF than Cyrene pretreatment. Thus, pyrolysis is currently more effective. However, Cyrene solvent pretreatment is yet to be optimized, and thus it remains a promising technique for PVDF removal.

4. Conclusions

The recycling of spent LIBs presents a significant challenge in dealing with PVDF binder, which hinders particle liberation and alters the surface properties of active materials. Methods like thermal treatment and solvent-based techniques are commonly used to remove the PVDF binder. However, accurately quantifying their efficiency for binder removal using conventional analytical tools has remained a challenge. In response to this, this study utilised ToF-SIMS to produce high spatial-resolution chemical maps of cathodic material, offering a detailed insight into the distribution of PVDF within CAM. The positive ToF-SIMS characteristic peaks for Cyrene and PVDF were also identified. Cyrene had two characteristic peaks, namely ketene (42 *m/z*) and 1,3-dioxole (86 *m/z*) along with intense $C_2H_3O^+$, $C_3H_3O^+$, $C_4H_7^+$ and $C_3H_5O^+$ peaks. While the characteristic peaks for PVDF were identified to be $C_3H_2F_5^+$ (133 *m/z*), $C_3H_2F_3^+$ (95 m/z) and $C_3HF_4^+$ (113 *m/z*).

The SEM and ToF-SIMS analysis concluded that both the Cyrene and pyrolysis pretreatments remove PVDF. However, both techniques do not fully remove the binder and lead to some degree of particle agglomeration. When using Cyrene pretreatment, the PVDF redeposited onto the CAM, resulting in thick layers of PVDF on the particle surfaces and large amounts of particle agglomeration. It was determined that the amount of PVDF redeposition could be minimized using hot filtration (>100 °C). After dissolution of pristine PVDF with Cyrene, both the recovered PVDF and spent Cyrene showed no residual PVDF in the Cyrene (post cooling). The recovered PVDF only contained some excess Cyrene which in theory can be removed through water rinsing. On the other hand when the CAM is pyrolyzed, a large amount of the PVDF decomposed (producing toxic halogenated hydrocarbons), leaving behind a carbonaceous material, that still binds the CAM particles to one another.

As of our current understanding, thermal treatment (i.e., pyrolysis) stands out as the most effective method for removing PVDF from CAMs. Nonetheless, this research has unveiled a promising alternative: the solvent-pretreatment of CAM with Cyrene, followed by a hot filtration process to separate the Cyrene-PVDF solution. It is crucial to underscore the substantial potential of the Cyrene pretreatment method. Through optimization of the operating parameters, Cyrene pretreatment could become a competitive alternative to pyrolysis, particularly when considering its ability to recover PVDF in an unaltered state and the elimination of harmful halogenated hydrocarbons as a waste product.

Supplementary Materials: The following supporting information can be downloaded at: https://www.mdpi.com/article/10.3390/recycling10020056/s1. Figure S1. Process flow diagrams for pre-treatment of PVDF (A), and NMC 111 and PVDF-coated NMC 811 (B). Figure S2. Comparison of positive ToF-SIMS spectra for NMC 811 post Cyrene and pyrolysis pretreatments (A). Enlarged images around characteristic PVDF peaks (B–D). Figure S3. TGA curve of pristine PVDF, 800 °C under Ar atmosphere at a heating rate of 10 K/min.

Author Contributions: Conceptualization, J.E. and A.V.; Methodology, M.S.H., A.M.S., W.D.A.R. and A.V.; Validation A.M.S. and M.S.H.; Formal Analysis, ToF-SIMS was performed by W.D.A.R.; Raman Spectroscopy was conducted by M.R. and SEM imaging was conducted by A.V., A.M.S. and M.S.H.; Investigation, M.S.H., A.M.S. and Á.J.R.M.; Resources, J.E. and M.R.; Data Curation, A.M.S. and M.S.H.; Writing—original draft preparation, M.S.H. and A.M.S. and A.V.; Writing—Review and Editing, M.S.H., A.M.S., A.V., C.C.B., J.E., M.R., E.A.O., W.D.A.R. and D.F.; Data Visualization, M.S.H., A.M.S., A.V. and W.D.A.R.; Supervision, A.V, J.E., M.R., E.A.O. and C.C.B.; Project Administration, A.V., J.E., M.R., E.A.O. and C.C.B.; Funding Acquisition, A.V., J.E. and M.R. All authors have read and agreed to the published version of the manuscript.

Funding: The RES-65109 RESTORE Project (Sustainable Recycling and Recovery of Electrode Materials from Spent lithium-ion Batteries) has been funded by the Future Battery Industries CRC (FBI-CRC, as part of the Australian Government Cooperative Research Centres Program) and the Universities Australia Deutscher Akademischer Austauschdienst (AU-DAAD). Curtin University and the Helmholtz Zentrum Dresden Rossendorf contributed to staff salaries and fee scholarships (for Mr Henderson). The IONTOF M6 ToF-SIMS at the JdLC, Curtin University, was obtained using funding from the Australian Research Council LIEF program (LE190100053).

Data Availability Statement: Data will be made available upon request to the corresponding author.

Conflicts of Interest: The authors declare no conflicts of interest.

References

1. Ali, H.; Khan, H.A.; Pecht, M. Preprocessing of spent lithium-ion batteries for recycling: Need, methods, and trends. *Renew. Sustain. Energy Rev.* **2022**, *168*, 112809.
2. Raj, T.; Chandrasekhar, K.; Kumar, A.N.; Sharma, P.; Pandey, A.; Jang, M.; Jeon, B.H.; Varjani, S.; Kim, S.H. Recycling of cathode material from spent lithium-ion batteries: Challenges and future perspectives. *J. Hazard. Mater.* **2022**, *429*, 128312.
3. Arshad, F.; Li, L.; Amin, K.; Fan, E.; Manurkar, N.; Ahmad, A.; Yang, J.; Wu, F.; Chen, R. A Comprehensive Review of the Advancement in Recycling the Anode and Electrolyte from Spent Lithium Ion Batteries. *ACS Sustain. Chem. Eng.* **2020**, *8*, 13527–13554. [CrossRef]
4. Mohanty, A.; Sahu, S.; Sukla, L.B.; Devi, N. Application of various processes to recycle lithium-ion batteries (LIBs): A brief review. *Mater. Today Proc.* **2021**, *47*, 1203–1212. [CrossRef]
5. Pinegar, H.; Smith, Y.R. Recycling of End-of-Life Lithium Ion Batteries, Part I: Commercial Processes. *J. Sustain. Metall.* **2019**, *5*, 402–416. [CrossRef]
6. Khodadadmahmoudi, G.; Tabar, K.J.; Homayouni, A.H.; Chelgani, S.C. Recycling spent lithium batteries—An overview of pretreatment flowsheet development based on metallurgical factors. *Environ. Technol. Rev.* **2023**, *12*, 2248559. [CrossRef]
7. Langdon, R.; Dominish, E.; Lara, H. *B-Cycle Benchmarking Program: Market Analysis & Fate Mapping, and Life Cycle Analysis*; Institute for Sustainable Futures: Sydney, Australia, 2023. Available online: https://bcycle.com.au/wp-content/uploads/2023/05/Battery-MA-Report-FINAL-20230927.pdf (accessed on 12 March 2025).
8. National Institute of Environmental Health Sciences. Perfluoroalkyl and Polyfluoroalkyl Substances (PFAS). 2023. Available online: https://www.niehs.nih.gov/health/topics/agents/pfc/index.cfm (accessed on 20 September 2023).
9. Fenton, S.E.; Ducatman, A.; Boobis, A.; Dewitt, J.C.; Lau, C.; Ng, C.; Smith, J.S.; Roberts, S.M. Per- and Polyfluoroalkyl Substance Toxicity and Human Health Review: Current State of Knowledge and Strategies for Informing Future Research. *Environ. Toxicol. Chem.* **2021**, *40*, 606–630.
10. Ehsan, M.N.; Riza, M.; Pervez, M.N.; Khyum, M.M.O.; Liang, Y.; Naddeo, V. Environmental and health impacts of PFAS: Sources, distribution and sustainable management in North Carolina (USA). *Sci. Total Environ.* **2023**, *878*, 163123.
11. Scott, A.; c&en. The Battle over PFAS in Europe. *Chemical and Engineering News: American Chemical Society*, 18 September 2023. Available online: https://cen.acs.org/policy/chemical-regulation/battle-over-PFAS-Europe/101/i31#:~:text=The%20European%20Commission%20(EC)%20is,a%20major%20subset%20of%20PFAS (accessed on 16 October 2023).
12. Kaspar, P.; Sobola, D.; Částková, K.; Knápek, A.; Burda, D.; Orudzhev, F.; Dallaev, R.; Tofel, P.; Trčka, T.; Grmela, L.; et al. Characterization of Polyvinylidene Fluoride (PVDF) Electrospun Fibers Doped by Carbon Flakes. *Polymers* **2020**, *12*, 2766. [CrossRef]
13. Medeiros, K.A.R.; Rangel, E.Q.; Sant'anna, A.R.; Louzada, D.R.; Barbosa, C.R.H.; D'almeida, J.R.M. Evaluation of the electromechanical behavior of polyvinylidene fluoride used as a component of risers in the offshore oil industry. *Oil Gas. Sci. Technol.-Rev. D'ifp Energ. Nouv.* **2018**, *73*, 48.
14. Versaci, D.; Nasi, R.; Zubair, U.; Amici, J.; Sgroi, M.; Dumitrescu, M.A.; Francia, C.; Bodoardo, S.; Penazzi, N. New eco-friendly low-cost binders for Li-ion anodes. *J. Solid. State Electrochem.* **2017**, *21*, 3429–3435.
15. Zhong, X.; Han, J.; Chen, L.; Liu, W.; Jiao, F.; Zhu, H.; Qin, W. Binding mechanisms of PVDF in lithium-ion batteries. *Appl. Surf. Sci.* **2021**, *553*, 149564.
16. Bresser, D.; Buchholz, D.; Moretti, A.; Varzi, A.; Passerini, S. Alternative binders for sustainable electrochemical energy storage—The transition to aqueous electrode processing and bio-derived polymers. *Energy Environ. Sci.* **2018**, *11*, 3096–3127. [CrossRef]
17. Rensmo, A.; Savvidou, E.K.; Cousins, I.T.; Hu, X.; Schellenberger, S.; Benskin, J.P. Lithium-ion battery recycling: A source of per- and polyfluoroalkyl substances (PFAS) to the environment? *Environ. Sci. Process Impacts* **2023**, *25*, 1015–1030.

18. Vanderbruggen, A.; Hayagan, N.; Bachmann, K.; Ferreira, A.; Werner, D.; Horn, D.; Peuker, U.; Serna-Guerrero, R.; Rudolph, M. Lithium-Ion Battery Recycling—Influence of Recycling Processes on Component Liberation and Flotation Separation Efficiency. *ACS EST Eng.* **2022**, *2*, 2130–2141. [CrossRef]
19. Vanderbruggen, A.; Salces, A.; Ferreira, A.; Rudolph, M.; Serna-Guerrero, R. Improving Separation Efficiency in End-of-Life Lithium-Ion Batteries Flotation Using Attrition Pre-Treatment. *Minerals* **2022**, *12*, 72. [CrossRef]
20. Salces, A.; Bremerstein, I.; Rudolph, M.; Vanderbruggen, A. Joint recovery of graphite and lithium metal oxides from spent lithium-ion batteries using froth flotation and investigation on process water re-use. *Miner. Eng.* **2022**, *184*, 107670. [CrossRef]
21. Golmohammadzadeh, R.; Dimachki, Z.; Bryant, W.; Zhang, J.; Biniaz, P.; Holl, M.M.B.; Pozo-Gonzalo, C.; Chakraborty Banerjee, P. Removal of polyvinylidene fluoride binder and other organics for enhancing the leaching efficiency of lithium and cobalt from black mass. *J. Environ. Manag.* **2023**, *343*, 118205.
22. Wang, M.; Liu, K.; Dutta, S.; Alessi, D.S.; Rinklebe, J.; Ok, Y.S.; Tsang, D.C.W. Recycling of lithium iron phosphate batteries: Status, technologies, challenges, and prospects. *Renew. Sustain. Energy Rev.* **2022**, *163*, 112515.
23. Kaya, M. State-of-the-art lithium-ion battery recycling technologies. *Circ. Econ.* **2022**, *1*, 100015.
24. Ji, Y.; Jafvert, C.T.; Zyaykina, N.N.; Zhao, F. Decomposition of PVDF to delaminate cathode materials from end-of-life lithium-ion battery cathodes. *J. Clean. Prod.* **2022**, *367*, 133112.
25. Jaleh, B.; Gavary, N.; Fakhri, P.; Muensit, N.; Taheri, S.M. Characteristics of PVDF Membranes Irradiated by Electron Beam. *Membranes* **2015**, *5*, 1–10. [CrossRef] [PubMed]
26. Zhang, G.; He, Y.; Feng, Y.; Wang, H.; Zhang, T.; Xie, W.; Zhu, X. Enhancement in liberation of electrode materials derived from spent lithium-ion battery by pyrolysis. *J. Clean. Prod.* **2018**, *199*, 62–68.
27. Huang, H.; Liu, C.; Sun, Z. In-situ pyrolysis based on alkaline medium removes fluorine-containing contaminants from spent lithium-ion batteries. *J. Hazard. Mater.* **2023**, *457*, 131782.
28. Armand, M.; Axmann, P.; Bresser, D.; Copley, M.; Edström, K.; Ekberg, C.; Guyomard, D.; Lestriez, B.; Novák, P.; Petranikova, M.; et al. Lithium-ion batteries–Current state of the art and anticipated developments. *J. Power Sources* **2020**, *479*, 228708.
29. Chen, Y.; Liu, N.; Jie, Y.; Hu, F.; Li, Y.; Wilson, B.P.; Xi, Y.; Lai, Y.; Yang, S. Toxicity Identification and Evolution Mechanism of Thermolysis-Driven Gas Emissions from Cathodes of Spent Lithium-Ion Batteries. *ACS Sustain. Chem. Eng.* **2019**, *7*, 18228–18235. [CrossRef]
30. Wang, M.; Liu, K.; Yu, J.; Zhang, Q.; Zhang, Y.; Valix, M.; Tsang, D.C.W. Challenges in Recycling Spent Lithium-Ion Batteries: Spotlight on Polyvinylidene Fluoride Removal. *Glob. Chall.* **2023**, *7*, 2200237. [CrossRef]
31. Citarella, A.; Amenta, A.; Passarella, D.; Micale, N. Cyrene: A Green Solvent for the Synthesis of Bioactive Molecules and Functional Biomaterials. *Int. J. Mol. Sci.* **2022**, *23*, 15960. [CrossRef]
32. De Bruyn, M.; Budarin, V.L.; Misefari, A.; Shimizu, S.; Fish, H.; Cockett, M.; Hunt, A.J.; Hofstetter, H.; Weckhuysen, B.M.; Clark, J.H.; et al. Geminal Diol of Dihydrolevoglucosenone as a Switchable Hydrotrope: A Continuum of Green Nanostructured Solvents. *ACS Sustain. Chem. Eng.* **2019**, *7*, 7878–7883. [CrossRef]
33. Kong, D.; Dolzhenko, A.V. Cyrene: A bio-based sustainable solvent for organic synthesis. *Sustain. Chem. Pharm.* **2022**, *25*, 100591. [CrossRef]
34. Sherwood, J.; Bruyn, M.D.; Constantinou, A.; Moity, L.; Mcelroy, C.R.; Farmer, T.J.; Duncan, T.; Raverty, W.; Hunt, A.J.; Clark, J.H. Dihydrolevoglucosenone (Cyrene) as a Bio-Based Alternative for Dipolar Aprotic Solvents. *Chem. Commun.* **2014**, *50*, 9650–9652.
35. Zhenova, A. Green Solvents for Polymer Applications. Ph.D. Thesis, University of York, York, UK, 2019.
36. Marshall, J.E.; Zhenova, A.; Roberts, S.; Petchey, T.; Zhu, P.; Dancer, C.E.J.; Mcelroy, C.R.; Kendrick, E.; Goodship, V. On the Solubility and Stability of Polyvinylidene Fluoride. *Polymers* **2021**, *13*, 1354. [CrossRef] [PubMed]
37. Sigma-Aldrich. Cyrene 99%, 1907/2006. Available online: https://www.sigmaaldrich.com/ZA/en/product/sial/807796?srsltid=AfmBOoquFtkhwM2PoU4R4AgqOMdtBMEObWSVJ-95bEEuPJRaLnFbmXQM (accessed on 13 March 2025).
38. Bai, Y.; Hawley, W.B.; Jafta, C.J.; Muralidharan, N.; Polzin, B.J.; Belharouak, I. Sustainable recycling of cathode scraps via Dihydrolevoglucosenone-based separation. *Sustain. Mater. Technol.* **2020**, *25*, e00202.
39. Zhou, H.; Pei, B.; Fan, Q.; Xin, F.; Whittingham, M.S. Can Greener Cyrene Replace NMP for Electrode Preparation of NMC 811 Cathodes? *J. Electrochem. Soc.* **2021**, *168*, 40536.
40. Sui, T.; Song, B.; Dluhos, J.; Lu, L.; Korsunsky, A.M. Nanoscale chemical mapping of Li-ion battery cathode material by FIB-SEM and ToF-SIMS multi-modal microscopy. *Nano Energy* **2015**, *17*, 254–260.
41. Vanderbruggen, A.; Gugala, E.; Blannin, R.; Bachmann, K.; Serna-Guerrero, R.; Rudolph, M. Automated mineralogy as a novel approach for the compositional and textural characterization of spent lithium-ion batteries. *Miner. Eng.* **2021**, *169*, 106924. [CrossRef]
42. Elashmawi, I.S.; Gaabour, L.H. Raman, morphology and electrical behavior of nanocomposites based on PEO/PVDF with multi-walled carbon nanotubes. *Results Phys.* **2015**, *5*, 105–110.
43. Pramod, K.; Gangineni, R.B. Influence of solvent evaporation rate on crystallization of polyvinylidene fluoride thin films. *Bull. Mater. Sci.* **2015**, *38*, 1093–1098.

44. Sun, A.C.; Kosar, W.; Zhang, Y.; Feng, X. A Study of Thermodynamics and Kinetics Pertinent to Formation of PVDF Membranes by Phase Inversion. *Desalination* **2013**, *309*, 156–164.
45. Feng, J.; Chag, C.-M.; Weng, L.-T. Influence of chain sequence structure of polymers on ToF-SIMS spectra. *Polym. Commun.* **2000**, *41*, 2695–2699. [CrossRef]
46. Liu, X.; Pollard, B.; Banwell, M.G.; Yu, L.-J.; Coote, M.L.; Gardiner, M.G.; van Vugt-Lussenburg, B.M.A.; van der Burg, B.; Grasset, F.L.; Campillo, E.; et al. Simple and modestly scalable synthesis of iso-Cyrene from levoglucosenone and its comparison to the bio-derived and polar aprotic solvent Cyrene. *Aust. J. Chem.* **2022**, *75*, 331–344.
47. Li, W.; Li, H.; Zhang, Y.-M. Preparation and investigation of PVDF/PMMA/TiO2 composite film. *J. Mater. Sci.* **2009**, *44*, 2977–2984.

recycling

MDPI

Article

Analyzing Organic Electrolyte Solvents from Spent Lithium-Ion Batteries as a Basis for Distillative Value Component Recovery

Martin Wolke [1], Kai Schröder [1], Konstantin Arnold [1], Pamina Mozumder [1], Till Beuerle [2], Katharina Jasch [1] and Stephan Scholl [1,*]

[1] Institute for Chemical and Thermal Process Engineering, Technische Universität Braunschweig, 38106 Braunschweig, Germany
[2] Institute for Pharmaceutical Biology, Technische Universität Braunschweig, 38106 Braunschweig, Germany
* Correspondence: s.scholl@tu-braunschweig.de

Abstract: The rapid expansion of lithium-ion batteries (LIBs), largely driven by the rising demand for electric vehicles, will lead to a significant increase in end-of-life (EOL) batteries, necessitating efficient recycling processes, which must be accompanied by equally efficient purification steps. This study addresses the challenge of reusing organic electrolyte solvents from spent LIBs, a key component often overlooked in existing recycling strategies. To address this issue, we developed a gas chromatography (GC) method. A variety of spent electrolyte samples of different origin, including mechanical-thermal pretreatment or direct cell recovery, were analyzed by quantification of common solvents and identified organic impurities. Results demonstrated that the composition of the recovered electrolytes was highly variable, with concentrations fluctuating. Impurities were identified, which may originate from various sources throughout the lifespan of an LIB and have the potential to reduce the performance of second-life LIBs by reusing the electrolyte without any purification. The findings highlight the necessity for advanced purification methods like a distillation process to remove these impurities and ensure the viability of recycled electrolytes in maintaining the performance and safety standards required for LIBs. This research contributes to the broader goal of enhancing the sustainability and reuse of battery materials.

Keywords: lithium-ion batteries; spent electrolyte; recycling; circular economy; analytics; gas chromatography; impurities; distillation; rectification

1. Introduction

As a consequence of the significant increase in the production of secondary batteries such as lithium-ion batteries (LIBs) and sodium-ion batteries, the global production capacity of over 1 TWh was reached for the first time in 2023 [1]. Based on the currently announced production facilities, a further increase in production is projected to reach approximately 9 TWh by 2030, with the majority of the market expected to be for battery electric vehicles (BEVs) [2]. Consequently, an increase in end-of-life (EOL) batteries can be expected with a time delay due to the lifespan of a LIB. Based on calculations from Fraunhofer ISI [3], it is estimated that approximately 420 kt of EOL batteries will be incurred in 2030, rising to approximately 2100 kt in 2040. It is anticipated that a large proportion of the battery waste will continue to originate from cell production. Assuming that the current forecasts materialize, the proportion of EOL LIBs from BEVs will increase from around 20% in 2035 to around 50% in 2040 [3]. Once a battery has reached EOL, it can either be downgraded for a second, less demanding use, e.g., as stationary storage, or recycled in order to recover the materials. The recycling of LIBs is regulated by the European Union, which has set a target recycling quota of 70% by average weight by the end of 2030 [4]. In order to achieve this recycling quota and close the material cycle, the recovery and processing of the electrolyte is becoming increasingly important. The electrolyte accounts for approximately 8–16 wt-% [5] and is therefore a substantial waste product on a kiloton scale.

Electrolytes used in LIBs usually consist of a lithium salt dissolved in an organic solvent mixture comprising a minimum of two components. Additionally, they are frequently combined with additives [6]. The electrolyte solvents are typically a combination of a cyclic carbonate, such as ethylene carbonate (EC) or propylene carbonate (PC), and at least one linear carbonate, such as dimethyl carbonate (DMC), ethyl methyl carbonate (EMC), or diethyl carbonate (DEC) [7–9]. Furthermore, other solvents, such as fluorobenzene (FB), are commonly used in commercial LIBs [10,11]. The advantages of FB include a positive influence on the formation of a stable solid electrolyte interface (SEI) between the anode and the electrolyte enriched in lithium fluoride (LiF) [12,13].

Additives are added to the electrolyte in order to enhance the characteristics of the LIB. Film-forming additives, such as vinylene carbonate (VC) or fluoroethylene carbonate (FEC), are utilized to optimize the interfaces between the electrodes and the electrolyte, thereby enhancing the lifespan and performance [8,14,15]. Cyclohexylbenzene (CHB) is a commonly employed overcharge protection additive [16].

Processes for recovering the solvents already exist. One option is a mechanical-thermal pre-treatment of the LIB. This process involves shredding the LIB, with the organic volatile electrolyte solvents subsequently separated from the solids through vacuum-drying in an inert gas atmosphere. This allows for the recovery of part of the electrolyte as condensate [17,18]. However, these solvents have not yet been considered for further processing; instead, they are thermally utilized. An alternative approach for the recovery of the electrolyte is sub- and supercritical CO_2 extraction [19–21]. In either case, the recovered electrolytes contain impurities that can be attributed to various sources. On the one hand, decomposition reactions within the cell can result in the liberation of impurities. These reactions are primarily observed during electrochemical ageing. Initially, this occurs during the formation and growth of the SEI within the cell [22–25] and subsequently during calendar ageing [19,22–24,26–30]. However, this phenomenon is also evident in thermal ageing, see, e.g., [22,23,27–29,31–34]. On the other hand, inadequate storage, particularly over an extended period of time in an atmosphere with a high water content or an excessively high temperature, can also result in the formation of contaminants. These include a reaction between the organic carbonates and water, as well as the transesterification of the organic carbonates to form impurities such as ethanol (EtOH) or methanol (MeOH), among others [24,35,36]. The recycling process itself may also be a potential source of contamination of the electrolyte. The degradation products of the electrolyte are analyzed during ageing and/or after removal from the electrolyte using suitable analytical methods, including gas chromatography-flame ionization detection (GC-FID), gas chromatography-mass spectrometry (GC-MS), high-performance liquid chromatography (HPLC), Fourier transform infrared spectroscopy (FT-IR), nuclear magnetic resonance spectroscopy (NMR), and others. However, the composition of the utilized electrolyte and cell material was clearly defined, and the experimental parameters are well established and contingent upon the specific field of investigation. Nevertheless, this is not the case in the context of industrial recycling of LIBs, where a significant number of different LIBs are processed, which differ in terms of their cell chemistry with various formulations for the solvent mixtures initially used. The recycling of LIBs from different manufacturers and battery generations with different life histories, recycled all together, can result in notable variation in the composition of the recovered electrolyte. Consequently, a diverse range of substances with varying compositions from all LIBs tend to accumulate in the recovered electrolyte.

Further studies have demonstrated that, in addition to the recovery of the initial electrolyte solvents, the removal of decomposition products and impurities, such as water, hydrofluoric acid (HF), and others, is crucial for the reuse of the recovered electrolyte. Liu et al. [37] recovered electrolytes through CO_2 extraction, and the extract was purified by anion exchange and molecular sieves to reduce the HF and water content. No further processing of the recovered electrolyte was carried out. In comparison to commercial electrolytes, the recovered electrolyte showed a reduced performance. It was postulated that additional purification steps for reclaimed electrolytes are necessary [37]. Xu et al. [38] pursued an approach for the reuse of spent electrolyte solvents with treatment by vacuum distillation. Compared to commercial electrolytes, the electrochemical performance of the recovered electrolyte was satisfactory in both charging and discharging, which substantiates the efficacy of the purification strategy for the recovered electrolyte, for instance, through distillation.

In order to illustrate the issue of recycling waste and to investigate the purification of recovered electrolyte solvents in more detail, a GC method was developed for analyzing organic electrolyte solvents from spent LIBs. So far, no standard GC method is available that allows for the detailed analysis of these complex mixtures. Due to the absence of a diluent, which is a rather unconventional technique for the analysis by GC, the newly developed method underwent a validation procedure. Samples were obtained directly from LIB cells or recycling processes at both laboratory and pilot plant scales. The aim was to quantify the frequently occurring components using GC-FID and to identify other substances, to the greatest possible extent, using GC-MS. The GC-MS analysis approach was used to generate a dataset of potential substances present in the recovered electrolyte solvents with the objective of aiding the design of future distillation approaches. These compounds are discussed and categorized in the context of their origin. This represents a pragmatic approach, focusing on the identification of potential impurities.

2. Materials and Methods

2.1. Chemicals and Reagents

DMC (≥99.8%), DEC (≥99.8%), and PC (≥99.7%) were supplied by Carl Roth (Karlsruhe, Germany). MeOH (≥ 99.8%), FB (99%), EC (99%), CHB (98%), chlorobenzene (CB, >99%), and acetonitrile (AcN, ≥99.8%) were supplied by Fisher Scientific (Schwerte, Germany). FEC (battery grade) and VC (battery grade) were supplied by BASF (Florham Park, NJ, USA). EMC (≥99.95%), EtOH (≥99.8%), acetone (≥99.8%), and sodium hydrogen carbonate ($NaHCO_3$, ≥99.7% Merck) were supplied by E-Lyte (Münster, Germany), Sigma-Aldrich (Taufkirchen, Germany), VWR International (Darmstadt, Germany), and Merck (Darmstadt, Germany). Commercial battery electrolyte (LP57) containing 1M lithium hexafluorophosphate ($LiPF_6$) dissolved in a 7:3 mass ratio of EMC:EC supplied by E-Lyte (Münster, Germany).

2.2. Spent Electrolytes

The recovered electrolyte samples from used LIBs were obtained by third parties and various battery sources, all in the context of commercial LIB recycling. Hence, detailed information on lifetime and life cycles was not always available. Presumably, Samples A and B originate from test cells with only a few cycles and Sample C_O from a mixture of EOL LIBs, test cells, and production scrap. The recovered samples were stored at 4 °C under atmospheric conditions.

Sample A, see Figure A1, was obtained from a prismatic cell with a nickel-manganese-cobalt (NMC) chemistry on the cathode side and a graphite chemistry on the anode side. Approximately 5 mL of surplus electrolyte directly accessible after opening the cell by cutting the housing under inert conditions in a glovebox was removed to obtain this sample.

Two electrolyte samples were obtained from the mechanical-thermal pre-treatment of the recycling route: Sample B, see Figure A1, was obtained from a pouch cell (NMC/graphite chemistry), which was shredded first and subsequently subjected to vacuum-drying in a laboratory scale under an inert atmosphere. The volatile compounds were recovered by means of condensation (approximately 10 mL).

A total of 5 L of Sample C_O, see Figure A1, was obtained from a pilot LIB recycling facility, and the cell chemistry was largely unknown; however, it can be assumed that most of the material originated from modules that had an NMC/graphite chemistry. All modules were shredded and treated in a similar manner to Sample B. Due to the vacuum-drying in an inert gas atmosphere followed by condensation, the recovered samples contained only volatile organic compounds, as the conduction salts remain in the solid fraction in this process.

Due to the absence of conduction salt and the large volume of accessible samples, approximately 400 mL of Sample C_O was distilled under atmospheric conditions in a laboratory setup located within a fume hood. This setup comprised a heating plate, a flat bottom flask containing the sample, a magnetic stirring bar, a condenser operated with a water cooler, and a flat bottom flask to collect the condensate. See Figure A2 for a visual representation of this setup. After distillation, approximately 390 mL were collected as distillate (Sample C_D, see Figure A1) and 8 mL as residue (Sample C_R, see Figure A1). All samples were prepared as outlined in Section 2.5 and analyzed by GC-FID and GC-MS (Sections 2.3 and 2.4, respectively).

2.3. GC-Method

In order to perform a quantitative analysis of the specified components, the following GC method was developed. GC SCION 45 (SCION Instruments, Goes, The Netherlands) equipped with an FID, combined with the autosampler Combi PAL-XT (CTC Analytics, Zwingen, Switzerland), was utilized via the SCION Instrument CDS software, version 3.0.1.16. A Rxi-624Sil MS Capillary column (30 m, 0.25 mm ID, 1.4 μm), provided by Restek (Bad Homburg, Germany), was used for this study. Prior to and after each injection, the 1.2 μL syringe was rinsed with acetone in order to prevent cross-contamination between samples. The 1177-type injector was set at a temperature of 280 °C using a split-liner (borosilicate) HP 78.5 × 6.3 × 4 mm and a split ratio of 1:200. The injection volume was 0.2 μL. Nitrogen was employed as the carrier gas with a flow rate of 0.63 $mL \cdot min^{-1}$. The temperature program was 60 °C (1 min)—20 $°C \cdot min^{-1}$—200 °C (0 min)—40 $°C \cdot min^{-1}$—260 °C (5 min). FID temperature was maintained at 350 °C, while gases were set at 300 $mL \cdot min^{-1}$, 30 $mL \cdot min^{-1}$, and 25 $mL \cdot min^{-1}$ for synthetic air, hydrogen, and nitrogen (makeup gas), respectively.

2.4. GC-MS Identification of LIBs Compounds

For the identification of unknown components, GC-MS measurements were conducted using an Agilent Technologies 6890 N (Agilent Technologies Deutschland GmbH, Böblingen, Germany) gas chromatograph coupled to an Agilent Technologies 5975B mass spectrometer (Agilent Technologies Deutschland GmbH, Böblingen, Germany). GC-column and temperature program was identical to the GC-FID method (Section 2.3), but helium was used as the carrier gas with a flow rate of 1.0 $mL \cdot min^{-1}$. Split ratio was set at 1:20, and the injection volume was 1 μL. The MS temperatures of the analyzer and source were set at 150 °C and 230 °C, respectively, and the mass range was recorded from 29 to 450 m/z.

The identification of separated components was achieved by comparing spectra using the NIST23 library program and spectra database.

2.5. Sample Preparation

For validation of the GC method (see Section 2.6), samples were prepared with synthetic substances in accordance with Section 2.1. To initiate the validation process, each pure substance (except for EC) was individually measured without further preparation by GC-FID. Due to the normal melting point of EC at 36.4 °C, EC was previously dissolved in EMC. Subsequently, a standard mixture comprising all components was prepared for the purpose of testing the selectivity of the method. This mixture was then combined with the internal standard (InSt) AcN in a 2:1 solvent-to-acetonitrile volumetric ratio and analyzed by GC-FID. Additionally, LP57 was also prepared and analyzed under atmospheric conditions in the same way to investigate whether the $LiPF_6$ contained therein had any influence on the analytic procedure and results.

In order to conduct calibration and linearity experiments without the use of a diluent, stock solutions (StSo) of the binary mixtures MeOH/EtOH, FB/DEC, and DMC/EMC were prepared in the range of 0.1–95 $g_i/g_{tot,StSo}$. Moreover, stock solutions of VC/FEC/EC/PC in the range of 0.1–50 $g_i/g_{tot,StSo}$ were prepared and, if necessary, filled with DMC. Additionally, stock solutions for CB/CHB in the range of CHB = 0.1–50 $g_{CHB}/g_{tot,StSo}$ were prepared. CB, which exhibited no reaction with the other compounds mentioned and is a well-known component in the field of distillation, was incorporated into the quantification process as a planned component for future processing steps in a distillation procedure. Subsequently, the stock solutions were prepared with the InSt AcN at a 2:1 solvent-to-acetonitrile volumetric ratio per vial. To extend the range of the calibration, pure MeOH, EtOH, CB, DMC, EMC, DEC, and FB were also prepared in a volumetric ratio of 2:1, 2.1:1, and 2.2:1, respectively, thus achieving a total range of these components in the vials of 0.067–0.6875 g_i/g_{tot}.

The masses of all stock solutions and samples in the vials were determined gravimetrically using a PRACTUM 224-1S balance (Sartorius Lab Instruments, Göttingen, Germany) and were prepared using a HandyStep touch repetitive pipettes (Brand, Wertheim, Germany).

As the recovered samples derived from spent LIBs may contain acids with the potential to cause damage to the GC system, the acids were neutralized prior to analysis by the addition of $NaHCO_3$. To ensure the neutralization with $NaHCO_3$, a series of preliminary tests were carried out. For the neutralization, a spatula tip of $NaHCO_3$ was added to approximately 5 mL of each sample, mentioned in Section 2.2. If a gas release were observed, this procedure was repeated until no more gas was released. This resulted in the formation of carbon dioxide, a sodium salt of the acid as a solid, and water as a reaction product. Excess sodium hydroxide carbonate and potential sodium salts were separated by centrifugation. The prepared samples were mixed in triplicates with the internal standard acetonitrile in a volumetric ratio of 2:1 and measured by GC-FID. In order to investigate the influence of neutralization on the analysis, a second measurement of the standard mixture was conducted. This was prepared in the same way by adding $NaHCO_3$ and then analyzed by GC-FID in comparison with the initial measurement.

Prior to switching to no diluent, GC-MS measurements were conducted. Consequently, the preparation of samples for GC-MS measurements differed from the previously described procedure. Instead, 100 µL of the prepared electrolyte samples were diluted in 1250 µL methanol by adding 150 µL of an InSt, which had a concentration of 30 mg/mL acetonitrile diluted in MeOH.

2.6. GC-Validation Procedure

The method was validated in accordance with ICH R2(Q2) [39] and FDA [40] guidelines. Therefore, selectivity, linearity, range, limit of quantification (*LOQ*), limit of detection (*LOD*), accuracy, and precision were determined.

Selectivity of the method was evaluated in conjunction with its system suitability. In order to achieve this, blank samples of pure acetone, pure AcN, as well as the samples of each substance and the sample containing all substances, see Section 2.5, were injected and measured in triplicate. The clear separation and unambiguous assignment of the peaks were assessed based on the resolution RS, with a satisfactory resolution defined as RS > 1.5 [41]. Additionally, the asymmetry factor and number of theoretical plates were determined according to USP 621 [42].

The linear dependence of the signal (ratio of analyte peak area to the peak area of InSt $A_i \cdot {A_{AcN}}^{-1}$) in relation to the ratio of the mass of the analytes to the mass of the InSt $m_i \cdot {m_{AcN}}^{-1}$ was evaluated. For the purpose of analysis, different mass ratios were determined by quintuplicate measurements, and Equation (1) was applied to these values.

$$A_i \cdot {A_{AcN}}^{-1} = a \cdot m_i \cdot {m_{AcN}}^{-1} + b \tag{1}$$

The slope a and y-intercept b are the regression parameters. In order to ascertain the presence of linearity, the coefficient of determination, R^2, is employed.

The range is defined by the upper and lower bounds of the calibration, which are determined by the linearity of the set mass ratios. Due to the wide concentration ranges, the calibration was divided into three categories: low, medium, and high. Each category was assigned its own calibration curve.

LOD and *LOQ* were calculated in accordance with the Guidance Document on the Estimation of *LOD* and *LOQ* for Measurements in the Field of Contaminants in Feed and Food [43]. The limits were established using a sample in which all solvents were dissolved in pure CB, with a mass lower than the minimum mass required for the calibration range. The stock solution was prepared once, and the samples for *LOD* determination were prepared 10 times individually, with each sample measured once. *LOD* and *LOQ* were calculated using Equations (2) and (3), with the slope a from the calibration and the standard deviation $s_{y,b}$ from these measurements.

$$LOD = 3.9 \cdot \frac{s_{y,b}}{b} \tag{2}$$

$$LOQ = 3.3 \cdot LOD \tag{3}$$

Accuracy and precision of the method were evaluated using the recovery percentage $\%R_i$, defined as the ratio of the measured value $m_{i,meas}$ to the theoretical value $m_{i,th.}$, Equation (4), and the relative standard deviation (RSD). $\%R_i$ and RSD were determined for nine separate solutions in which the components were mixed together in different mass ratios and combinations. Additionally, the sample, which contains all the aforementioned components, was used for this purpose. From these solutions, five separately applied samples were each measured.

$$\%R_i = \frac{m_{i,meas.}}{m_{i,th.}} \tag{4}$$

3. Results and Discussion

3.1. Method Development

In the course of method development, a number of different solvents was tested for sample dilution, including methanol, ethanol, chlorobenzene, toluene, and others. The GC-separation of most of these solvents would have been suitable because there was no chromatographic overlap between the solvent and the target analytes. However, most of the solvents led to highly asymmetric peaks of the target components. Only the use of methanol or ethanol as a solvent, which eluted earlier than the target compounds, resulted in symmetrical and reproducible target peaks.

However, it was noticed that methanol-diluted samples with a high EMC and low DMC content (stored for extended periods of time) contained ethanol. This observation can be explained by the transesterification of EMC with MeOH to form DMC and EtOH, Equation (5) [35].

$$C_4H_8O_3 + CH_3OH \rightleftharpoons C_3H_6O_3 + C_2H_5OH \quad (5)$$

Pilot experiments confirmed that samples that had been dissolved in methanol showed an ethanol content of <1.0 g_{EtOH}/g_{tot}, whereas the samples that had not been dissolved in a solvent exhibited no or significantly lower DMC and EtOH. To avoid such solvent-derived alterations of the samples, the method was revised and changed to direct injection of undiluted samples. Since GC-FID is an analytical technique for trace analysis, direct injection of undiluted samples is rather unusual. However, by decreasing the injection volume to 0.2 µL and increasing the split ratio to 1:200, peak shape and peak size could be obtained in the usual dimensions. Furthermore, the only sample preparation step was the addition of the InSt AcN. Utilization of nitrogen as a mobile phase keeps costs within a manageable range. The method has undergone a validation procedure due to the absence of a solvent, which is a rather unconventional technique for GC analysis.

3.2. *Method Validation*

3.2.1. Selectivity and System Suitability

The chromatogram, see Figure A3, of the sample, which was composed entirely of all synthetic components, indicated the absence of degradation products and demonstrated that the resolution of all components was at least RS > 3 and that the number of theoretical plates (NTP) was >170,000. No effect of $NaHCO_3$ addition was observed (Figure A3). During the linearity measurements, the asymmetry factors for all components were found to range between 0.73 and 1.52.

As shown in Figure A4, the chromatogram of LP57 revealed the presence of multiple peaks, of which 99.5% of the peak area is attributed to AcN, EMC, and EC. The presence of trace peaks has also been observed, and it has been theorized that these are generated during preparation under atmospheric conditions or in the GC itself due to decomposition reactions involving $LiPF_6$. As this method only leads to a small proportion of decomposition products, as can be seen from the peak area fraction of the known components, it is also capable of analyzing samples containing conduction salt. However, if decomposition reactions are to be avoided completely, it is necessary to handle the samples in an inert atmosphere and remove the dissolved conductive salt prior to analysis.

3.2.2. Linearity, Range, LOQ, and LOD

Given the large calibration range, the process was subdivided into two or more calibration curves, with each calibration line comprising a minimum of five calibration points. Table A1 presents the range of each calibration, the calibration parameters, the coefficient of determination R^2, and the *LOQ* and *LOD*. For each individual calibration, R^2 exceeds the recommended threshold value of 0.99, as suggested by [41]. Moreover, Figure 1 illustrates the calibration results as example for the solvents EMC and DMC. The highest *LOD* were identified for DMC with a value of 0.4189 $mg_{DMC} \cdot g_{AcN}{}^{-1}$.

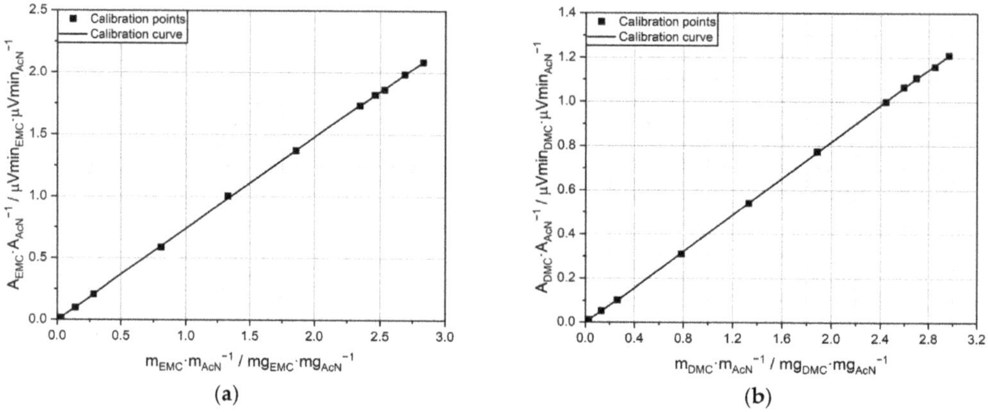

Figure 1. Calibration points and linear regressions for (**a**) EMC and (**b**) DMC.

3.2.3. Accuracy and Precision

The accuracy and precision for the low and medium calibration curves were checked at least once for all components. This resulted in 54 measurement results, which are presented in Table A2. Of these measurements, which also include the sample of the standard mixture prepared with $NaHCO_3$, 52 show a %R in the range 90–110, of which 44 are in the range 95–105. Only two measurements showed a %R > 110. The lowest recovery rate was observed for low calibrations of FB, with a value of %R = 90.39. In contrast, the highest recovery rate was observed for the medium calibration of EC, with an approximate value of %R = 116. The RSD for all components was ≤6.01%, except for two measurements for EC and CHB, where the RSD was <6.70%.

3.3. Composition and Impurities of Spent Electrolytes

A total of 106 peaks were identified across all samples, of which 42 could be unambiguously assigned based on the MS spectra and library spectra (NIST23) comparison using a match factor >800 as the criterion for identification. In general, it should be noted that in 14 cases, the assignment of certain peaks did not meet this level of certainty. However, these peaks could be adequately proposed based on literature data. A total of 50 peaks could only be characterized insufficiently and were registered as unknowns. These unknowns were only observed as trace components. If it turns out that one or more of these components play an important role in the recycling process, they would have to be specifically enriched in order to elucidate the structure of the component. Currently, there is no evidence for this.

The analysis of the samples revealed similarities in their composition and the identified substances. Upon neutralization of the samples with $NaHCO_3$, no visible gas formation was observed in any sample, which leads to the conclusion that the samples have no or very low acid content. It is evident that the procedure described above does not allow for the ascertainment of acidity levels. Consequently, any attempt to determine acidity levels would have to be conducted in a supplementary manner using a titration system, for example. Furthermore, all samples were found to contain EtOH, MeOH, DMC, EMC, DEC, and EC. The presence of carbonates can be attributed to their utilization as a solvent for the conducting salt. It seems feasible that the low concentration of some solvents, particularly DMC and DEC, may not have originated from their initial use as solvents but rather as a decomposition product within the cell during the recycling process or as a consequence of storage of the recovered electrolyte under ambient conditions. Matadi et al. [44] observed the presence of DMC and DEC during calendar ageing where initially the electrolyte solvent EMC/EC was used. This phenomenon can be explained by the reduction of EMC on lithiated graphite surface [44,45]. The aforementioned reactions (Equations (6) and (7)) were published by Kim et al. [45], which demonstrated that EMC decomposes to an alkoxide like (MeO^- or EtO^-) and an alkyl formate radical. This can lead to a reaction between the alkoxides and EMC to form DMC or DEC and the other alkoxide (EtO^- or MeO^-).

$$C_4H_8O_3 + e^- \rightarrow CH_3O^- + C_3O_2H_5 \quad (6)$$

$$C_4H_8O_3 + e^- \rightarrow C_2H_5O^- + C_2O_2H_3 \quad (7)$$

This provides an explanation for the low content of DMC and DEC, as well as the presence of MeOH and EtOH in the samples. Another route can be the disproportionation of EMC to DMC and DEC or the transesterification of EMC and MeOH or EtOH to DMC and EtOH or DEC and MeOH. See Equations (5), (8), and (9) [35]. Moreover, water, for instance, from the ambient environment can react with EMC due to inadequate storage conditions, resulting in the contamination with degradation products, including DMC, DEC, MeOH, and EtOH [46].

$$2\, C_4H_8O_3 \rightleftharpoons C_3H_6O_3 + C_5H_{10}O_3 \quad (8)$$

$$C_4H_8O_3 + C_2H_5OH \rightleftharpoons C_5H_{10}O_3 + CH_3OH \quad (9)$$

The two-phase formation in Sample B indicates the inclusion of water. The carbonates exhibit a mixing gap with water [47]. The aqueous droplet, which measured approximately 5 mm in diameter, was too small to be analyzed. To determine the water content in each sample, Karl Fischer titration (KFT) could be performed.

The calibrated components PC, FEC, and VC were suspected but not identified in any of the samples. This can be attributed to the fact that they were not initially utilized for the electrolyte or due to their decomposition during SEI formation. The former option is more plausible in the case of PC, given that PC has the potential to cause the exfoliation of graphene on the anode side. In contrast, EC has been demonstrated to form a more stable SEI protective film on graphitic anodes in comparison to PC [6,7]. Consequently, the detection of EC in all samples provides compelling evidence that no additional PC was utilized. In the case of VC and FEC, which are used as SEI building additives, it is more probable that if they were initially used, these additives were fully decomposed during the formation process.

3.3.1. Sample A—Electrolyte Directly Out of Prismatic Cell

The electrolyte sample exhibited a brown hue (see Figure A1) and displayed a total of 25 peaks combined in the GC-MS and GC-FID measurements, as illustrated in Figure 2 in partial scale as well as in Figure A5 in full scale and listed in Table A3. Of these, 11 substances could not be further identified. It should be noted that the chromatograms of the GC-MS and GC-FID measurements can be aligned, allowing for the clear assignment of individual peaks in the unaltered (undiluted) GC-FID run using the identification data from the GC-MS analysis (Figure 2). Nevertheless, there are isolated cases in which peaks are observed in only one of the two analytical approaches. To illustrate, substances with a retention time of less than 3.4 min in the FID are undetectable in the MS because of the solvent delay feature used in GC-MS to protect the filament and the detector from damage due to the excessive solvent content. Additionally, acetone can only be detected in the FID since it was only employed for GC-FID as a rinsing solvent. Moreover, Sample A displays Peak 7 in the GC-FID measurement, whereas Peaks 15 and 24 are only present in the MS measurement, neither of which are observed in the other chromatograms. As a result, those remain unassigned.

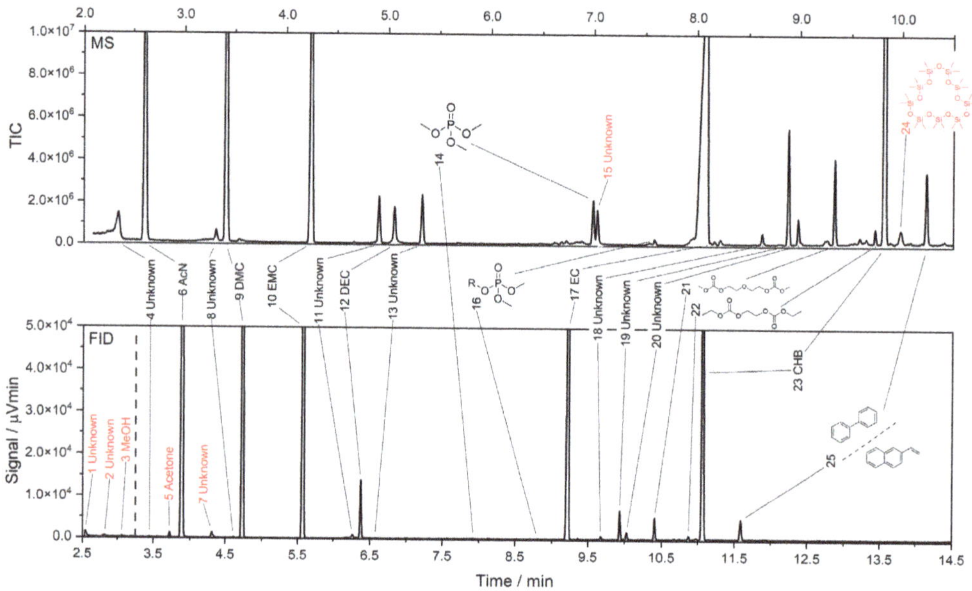

Figure 2. Comparison of gas chromatograms detected with GC-MS (top) and GC-FID (bottom) for Sample A.

The GC-FID measurements demonstrate a peak area fraction of the quantifiable components (knowns) of 97.8% (including AcN). This finding, at least qualitatively, suggests that the quantity of non-quantifiable components is, for the moment, negligible, with a peak area fraction of 2.2%. The analysis of Sample A, conducted using the calibrated GC-FID method, revealed the presence of DMC (375.57 ± 1.78 mg·mL^{-1}), EMC (312.64 ± 2.60 mg·mL^{-1}), DEC (9.05 ± 0.10 mg·mL^{-1}), EC (322.11 ± 11.07 mg·mL^{-1}), and CHB (39.35 ± 0.86 mg·mL^{-1}). The elevated concentration of DMC, EMC, and EC indicates that a solvent mixture of these three components was utilized within the cell, with a probable ratio of DMC:EMC:EC of 1:1:1. It seems reasonable to postulate that CHB was added as an additive and that the low proportion of DEC was the result of the aforementioned reaction of DMC or EMC during the ageing of the cell. This also provides an explanation for the low concentration of MeOH.

In the case of Sample A, the presence of a dissolved conducting salt, which is expected to remain in the sample due to the nature of the recovery process, suggests that reactions during storage and handling under atmospheric conditions or in the GC with the setup method may influence the chromatogram. This is evident in the comparison with the chromatograph of LP57 (Figure A4). The occurrence of Peaks 1, 2, 7, 11, 14, 18, and 20 in Sample A is consistent with the peaks observed in LP57, suggesting they may be attributed to decomposition products resulting from sample handling or GC-analysis. While the treatment presented here of samples containing conduction salt leads to an increase in conversion components, these only contribute a small proportion to the total peak area, as already explained in Section 3.2.1. Thus, it can be deduced that the decomposition products occurring in traces only have a minor influence on the concentrations. However, in instances where a comprehensive characterization of a sample containing conduction salts is required, the aforementioned changes in sample preparation and treatment (see Section 3.2.1) and the use of ion chromatography (IC) are nevertheless valuable.

Furthermore, analysis identified the presence of trimethyl phosphate (TMP) and an R-dimethyl phosphate. The most probable match in the database for the residue was identified as an R=ethyl or butyl. TMP is employed in LIB as a flame-retardant additive [48,49] and also occurs as an ageing product. This is demonstrated by Weber et al. [34] in the ageing of LIB cells under defined conditions and thermal treatment at 95 °C. However, as TMP has the same retention time as a peak in the LP57 analysis, it is possible that this peak is also TMP and the thermal treatment in the GC could lead to the formation of TMP and others. The formation of TMP and other phosphates, such as R-dimethyl phosphate with two methyl groups and one residue, can be explained by successive reactions of phosphorus fluoridates with organic carbonates. Alternatively, TMP may be formed in the reaction of phosphorus fluoridates with MeOLi, which results in the release of LiF and the subsequent formation of TMP [22,23].

Similarly, Lemordant et al. [50] identified the presence of oligoethylene glycols in the analysis of the electrolyte following its formation. One of the mass spectra presented in the study is analogous to the spectrum of Peak 21 and was also attributed to dimethyl diglycol carbonate. However, Lemordant et al. posit that this spectrum is more consistent with oligocarbonates. Nevertheless, the occurrence of diglycolcarbonates cannot be discounted. This is because diglycolcarbonates are well documented in the context of electrolyte ageing, as evidenced by the studies of Takeda et al. [29] and Gachot et al. [23]. Takeda et al. [29] identified the presence of diethyl diglycolcarbonate in the cell ageing of a DEC/EC electrolyte. In contrast, Gachot et al. [23] identified dimethyl diglycol carbonate in an EC/DMC electrolyte derived with a reaction scheme based on oligocarbonates, including diethyl-2,5-dioxahexane carboxylate (DEDOHC). It should be noted that oligocarbonates, including dimethyl-2,5-dioxahexane carboxylate (DMDOHC), ethylmethyl-2,5-dioxahexane carboxylate (EMDOHC), and DEDOHC, have been identified as ageing products in LIB electrolytes [19,23–25,27,29,51]. This hypothesis is also supported by the occurrence of Peak 22, DEDOHC. This is due to the fact that DEDOHC can be formed from dioxahexane carboxylate compounds in conjunction with the action of lithium alkoxides and the presence of linear organic carbonates [23]. Additionally, a nucleophilic attack of a lithium alkoxide on EC and a linear carbonate may occur [51]. In a recent study, Hofmann et al. (2023) [24] examined the formation of dicarbonates, including DMDOHC, DEDOHC, and EMDOHC. The researchers investigated the formation of these dicarbonate products both with and without the presence of conductive salts, such as $LiPF_6$. Their findings suggest that the formation of these dicarbonates is enhanced in the presence of conductive salts.

A comparison with the database for Peak 25 did not permit a clear assignment to one substance while indicating a match factor >900 for two possible compounds. The comparison with 2-vinylnaphthalenes, as well as with biphylene (BP), demonstrated a high degree of spectral agreement. Nevertheless, a further occurrence of 2-vinylnaphthalenes from spent LIBs is not documented in the existing literature. Conversely, the utilization of BP as a flame-retardant additive in LIBs is well known. Consequently, the impact of BP has been subjected to comprehensive examination within the existing literature [44,52–55]. For example, Lee et al. [52] demonstrate that BP can be employed as a stand-alone agent or in conjunction with other additives, such as CHB. The combination of BP and CHB has been demonstrated to offer enhanced overcharge protection compared to the use of either compound alone. Consequently, BP is frequently employed in LIBs and is regarded as a more plausible candidate for Peak 25. Therefore, it can be assumed that a combination of BP and CHB was initially utilized in this cell.

Additionally, octadecamethylcyclononasiloxane (Peak 24) was identified, yet the match (793) is insufficient for a definitive assignment, and this substance is not documented in the literature in relation to LIBs. Nevertheless, it appears to be a siloxane, which was also identified in Sample B and is discussed in greater detail there.

3.3.2. Sample B—Condensate from Mechanical-Thermal Recycling Route in Laboratory Scale

The sample resulting from the mechanical-thermal drying step during the recycling of a LIB on a laboratory scale exhibits 13 peaks, displayed in Figure 3 in partial scale as well as in Figure A6 in full scale and Table A4. Nine of these can be identified. These include DMC (2.67 ± 0.03 mg·mL^{-1}), EMC (964.62 ± 12.67 mg·mL^{-1}), DEC (10.75 ± 0.19 mg·mL^{-1}), and EC (10.40 ± 7.77 mg·mL^{-1}). In this instance, the peak area fraction of quantifiable components was found to be 99.97%, a value that can be attributed to three factors. Firstly, there was a low number of non-quantifiable peaks. Secondly, there was the method of recovery, where the conduction salt and components with low saturation vapor pressure and low concentration remain in the solid fraction of the shredded LIB. The absence of conduction salt has been identified as a factor that reduces the likelihood of decomposition products. Additionally, the utilization of the LIB may be restricted to a test level, with a reduced number of cycles, as outlined in Section 2.2, thereby leading to a reduced generation of decomposition products.

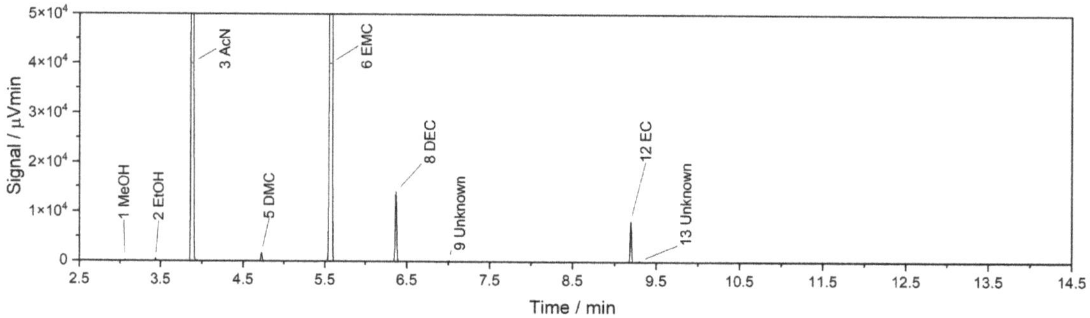

Figure 3. Chromatogram detected by GC-FID from Sample B recovered through mechanical disclosure combined with thermal drying.

In comparison to Sample A, the EC content is notably reduced. This is due to the fact that the recovery via drying enables low-boiling components, in particular, and heavier-boiling components, such as EC or CHB, to remain in the solid fraction of the crushed LIB [18]. Based on the available information, it can be concluded that a solvent mixture comprising EMC and EC was employed for the cell. The low proportions of DMC and DEC suggest that the mentioned transesterification of EMC occurred.

Moreover, ethylene glycol (MEG) was identified in the MS sample. This can be attributed to inadequate storage of the sample, resulting in water retention and subsequent hydrolysis of EC [36,46,56], or to the addition of methanol and the transformation of EC to EG and DMC [57,58].

In addition, octamethylcyclotetrasiloxane has also been identified in the MS sample. In general, siloxanes can be added to the electrolyte to facilitate the formation of interfaces between the electrolyte and the cathode and anode surfaces, among other functions [59]. Research is also being conducted on octamethylcyclotetrasiloxanes, which Wang et al. [60] added as an additive in an EC:EMC:DMC electrolyte. In addition to their use in additives, siloxanes such as dodecamethylcyclohexasiloxanes have also been detected in investigations of thermal runaways on LIBs in particulate recoils. It is assumed that this is due to the used instruments. In this instance, the identified siloxanes may also manifest in the chromatogram through column bleeding. Both are possible explanations for the occurrence of siloxanes in Sample A.

3.3.3. Sample C—Condensate from Mechanical-Thermal Recycling Route in Pilot Scale

A total of 68 peaks were identified in the chromatograms of the three samples C_O, C_D, and C_R, as illustrated in Figure 4 in partial scale as well as in Figure A7 in full scale. Of these, 25 could be assigned using the NIST database with good agreement; 8 showed moderate agreement with the database. However, the proposed structures could be justified with additional literature data, and 33 peaks could not be assigned (see Table A5).

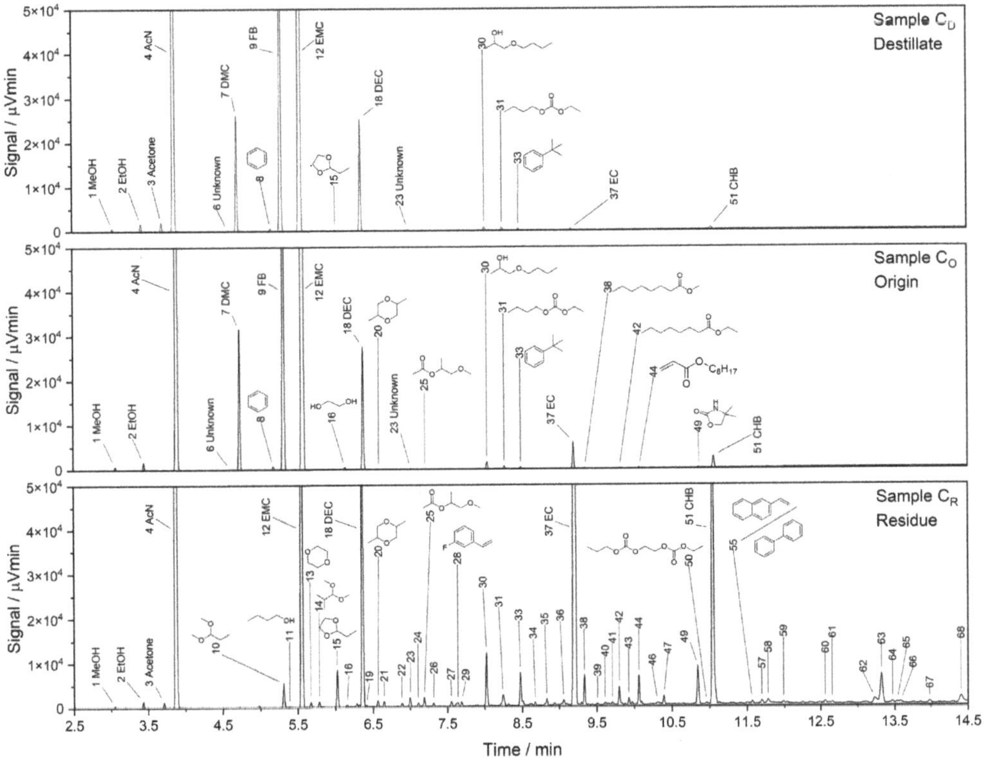

Figure 4. GC-FID chromatograms of Sample C_O (origin, center), Sample C_D (distillate, top), and Sample C_R (residue, bottom).

The original Sample C_O exhibited 23 peaks, in addition to the known quantifiable components MeOH (1.23 ± 0.02 mg·mL^{-1}), EtOH (1.03 ± 0.01 mg·mL^{-1}), and DMC (52.48 ± 0.23 mg·mL^{-1}), FB (48.08 ± 0.22 mg·mL^{-1}), EMC (831.45 ± 6.06 mg·mL^{-1}), DEC (21. 58 ± 0.21 mg·mL^{-1}), EC (10.87 ± 0.09 mg·mL^{-1}), and CHB (1.07 ± 0.01 mg·mL^{-1}). The analysis also encompassed 3 unknowns and 12 other substances.

In comparison to Sample B, a reduced concentration of EMC and a comparable EC concentration can be recorded. The higher concentrations of MeOH, EtOH, DMC, and DEC in Sample C_O can be attributed to two reasons. Firstly, the use of a different initial electrolyte is likely due to the appearance of FB. Secondly, the recycling of multiple LIBs, which may differ in initially used electrolytes as well in lifetime and life cycles, can also explain the higher number of impurities. However, this sample probably represents a more realistic scenario, as it will be observed in the future for the recycling of LIBs of different lifetimes and manufacturing technologies.

After distillation of the original Sample C_O leading to the distillate, Sample C_D, and the residue, Sample C_R, alterations were observed in both samples. These are demonstrated in the material balance presented in Figure 5. The distillate exhibited an elevated concentration of low-boiling components (e.g., EMC, reaching 866.72 ± 9.61 mg·mL^{-1}), along with a complete transfer of light-boiling components (e.g., FB, benzene, or DMC) to the distillate. Conversely, the residue C_R demonstrated an enrichment of high-boiling components (e.g., EC or CHB). This also resulted in a significant number of components that were not previously detected because they were below the *LOD* in Sample C_O but became quantifiable after distillation due to the concentration increase in Sample C_R. While the proportion of quantifiable components in Sample C_O is 99.7%, there was a slight increase to 99.8% for Sample C_D and a decrease to 90.6% for the residue, Sample C_R, suggesting an accumulation of impurities in the residue, as previously mentioned. Consequently, 63 peaks were identified in the residue, Sample C_R, and peaks that had previously been unclassifiable in Sample C_O could be assigned to individual substances in Sample C_R due to the presence of clear mass spectra and identical retention times, as depicted with the chromatograms displayed one above the other in Figure 4 in partial scale as well as in Figure A7 in full scale.

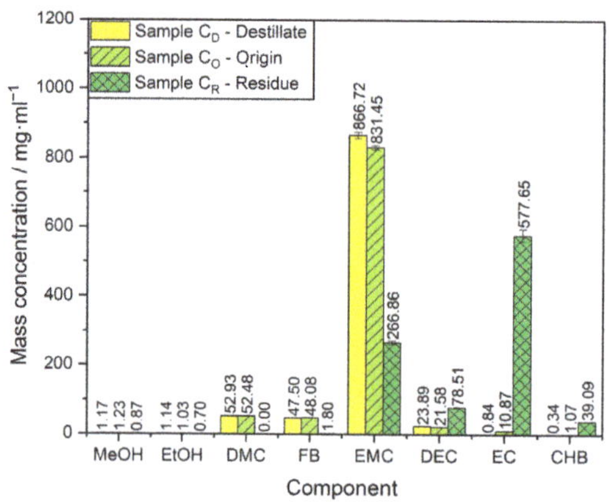

Figure 5. Mass concentrations of Samples C_O, C_D, and C_R.

The elevated number of peaks observed in Sample C_R is attributable to the fact that this sample originated from a pilot plant where a substantial quantity of spent LIBs were recycled in batch, and the impurities were subsequently concentrated through distillation. It can be assumed that the additional heat input during distillation led to further decomposition of the solvents and an increase in decomposition products. To avoid the accumulation of decomposition products during the electrolyte purification process, a more thermally gentle treatment, such as distillation or evaporation under vacuum conditions, would be preferable.

Examination of the identified components reveals the presence of trimethoxyborane (TMB) and tert-butylbenzene, which both can be added to the electrolyte as additives. TMB has been shown to enhance the stability of the interface between the electrode and electrolyte at high voltages, while the formation of a protective layer on the cathode surface inhibits the decomposition of the electrolyte and prevents the structural degradation of the electrode [61,62]. In general, tert-alkylbenzenes can be employed in combination with biphenyl, resulting in enhanced overcharge inhibition [63]. This is consistent with the observation of a peak in Sample C_R, which could be attributed to both biphenyl and 2-vinylnaphthalene, as discussed in Section 3.3.1.

In addition to tert-alkylbenzene, further substances containing benzene rings were identified in the samples, including benzene, fluorostyrene, BP, 2-vinylnaphthalene, and

5-methylquinoline. Benzene can be introduced either through the addition of an electrolyte [10] or as a result of a decomposition reaction in the electrolytes themselves [64]. Benzene was identified in the investigation of thermal decomposition by Chen et al. [64]. The occurrence of benzene can be explained by a decomposition reaction from EC, PC, and DEC to ethene and methyl-ethene via ethyne and methyl-ethyne, which then undergoes further decomposition to form benzene and methyl-benzene. An alternative hypothesis is that FB releases fluorine through cell ageing reactions during SEI formation, resulting in the incorporation of LiF into the SEI and the formation of benzene in the electrolyte. The occurrence of fluorostyrene is likely to be derived from a reaction based on FB, although this has not yet been documented in the literature, as little information on the influence of FB is available in the public domain.

Additionally, the presence of dioxanes and dioxalanes was identified by mass spectrometry. The presence of 1,4-dioxanes is identified in two distinct contexts in literature: during the thermal ageing of EMC/EC (1 mol $LiPF_6$) electrolytes [32] and on the anode and cathode sides, where the electrolyte was to be washed out in both cases [30]. Furthermore, Gachot et al. [22] identified the presence of 2-methyl-1,3-dioxalane in addition to 1,4-dioxanes, even after the cycling of cells. Although the latter was not detected in the present study, it only shows a longer alkyl residue up to 2-ethyl-1,3-dioxalane. This indicates that comparable reaction mechanisms for the formation of dioxalanes, as well as dioxanes such as 2,5-dimethyl-1,4-dioxanes, may occur within the cell.

The electrolyte samples were found to contain several other ester compounds, including methyl octanoate, ethyl octanoate, an isomer of octyl cyclohexane, 1-methoxy-2-propyl acetate, 2-hydroxyethyl octanoate, and butyl caprylate. The group of esters is well documented in the context of aged electrolytes in LIBs, with the majority of research focusing on short-chain ester compounds. A reaction of linear carbonates, such as DMC, with a lithium cation can result in the formation of radicals, including the methyl radical and the methyl carboxyl radical [22]. Similarly, it is conceivable that ethyl radicals and ethyl carboxy radicals are formed by the living reaction mechanism with EMC or DEC. Consequently, alkyl radicals, hydrogen ions (hydronium ions), and carboxy radicals may undergo a reaction to form acetate or (methyl) formate. It can be assumed that ethene, which can be formed during the reaction of EC with Li+ [23,26], is attacked by the radicals. This could result in a chain elongation of the carboxy radical, which can be terminated by the reaction with an alkyl radical or H+. This would explain the occurrence of the majority of the listed esters [65]. In addition, long-chain ester compounds were identified, which were attributed to a similar rationale and a reaction with ethene.

Although there is considerable overlap between the components listed here and those identified in the literature, the origin of certain components, such as 1,1-dimethoxypropane, 1-butanol, 1,1-dimethoxy-2-methylpropane, 1-butoxy-2-propanol, ethyl butyl carbonate, and 4,4-dimethyl-2-oxazolidinone, remains unexplained. Despite an extensive literature search, no evidence of corresponding ageing processes could be found.

The analysis of Sample C_O illustrates the difficulties associated with the preparation of recycled electrolytes. The impact of individual impurities remains unknown; therefore, the purification of target solvents, such as linear carbonates or FB, requires further investigation.

4. Conclusions

This contribution presents a new GC-FID and GC-MS method capable of identifying the organic components of recovered electrolytes originating from spent LIBs.

The method was applied to three samples of spent electrolytes, of which Sample A was recovered directly from the cell and Sample B and C_O were recovered from shredded LIB scrap through vacuum-drying in an inert gas atmosphere. In addition, Sample C_O

from a recycling plant in pilot scale was distilled, resulting in a distillate (Sample C_D) and a residue fraction (Sample C_R).

The analyses enabled the formulation of conclusions regarding the solvents and additives employed. Quantification and measurement of some of the most-used solvents for LIB electrolytes, EMC, DEC, and DMC and EC were conducted in all samples. The analysis revealed that EMC was present in the highest concentrations in almost all original samples. Although DMC and DEC were detected in the samples, this can be partly attributed to a decomposition reaction due to their low concentrations in Samples B and C_O. The presence of MeOH and EtOH, which are also present in traces in all samples, can be explained in a similar manner. In summary, conclusions can be drawn about the solvents used, but these are also strongly influenced by the recovery method of the electrolyte samples. For example, the comparatively low EC content in Samples B and C_O can be explained by the fact that, due to the low saturation vapor pressure of EC, the drying process during vacuum-drying is insufficient; therefore, less EC is present in the recovered electrolyte sample. The presence of common additives such as CHB or BP, along with the occurrence of decomposition products, which are formed during cell ageing, including methyl octanoate, ethyl octanoate, and DEDOHC, has been proven. Notably, the presence of FB and benzene-ring-containing substances in Sample C_O, which are absent in all other samples, is an important finding.

Additionally, some components remain unidentified, and their provenance may be traced back to formed components inside the LIB, alternatively attributed to the handling of the samples and the conditions in the GC, or perhaps have been formed during the distillation process. While the analytics employed offer insight into the composition of the recovered electrolyte samples and draw attention to potential processing challenges, it should be noted that further analyses and, if necessary, further analysis techniques are necessary for a comprehensive characterization of samples from spent LIB. These may include IC to determine the salt content, KFT to determine the water content, or an enrichment for specific, as yet unknown, contaminations to elucidate the structure of the components. These investigations are an important step to learn how complex these recovered electrolyte mixtures from spent LIBs are, elucidate the structures of the main components, and determine what processes might be involved in the formation of degradation products and trace components. This was particularly evident for Sample C_D obtained by distillation. The sample exhibited a distinct chromatogram, indicating that the process of distillation or rectification may facilitate the recovery and reuse of electrolyte solvents. Moreover, the retention times of certain components, including high-boiling DEC, EC, and CHB, provide further support for this conclusion. To reuse the target solvents, it is necessary to separate alcohols such MEOH and EtOH from decomposition reactions in addition to other impurities. The distillation of Sample C_O enabled the identification of some of these components, with most of these components being primarily impurities and consistent with the findings of existing literature.

Finally, a comparison of all samples revealed considerable variations in their composition and the components present, which provides an outlook of what can be expected from future large-scale industrial LIB recycling processes. Studies like those presented here will help to find out whether a single general process can be carried out to enable sustainable reuse of electrolyte materials in new battery cells. The analytical techniques presented here will not only help in the development and optimization of such processes but will also be essential in ensuring the quality of the recycled materials.

Author Contributions: Conceptualization, M.W., K.J., T.B. and S.S.; Data curation, M.W.; Formal analysis, M.W. and K.S.; Funding acquisition, K.J. and S.S.; Investigation, M.W., K.S., K.A., P.M. and T.B.; Methodology, M.W., K.S., K.A. and P.M.; Project administration, K.J.; Resources, M.W.;

Supervision, K.J. and S.S.; Validation, M.W., K.S. and P.M.; Visualization, M.W.; Writing—original draft, M.W.; Writing—review & editing, M.W., K.J., T.B. and S.S. All authors have read and agreed to the published version of the manuscript.

Funding: This research was funded by Federal Ministry for Economic Affairs and Climate Action, grant number 16BZF342F.

Data Availability Statement: The original contributions presented in the study are included in the article; further inquiries can be directed to the corresponding author.

Acknowledgments: Special thanks to Steffen Fischer of the TU Braunschweig, Institute of Particle Technology, for providing the samples.

Conflicts of Interest: The authors declare no conflicts of interest.

Nomenclature

Symbol	**Description**
A_i	peak area from component i
a	slope
b	y-intercept
m_i	mass of component i
$m_{i,meas}$	measured value (mass) of component i
$m_{i,th}$	theoretical value (mass) of component i
R^2	coefficient of determination
RS	resolution
$s_{y,b}$	standard deviation
$\%R_i$	recovery percentage of component i
Abbreviations	**Description**
AcN	acetonitrile
BEVs	battery electric vehicles
BP	biphylene
CB	chlorobenzene
CHB	cyclohexylbenzene
DEC	diethyl carbonate
DEDOHC	diethyl-2,5-dioxahexane carboxylate
DMC	dimethyl carbonate
DMDOHC	dimethyl-2,5-dioxahexane carboxylate
EC	ethylene carbonate
EMC	ethyl methyl carbonate
EMDOHC	ethylmethyl-2,5-dioxahexane carboxylate
EOL	end-of-life
EtOH	ethanol
FB	fluorobenzene
FEC	fluoroethylene carbonate
FID	flame ionization detection
GC	gas chromatography
HF	hydrofluoric acid
IC	ion chromatography
InSt	internal standard
KFT	Karl Fischer titration
LIB	lithium-ion battery
LiF	lithium fluoride
LiPF6	lithium hexafluorophosphate
LOD	limit of detection
LOQ	limit of quantification
MEG	ethylene glycol

MeOH	methanol
MS	mass spectrometry
$NaHCO_3$	sodium hydrogen carbonate
NMC	nickel-manganese-cobalt
PC	propylene carbonate
RS	resolution
RSD	relative standard deviation
SEI	solid electrolyte interface
StSo	stock solution
TMB	trimethoxyborane
TMP	trimethyl phosphate
VC	vinylene carbonate

Appendix A

Appendix A.1. Samples and Setup

(**a**) Sample A (**b**) Sample B (**c**) Sample C_O (**d**) Sample C_D (**e**) Sample C_R

Figure A1. Photo of each sample. (**a**) Sample A after sample preparation. (**b**) Sample B after addition of $NaHCO_3$ (right) and after sample preparation (left). (**c**) Sample C_O (right) and after sample preparation (left). (**d**) Sample C_D after addition of $NaHCO_3$ (right) after sample preparation (left). (**e**) Sample C_R (right) and after sample preparation (left).

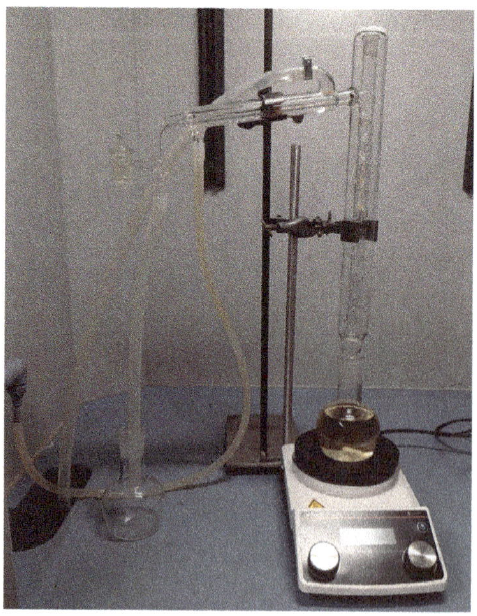

Figure A2. Experimental setup for the distillation of sample C_O.

Appendix A.2. Additional Information for Method Validation

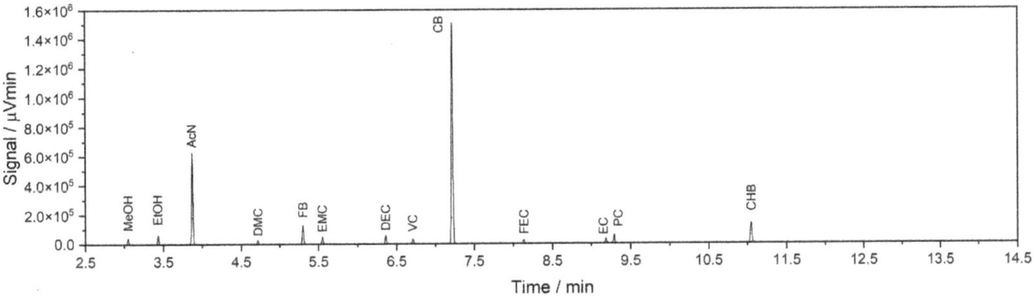

(**a**) GC-FID chromatogram of the standard mixture.

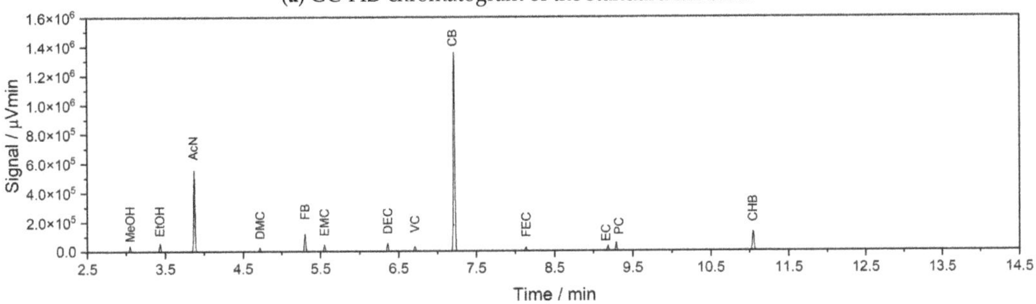

(**b**) GC-FID chromatogram of the standard mixture after preparation with $NaHCO_3$.

Figure A3. GC-FID chromatograms of standard mixture comprising all components for validating the selectivity of the method (**a**) without and (**b**) with $NaHCO_3$ preparation.

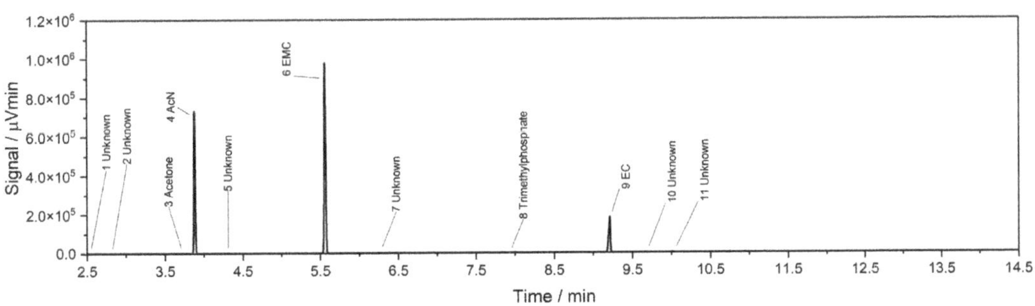

(**a**) GC-FID chromatogram of LP57 in full scale.

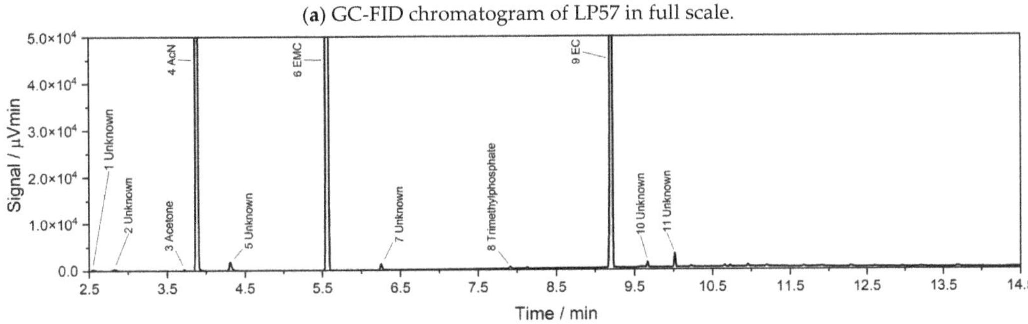

(**b**) GC-FID chromatogram of LP57 in partial scale.

Figure A4. GC-FID chromatograms of LP57 in (**a**) full scale and (**b**) partial scale.

Table A1. Range, slope a, intercept b, and coefficient of determination R^2 for each calibration curve and LOD and LOQ for each component.

Substance	Upper Limit	Lower Limit	a	b	R^2	LOD	LOQ
	$\frac{g_i}{g_{AcN}}$	$\frac{g_i}{g_{AcN}}$	$\frac{\mu Vmin_i \cdot g_{AcN}}{\mu Vmin_{AcN} \cdot g_i}$	$\frac{\mu Vmin_i}{\mu Vmin_{AcN}}$	-	$\frac{mg_i}{g_{AcN}}$	$\frac{mg_i}{g_{AcN}}$
MeOH	0.2131	0.0031	0.7419	-1.1167×10^{-3}	0.9994	0.0877	0.2893
	1.8247	0.2131	0.7519	2.2946×10^{-3}	0.9998		
	2.2128	1.8247	0.7735	-3.8319×10^{-2}	0.9987		
EtOH	0.2028	0.0031	1.0587	1.5370×10^{-5}	0.9995	0.0917	0.3027
	1.8136	0.2028	1.1243	-1.5579×10^{-2}	0.9999		
	2.2143	1.8136	0.9866	2.4923×10^{-1}	0.9968		
FB	0.2486	0.0037	2.2373	-1.3387×10^{-3}	0.9993	0.0948	0.3129
	2.3244	0.2486	2.0201	2.7882×10^{-2}	0.9999		
	2.8624	2.3244	2.1570	-2.7103×10^{-1}	0.9948		
DMC	0.2613	0.0047	0.3897	1.0010×10^{-4}	0.9998	0.4189	1.3823
	2.4424	0.2613	0.4135	-8.8866×10^{-3}	0.9999		
	2.9646	2.4424	0.3965	3.4019×10^{-2}	0.9974		
EMC	0.2823	0.0043	0.7326	-8.5645×10^{-4}	0.9998	0.1101	0.3635
	2.3473	0.2823	0.7412	-1.3123×10^{-3}	0.9997		
	2.8326	2.3473	0.7296	2.1402×10^{-2}	0.9991		
DEC	0.2731	0.0040	0.8916	-8.3882×10^{-4}	0.9968	0.0460	0.1518
	2.2376	0.2731	0.8911	1.6395×10^{-2}	0.9997		
	2.7289	2.2376	0.8777	3.1273×10^{-2}	0.9951		
VC	0.2554	0.0065	0.3563	2.8799×10^{-4}	1.0000	0.3263	1.0767
	1.7008	0.2554	0.3585	-2.7006×10^{-4}	0.9995		
FEC	0.2461	0.0178	0.2448	-4.8183×10^{-4}	1.0000	0.2618	0.8639
	1.7090	0.2461	0.2203	-2.0051×10^{-3}	0.9943		
PC	0.2577	0.0069	0.6279	-2.9934×10^{-4}	0.9995	0.1526	0.5037
	1.7050	0.2577	0.5700	-1.4452×10^{-3}	0.9965		
EC	0.2415	0.0032	0.3564	-2.1831×10^{-4}	0.9999	0.0999	0.3296
	1.6764	0.2415	0.3148	1.8939×10^{-3}	0.9976		
CHB	0.2597	0.0061	2.4046	-1.1336×10^{-3}	1.0000	0.2924	0.9649
	1.3052	0.2597	2.5992	-5.7071×10^{-2}	0.9994		
CB	2.5251	3.1126	1.6368	2.5157×10^{-1}	0.9908	-	-
	3.1126	2.5251	2.0577	-8.9418×10^{-1}	0.9908		

Table A2. %R and RSD of all measurements for checking accuracy and precision for all components.

$m_{i,th}$	$m_{i,meas.}$	%R	RSD	$m_{i,th}$	$m_{i,meas.}$	%R	RSD
mg	mg	%	%	mg	mg	%	%
	MeOH					EC	
82.47	80.54	97.65	1.93	53.21	51.68	97.13	3.9
358.38	360.41	100.57	0.87	538.85	626.93	116.35	6.5
31.31	29.35	93.74	5.22	54.44	57.1	104.88	3.53
31.42	28.76	91.53	5.23	54.63	55.63	101.83	5.34
	EtOH					FB	
85.45	85.55	100.12	0.78	804.28	813.56	101.15	1.37
396.57	393.14	99.14	0.42	39.95	36.11	90.39	1.7
33.33	31.97	95.93	1.98	42.28	38.48	91.02	3.12
33.45	32.17	96.19	1.39	42.43	39.51	93.12	2.36

Table A2. *Cont.*

$m_{i,th}$	$m_{i,meas.}$	%R	RSD	$m_{i,th}$	$m_{i,meas.}$	%R	RSD
mg	mg	%	%	mg	mg	%	%
	DEC				FEC		
845.01	852.75	100.92	1.88	50.56	47.78	94.5	2.39
91.57	88.44	96.58	1.05	567.94	562.48	99.04	1.8
42.62	41.6	97.61	5.98	61.17	63.43	103.7	4.24
856.04	860.31	100.5	2.13	83.51	84.76	101.51	2.6
42.76	42.13	98.51	3.15	61.39	62.54	101.89	4.6
	PC				CHB		
52.11	55.59	106.69	3.13	51.52	53.96	104.73	2.68
629.39	639.28	101.57	2.52	41.18	42.31	102.74	6.43
53.47	55.58	103.96	3.98	53.98	54.07	100.18	6.69
78.76	87.06	110.54	3.61	211.66	210.41	99.41	3.64
53.65	54.55	101.68	5.12	41.32	42.59	103.07	4.94
	DMC				VC		
480.45	480.85	100.08	1.69	53.63	54.03	100.75	4.12
117.08	121.72	103.96	2.60	539.12	576.81	106.99	3.94
46.04	45.22	98.23	3.49	59.84	59.85	100.02	4.25
46.2	46.2	100.00	2.12	60.04	59.25	98.69	3.63
	EMC				CB		
477.21	483.47	101.31	2.92	953.32	960.49	100.75	3.59
96.35	97.29	100.97	3.11	627.96	623.17	99.24	2.82
43.49	42.18	97.00	5.03	584.84	582.65	99.63	5.02
79.51	77.69	97.71	3.04	586.87	590.3	100.58	2.76
209.08	207.23	99.11	2.43				
43.64	42.93	98.39	2.73				

Appendix A.3. Full-Scale Chromatograms and Overview of Appeared Peaks

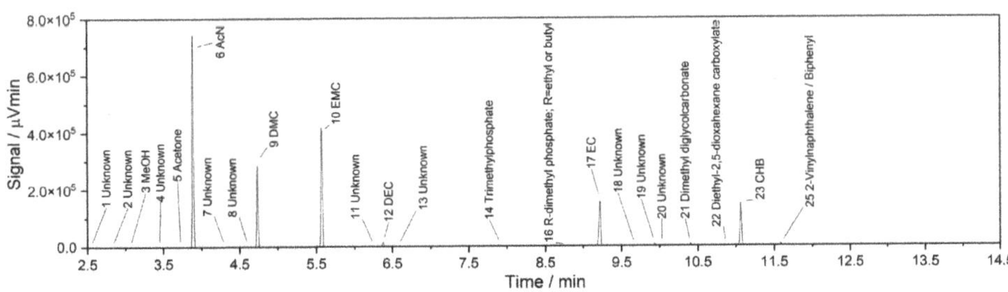

(**a**) GC-FID chromatogram of Sample A in full scale.

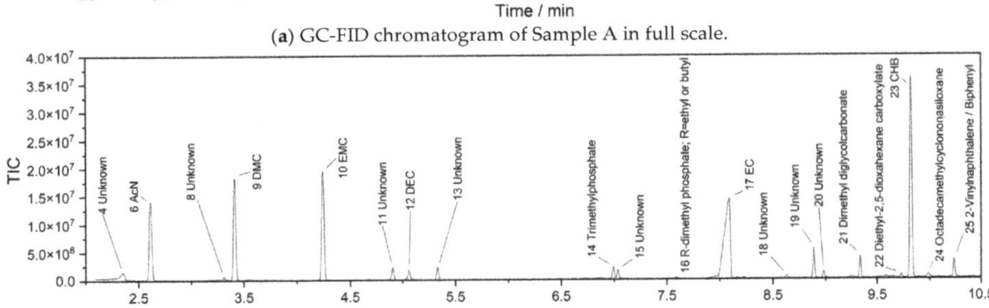

(**b**) GC-MS chromatogram of Sample A in full scale.

Figure A5. Chromatogram of Sample A in full-scale for (**a**) GC-FID and (**b**) GC-MS.

Table A3. List of peaks and the assignment for Sample A.

Peak-ID	Substance	Match Factor	Concentration
			$mg \cdot mL^{-1}$
1 [FID]	Unknown		Tr
2 [FID]	Unknown		Tr
3 [FID,p]	MeOH		0.80 ± 0.09
4	Unknown	701	Tr
5 [FID]	Acetone		Tr
6	AcN	912	Tr
7 [FID]	Unknown		Tr
8	Unknown	557	Tr
9	DMC	894	375.57 ± 1.78
10	EMC	806	312.64 ± 2.59
11	Unknown	736	Tr
12	DEC	817	9.05 ± 0.10
13	Unknown	677	Tr
14	Trimethylphosphate	903	Tr
15 [MS]	Unknown	710	Tr
16 [t,p]	R-Dimethylphosphat with R: Ethyl or Butyl	757–780	Tr
17	EC	888	322.11 ± 11.07
18	Unknown	552	Tr
19	Unknown	582	Tr
20	Unknown	687	Tr
21 [p]	Dimethyl diglycolcarbonate	613	Tr
22 [p]	Diethyl-2,5-dioxahexane carboxylate	733	Tr
23	CHB	944	39.35 ± 0.86
24 [MS,p]	Octadecamethylcyclononasiloxane	793	Tr
25 [t,p]	2-Vinylnaphthalene	945	Tr
25 [t,p]	Biphenyl	941	Tr

[MS] only detected via MS, [FID] only detected via FID, [p] putative, [t] tentatively.

Table A4. List of peaks and the assignment for Sample B.

Peak-ID	Substance	Match Factor	Concentration
			$mg \cdot mL^{-1}$
1 [FID]	MeOH		0.73 ± 0.01
2	EtOH	807	0.22 ± 0.01
3	AcN	876	Tr
4 [MS]	Unknown	785	Tr
5	DMC	877	2.67 ± 0.03
6	EMC	820	964.62 ± 12.67
7 [MS]	MEG	813	Tr
8	DEC	857	10.75 ± 0.19
9	Unknown	507	Tr
10 [MS]	Octamethylcyclotetrasiloxane	811	Tr
11 [MS]	Unknown	596	Tr
12 [p]	EC	764	10.40 ± 7.77
13	Unknown		Tr

[MS] only detected via MS, [FID] only detected via FID, [p] putative, [t] tentatively.

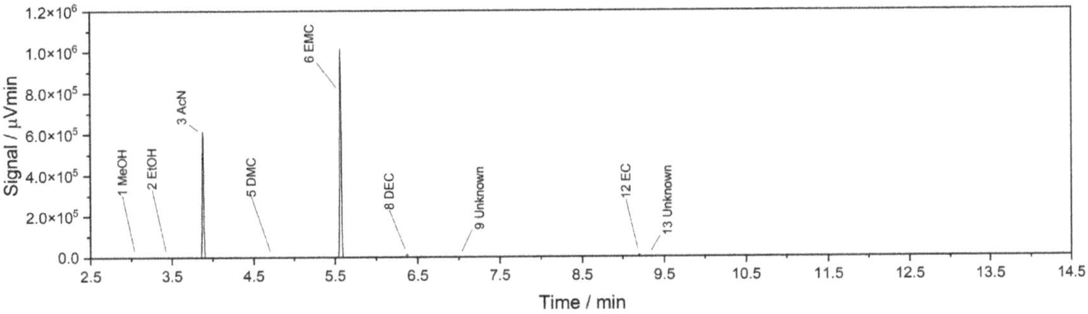

Figure A6. GC-FID chromatogram of Sample B in full scale.

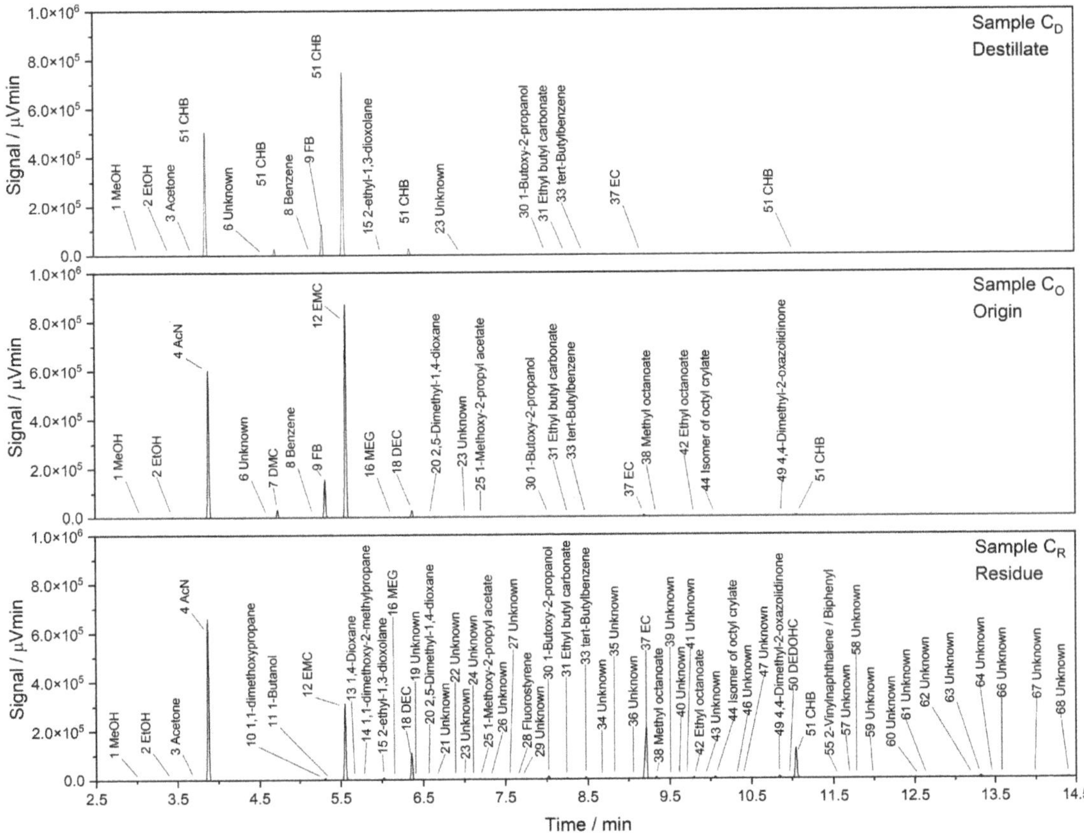

Figure A7. Full-scale GC-FID chromatograms of Sample C_O (origin, (**center**)), Sample C_D (distillate, (**top**)), and Sample C_R (residue, (**bottom**)).

Table A5. List of peaks and the assignment for sample C_O, C_D, and C_R.

Peak-ID	Substance	Concentration in Sample mg·mL^{-1}		
		C_O	C_D	C_R
1 ^{FID}	MeOH	1.23 ± 0.02	1.17 ± 0.02	0.87 ± 0.07
2	EtOH	1.03 ± 0.01	1.14 ± 0.01	0.70 ± 0.081
3 ^{FID}	Acetone		Tr	Tr
4	can	Tr	Tr	Tr
5 ^{MS}	Trimethoxyborane			Tr
6	Unknown	Tr	Tr	
7	DMC	52.48 ± 0.23	52.93 ± 0.45	
8 ^{P}	Benzene	Tr	Tr	
9	FB	48.08 ± 0.22	47.50 ± 0.39	
10	1,1-dimethoxypropane			Tr
11	1-Butanol			Tr
12	EMC	831.45 ± 6.06	866.72 ± 9.61	266.86 ± 4.78
13 ^{P}	1,4-Dioxane			Tr
14	1,1-dimethoxy-2-methylpropane			Tr
15	2-ethyl-1,3-dioxolane		Tr	Tr
16 ^{P}	MEG	Tr		Tr
17 ^{MS}	Unknown		Tr	
18	DEC	21.58 ± 0.21	23.89 ± 0.27	78.51 ± 1.62
19	Unknown			Tr
20 ^{P}	2,5-Dimethyl-1,4-dioxane	Tr		Tr
21	Unknown			Tr
22	Unknown			Tr
23	Unknown	Tr	Tr	Tr
24 ^{FID}	Unknown			Tr
25	1-Methoxy-2-propyl acetate	Tr		Tr
26 ^{FID}	Unknown			Tr
27	Unknown			Tr
28 ^{P}	Fluorostyrene			Tr
29	Unknown			Tr
30	1-Butoxy-2-propanol	Tr	Tr	Tr
31 ^{P}	Ethyl butyl carbonate	Tr	Tr	Tr
32 ^{MS}	Unknown	Tr		Tr
33	tert-Butylbenzene	Tr	Tr	Tr
34	Unknown			Tr
35	Unknown			Tr
36	Unknown			Tr
37	EC	10.87 ± 0.09	0.84 ± 0.02	577.65 ± 18.11
38	Methyl octanoate	Tr		Tr
39	Unknown			Tr
40	Unknown			Tr
41	Unknown			Tr
42	Ethyl octanoate	Tr		Tr
43	Unknown			Tr
44 ^{t}	Isomer of octyl crylate	Tr		Tr
45 ^{MS}	Unknown			Tr
46	Unknown			Tr
47	Unknown			Tr
48 ^{MS}	Unknown			Tr
49	4,4-Dimethyl-2-oxazolidinone	Tr		Tr
50	diethyl-2,5-dioxahexane dicarboxylate			Tr
51	CHB	1.07 ± 0.01	0.34 ± 0.01	39.09 ± 1.08
52 ^{MS}	Butyl caprylate			Tr
53 ^{MS}	5-Methylquinoxaline			Tr

Table A5. *Cont.*

Peak-ID	Substance	Concentration in Sample mg·mL^{-1}		
		C_O	C_D	C_R
54 ^{MS}	Unknown			Tr
55 ^{t,p}	Biphenyl			Tr
55 ^{t,p}	2-Vinylnaphthalene			Tr
56 ^{MS}	2-Hydroxyethyl octanoate			Tr
57 ^{FID}	Unknown			Tr
58 ^{FID}	Unknown			Tr
59 ^{FID}	Unknown			Tr
60 ^{FID}	Unknown			Tr
61 ^{FID}	Unknown			Tr
62 ^{FID}	Unknown			Tr
63 ^{FID}	Unknown			Tr
64 ^{FID}	Unknown			Tr
65 ^{FID}	Unknown			Tr
66 ^{FID}	Unknown			Tr
67 ^{FID}	Unknown			Tr
68 ^{FID}	Unknown			Tr

^{MS} only detected via MS, ^{FID} only detected via FID, ^{p} putative, ^{t} tentatively.

References

1. Hettesheimer, T.; Neef, C.; Rosellón Inclán, I.; Link, S.; Schmaltz, T.; Schuckert, F.; Stephan, A.; Stephan, M.; Thielmann, A.; Weymann, L.; et al. *Lithium-Ion Battery Roadmap—Industrialization Perspectives Toward 2030*; Fraunhofer ISI: Karlsruhe, Germany, 2023. [CrossRef]
2. Kampker, A.; Heimes, H.; Dorn, B. Battery Monitor 2023. Available online: https://content.rolandberger.com/hubfs/07_presse/Batteriemonitor_2023_digital_final.pdf (accessed on 29 November 2024).
3. Schmaltz, T. Recycling of Lithium-Ion Batteries Will Increase Strongly in Europe. Available online: https://www.isi.fraunhofer.de/en/blog/themen/batterie-update/recycling-lithium-ionen-batterien-europa-starke-zunahme-2030-2040.html (accessed on 9 August 2024).
4. European Union. Regulation (EU) 2023/1542 of the European Parliament and of the Council of 12 July 2023 concerning batteries and waste batteries, amending Directive 2008/98/EC and Regulation (EU) 2019/1020 and repealing Directive 2006/66/EC. *Off. J. Eur. Union* **2023**. Available online: https://eur-lex.europa.eu/eli/reg/2023/1542/oj/eng (accessed on 24 August 2024).
5. Arnberger, A.; Coskun, E.; Rutrecht, B. Recycling von Lithium-Ionen-Batterien. In *Recycling und Rohstoffe*; Thomé-Kozmiensky Verlag GmbH: Neuruppin, Germany, 2018; Volume 11, pp. 583–599.
6. Xu, K. Nonaqueous Liquid Electrolytes for Lithium-Based Rechargeable Batteries. *ChemInform* **2004**, *104*, 4303–4418. [CrossRef]
7. Ehrlich, G.M. LITHIUM-ION BATTERIES. In *Handbook of Batteries*; McGraw-Hill handbooks; McGraw-Hill: New York, NY, USA, 2002; ISBN 978-0-07-135978-8.
8. Hartnig, C.; Schmidt, M. Elektrolyte und Leitsalze. In *Handbuch Lithium-Ionen-Batterien*; Springer: Berlin/Heidelberg, Germany, 2013; pp. 61–77, ISBN 978-3-642-30652-5.
9. Julien, C.; Mauger, A.; Vijh, A.; Zaghib, K. *Lithium Batteries: Science and Technology*; Springer International Publishing: Cham, Switzerland, 2016; ISBN 978-3-319-19107-2.
10. Yew, K.-H.; Song; Jung, C.-S.; Lee, Y.-B. Elektrolyt für Eine Lithiumbatterie und Lithiumbatterie Denselben Enthaltend. DE602006000279T2, 27 November 2008.
11. Pražanová, A.; Kočí, J.; Míka, M.H.; Pilnaj, D.; Plachý, Z.; Knap, V. Pre-Recycling Material Analysis of NMC Lithium-Ion Battery Cells from Electric Vehicles. *Crystals* **2023**, *13*, 214. [CrossRef]
12. Jiang, Z.; Zeng, Z.; Liang, X.; Yang, L.; Hu, W.; Zhang, C.; Han, Z.; Feng, J.; Xie, J. Fluorobenzene, A Low-Density, Economical, and Bifunctional Hydrocarbon Cosolvent for Practical Lithium Metal Batteries. *Adv. Funct. Mater.* **2021**, *31*, 2005991. [CrossRef]
13. Jiang, Z.; Zeng, Z.; Zhai, B.; Li, X.; Hu, W.; Zhang, H.; Cheng, S.; Xie, J. Fluorobenzene-Based Diluted Highly Concentrated Carbonate Electrolyte for Practical High-Voltage Lithium Metal Batteries. *J. Power Sources* **2021**, *506*, 230086. [CrossRef]
14. Markevich, E.; Salitra, G.; Aurbach, D. Fluoroethylene Carbonate as an Important Component for the Formation of an Effective Solid Electrolyte Interphase on Anodes and Cathodes for Advanced Li-Ion Batteries. *ACS Energy Lett.* **2017**, *2*, 1337–1345. [CrossRef]

15. Wang, D.Y.; Sinha, N.N.; Burns, J.C.; Aiken, C.P.; Petibon, R.; Dahn, J.R. A Comparative Study of Vinylene Carbonate and Fluoroethylene Carbonate Additives for LiCoO2/Graphite Pouch Cells. *J. Electrochem. Soc.* **2014**, *161*, A467–A472. [CrossRef]
16. Tian, X.; Yi, Y.; Fang, B.; Yang, P.; Wang, T.; Liu, P.; Qu, L.; Li, M.; Zhang, S. Design Strategies of Safe Electrolytes for Preventing Thermal Runaway in Lithium Ion Batteries. *Chem. Mater.* **2020**, *32*, 9821–9848. [CrossRef]
17. Hanisch, C.; Westphal, B.; Haselrieder, W.; Schoenitz, M. Method for the Treatment of Used Batteries, in Particular Rechargeable Batteries, and Battery Processing Installation. US11050097B2, 29 June 2021.
18. Stehmann, F.; Bradtmöller, C.; Scholl, S. Separation of the Electrolyte—Thermal Drying. In *Recycling of Lithium-Ion Batteries*; Sustainable Production, Life Cycle Engineering and Management; Springer International Publishing: Cham, Switzerland, 2018; pp. 139–153, ISBN 978-3-319-70571-2.
19. Grützke, M.; Mönnighoff, X.; Horsthemke, F.; Kraft, V.; Winter, M.; Nowak, S. Extraction of Lithium-Ion Battery Electrolytes with Liquid and Supercritical Carbon Dioxide and Additional Solvents. *RSC Adv.* **2015**, *5*, 43209–43217. [CrossRef]
20. Nowak, S.; Winter, M. The Role of Sub- and Supercritical CO_2 as "Processing Solvent" for the Recycling and Sample Preparation of Lithium Ion Battery Electrolytes. *Molecules* **2017**, *22*, 403. [CrossRef]
21. Zachmann, N.; Fox, R.V.; Petranikova, M.; Ebin, B. Implementation of a Sub-and Supercritical Carbon Dioxide Process for the Selective Recycling of the Electrolyte from Spent Li-Ion Battery. *J. CO2 Util.* **2024**, *81*, 102703. [CrossRef]
22. Gachot, G.; Ribière, P.; Mathiron, D.; Grugeon, S.; Armand, M.; Leriche, J.-B.; Pilard, S.; Laruelle, S. Gas Chromatography/Mass Spectrometry As a Suitable Tool for the Li-Ion Battery Electrolyte Degradation Mechanisms Study. *Anal. Chem.* **2011**, *83*, 478–485. [CrossRef]
23. Gachot, G.; Grugeon, S.; Armand, M.; Pilard, S.; Guenot, P.; Tarascon, J.-M.; Laruelle, S. Deciphering the Multi-Step Degradation Mechanisms of Carbonate-Based Electrolyte in Li Batteries. *J. Power Sources* **2008**, *178*, 409–421. [CrossRef]
24. Hofmann, A.; Müller, F.; Schöner, S.; Jeschull, F. Revealing the Formation of Dialkyl Dioxahexane Dioate Products from Ethylene Carbonate Based Electrolytes on Lithium and Potassium Surfaces. *Batter. Supercaps* **2023**, *6*, e202300325. [CrossRef]
25. Yoshida, H.; Fukunaga, T.; Hazama, T.; Terasaki, M.; Mizutani, M.; Yamachi, M. Degradation Mechanism of Alkyl Carbonate Solvents Used in Lithium-Ion Cells during Initial Charging. *J. Power Sources* **1997**, *68*, 311–315. [CrossRef]
26. Laruelle, S.; Pilard, S.; Guenot, P.; Grugeon, S.; Tarascon, J.-M. Identification of Li-Based Electrolyte Degradation Products Through DEI and ESI High-Resolution Mass Spectrometry. *J. Electrochem. Soc.* **2004**, *151*, A1202–A1209. [CrossRef]
27. Schultz, C.; Vedder, S.; Streipert, B.; Winter, M.; Nowak, S. Quantitative Investigation of the Decomposition of Organic Lithium Ion Battery Electrolytes with LC-MS/MS. *RSC Adv.* **2017**, *7*, 27853–27862. [CrossRef]
28. Schultz, C.; Kraft, V.; Pyschik, M.; Weber, S.; Schappacher, F.; Winter, M.; Nowak, S. Separation and Quantification of Organic Electrolyte Components in Lithium-Ion Batteries via a Developed HPLC Method. *J. Electrochem. Soc.* **2015**, *162*, A629–A634. [CrossRef]
29. Takeda, S.; Morimura, W.; Liu, Y.; Sakai, T.; Saito, Y. Identification and Formation Mechanism of Individual Degradation Products in Lithium-ion Batteries Studied by Liquid Chromatography/Electrospray Ionization Mass Spectrometry and Atmospheric Solid Analysis Probe Mass Spectrometry. *Rapid Commun. Mass Spectrom.* **2016**, *30*, 1754–1762. [CrossRef]
30. Terborg, L.; Weber, S.; Passerini, S.; Winter, M.; Karst, U.; Nowak, S. Development of Gas Chromatographic Methods for the Analyses of Organic Carbonate-Based Electrolytes. *J. Power Sources* **2014**, *245*, 836–840. [CrossRef]
31. Campion, C.L.; Li, W.; Euler, W.B.; Lucht, B.L.; Ravdel, B.; DiCarlo, J.F.; Gitzendanner, R.; Abraham, K.M. Suppression of Toxic Compounds Produced in the Decomposition of Lithium-Ion Battery Electrolytes. *Electrochem. Solid-State Lett.* **2004**, *7*, A194. [CrossRef]
32. Grützke, M.; Weber, W.; Winter, M.; Nowak, S. Structure Determination of Organic Aging Products in Lithium-Ion Battery Electrolytes with Gas Chromatography Chemical Ionization Mass Spectrometry (GC-CI-MS). *RSC Adv.* **2016**, *6*, 57253–57260. [CrossRef]
33. Kraft, V.; Weber, W.; Grützke, M.; Winter, M.; Nowak, S. Study of Decomposition Products by Gas Chromatography-Mass Spectrometry and Ion Chromatography-Electrospray Ionization-Mass Spectrometry in Thermally Decomposed Lithium Hexafluorophosphate-Based Lithium Ion Battery Electrolytes. *RSC Adv.* **2015**, *5*, 80150–80157. [CrossRef]
34. Weber, W.; Kraft, V.; Grützke, M.; Wagner, R.; Winter, M.; Nowak, S. Identification of Alkylated Phosphates by Gas Chromatography–Mass Spectrometric Investigations with Different Ionization Principles of a Thermally Aged Commercial Lithium Ion Battery Electrolyte. *J. Chromatogr. A* **2015**, *1394*, 128–136. [CrossRef] [PubMed]
35. Keller, T.; Holtbruegge, J.; Niesbach, A.; Górak, A. Transesterification of Dimethyl Carbonate with Ethanol To Form Ethyl Methyl Carbonate and Diethyl Carbonate: A Comprehensive Study on Chemical Equilibrium and Reaction Kinetics. *Ind. Eng. Chem. Res.* **2011**, *50*, 11073–11086. [CrossRef]
36. Metzger, M.; Strehle, B.; Solchenbach, S.; Gasteiger, H.A. Hydrolysis of Ethylene Carbonate with Water and Hydroxide under Battery Operating Conditions. *J. Electrochem. Soc.* **2016**, *163*, A1219–A1225. [CrossRef]
37. Liu, Y.; Mu, D.; Li, R.; Ma, Q.; Zheng, R.; Dai, C. Purification and Characterization of Reclaimed Electrolytes from Spent Lithium-Ion Batteries. *J. Phys. Chem. C* **2017**, *121*, 4181–4187. [CrossRef]

38. Xu, R.; Lei, S.; Wang, T.; Yi, C.; Sun, W.; Yang, Y. Lithium Recovery and Solvent Reuse from Electrolyte of Spent Lithium-Ion Battery. *Waste Manag.* **2023**, *167*, 135–140. [CrossRef]
39. *Q2(R2) Validation of Analytical Procedures*; Center for Drug Evaluation and Research, Office of Regulatory Policy: Silver Spring, MD, USA, 2024.
40. *Analytical Procedures and Methods Validation for Drugs and Biologics*; FDA: Silver Spring, MD, USA, 2015.
41. Kromidas, S. *Handbuch Validierung in der Analytik*, 1st ed.; Wiley: Hoboken, NJ, USA, 2000; ISBN 978-3-527-29811-2.
42. <621> Chromatography. Available online: https://www.usp.org/harmonization-standards/pdg/excipients/chromatography (accessed on 29 August 2024).
43. Wenzl, T.; Haedrich, J.; Schaechtele, A.; Robouch, P.; Stroka, J. *Guidance Document on the Estimation of LOD and LOQ for Measurements in the Field of Contaminants in Feed and Food*; Publications Office: Luxembourg, 2016; ISBN 978-92-79-61768-3.
44. Matadi, B.P.; Geniès, S.; Delaille, A.; Waldmann, T.; Kasper, M.; Wohlfahrt-Mehrens, M.; Aguesse, F.; Bekaert, E.; Jiménez-Gordon, I.; Daniel, L.; et al. Effects of Biphenyl Polymerization on Lithium Deposition in Commercial Graphite/NMC Lithium-Ion Pouch-Cells during Calendar Aging at High Temperature. *J. Electrochem. Soc.* **2017**, *164*, A1089–A1097. [CrossRef]
45. Kim, H.; Grugeon, S.; Gachot, G.; Armand, M.; Sannier, L.; Laruelle, S. Ethylene Bis-Carbonates as Telltales of SEI and Electrolyte Health, Role of Carbonate Type and New Additives. *Electrochim. Acta* **2014**, *136*, 157–165. [CrossRef]
46. Heider, U.; Oesten, R.; Jungnitz, M. Challenge in Manufacturing Electrolyte Solutions for Lithium and Lithium Ion Batteries Quality Control and Minimizing Contamination Level. *J. Power Sources* **1999**, *81–82*, 119–122. [CrossRef]
47. Cheng, K.; Feng, W.; Wang, L.; Chen, Y. Phase Equilibria for Quaternary Systems of Water + Methanol or Ethanol + Ethyl Methyl Carbonate + Octane at T = 298.15 K. *J Solut. Chem* **2017**, *46*, 1337–1348. [CrossRef]
48. Wang, X.; Yasukawa, E.; Kasuya, S. Nonflammable Trimethyl Phosphate Solvent-Containing Electrolytes for Lithium-Ion Batteries: I. Fundamental Properties. *J. Electrochem. Soc.* **2001**, *148*, A1058–A1065. [CrossRef]
49. Xu, F.; Cai, X.; Zhang, J.; Zhang, L.; Wang, J.; Zhang, N.; Li, S. Improving Compatibility between Trimethyl Phosphate and Graphite Anodes by Preconstructing a Stable Solid Electrolyte Interphase Film. *ACS Appl. Energy Mater.* **2022**, *5*, 11370–11378. [CrossRef]
50. Lemordant, D.; Zhang, W.; Ghamouss, F.; Farhat, D.; Darwiche, A.; Monconduit, L.; Dedryvère, R.; Martinez, H.; Cadra, S.; Lestriez, B. Artificial SEI for Lithium-Ion Battery Anodes. In *Advanced Fluoride-Based Materials for Energy Conversion*; Elsevier: Amsterdam, The Netherlands, 2015; pp. 173–202, ISBN 978-0-12-800679-5.
51. Sasaki, T.; Abe, T.; Iriyama, Y.; Inaba, M.; Ogumi, Z. Formation Mechanism of Alkyl Dicarbonates in Li-Ion Cells. *J. Power Sources* **2005**, *150*, 208–215. [CrossRef]
52. Lee, H.; Lee, J.H.; Ahn, S.; Kim, H.-J.; Cho, J.-J. Co-Use of Cyclohexyl Benzene and Biphenyl for Overcharge Protection of Lithium-Ion Batteries. *Electrochem. Solid-State Lett.* **2006**, *9*, A307–A310. [CrossRef]
53. Shim, E.-G.; Nam, T.-H.; Kim, J.-G.; Kim, H.-S.; Moon, S.-I. Effect of Vinyl Acetate plus Vinylene Carbonate and Vinyl Ethylene Carbonate plus Biphenyl as Electrolyte Additives on the Electrochemical Performance of Li-Ion Batteries. *Electrochim. Acta* **2007**, *53*, 650–656. [CrossRef]
54. Tobishima, S.; Ogino, Y.; Watanabe, Y. Influence of Electrolyte Additives on Safety and Cycle Life of Rechargeable Lithium Cells. *J. Appl. Electrochem.* **2003**, *33*, 143–150. [CrossRef]
55. Xiao, L.; Ai, X.; Cao, Y.; Yang, H. Electrochemical Behavior of Biphenyl as Polymerizable Additive for Overcharge Protection of Lithium Ion Batteries. *Electrochim. Acta* **2004**, *49*, 4189–4196. [CrossRef]
56. Yang, Y.-L.; Ramaswamy, S.G.; Jakoby, W.B. Enzymatic Hydrolysis of Organic Cyclic Carbonates. *J. Biol. Chem.* **1998**, *273*, 7814–7817. [CrossRef]
57. Bhanage, B.M.; Fujita, S.; Ikushima, Y.; Arai, M. Transesterification of Urea and Ethylene Glycol to Ethylene Carbonate as an Important Step for Urea Based Dimethyl Carbonate Synthesis. *Green Chem.* **2003**, *5*, 429–432. [CrossRef]
58. Fang, Y.-J.; Xiao, W.-D. Experimental and Modeling Studies on a Homogeneous Reactive Distillation System for Dimethyl Carbonate Synthesis by Transesterification. *Sep. Purif. Technol.* **2004**, *34*, 255–263. [CrossRef]
59. Wang, H.; Chen, S.; Li, Y.; Liu, Y.; Jing, Q.; Liu, X.; Liu, Z.; Zhang, X. Organosilicon-Based Functional Electrolytes for High-Performance Lithium Batteries. *Adv. Energy Mater.* **2021**, *11*, 2101057. [CrossRef]
60. Wang, H.; Sun, D.; Li, X.; Ge, W.; Deng, B.; Qu, M.; Peng, G. Alternative Multifunctional Cyclic Organosilicon as an Efficient Electrolyte Additive for High Performance Lithium-Ion Batteries. *Electrochim. Acta* **2017**, *254*, 112–122. [CrossRef]
61. Ding, J.; Wen, Y.; Lan, X.; Hu, R. Roles of Trimethyl Borate in Constructing an Interphase on Li Anode: Angel or Demon? *ACS Appl. Mater. Interfaces* **2023**, *15*, 6768–6776. [CrossRef]
62. Liu, Q.; Yang, G.; Liu, S.; Han, M.; Wang, Z.; Chen, L. Trimethyl Borate as Film-Forming Electrolyte Additive To Improve High-Voltage Performances. *ACS Appl. Mater. Interfaces* **2019**, *11*, 17435–17443. [CrossRef] [PubMed]
63. ABE, K.; Matsumori, Y.; Ueki, A. Nonaqueouselectrolytic Solution and Lithium Secondary Batteries. US 2008/0050658 A1, 28 February 2008.

64. Chen, S.; Wang, Z.; Wang, J.; Tong, X.; Yan, W. Lower Explosion Limit of the Vented Gases from Li-Ion Batteries Thermal Runaway in High Temperature Condition. *J. Loss Prev. Process Ind.* **2020**, *63*, 103992. [CrossRef]
65. Chen, S.; Wang, Z.; Yan, W. Identification and Characteristic Analysis of Powder Ejected from a Lithium Ion Battery during Thermal Runaway at Elevated Temperatures. *J. Hazard. Mater* **2020**, *400*, 123169. [CrossRef]

recycling

MDPI

Article

Global Supply of Secondary Lithium from Lithium-Ion Battery Recycling

Carolin Kresse [1,*], Britta Bookhagen [1], Laura Buarque Andrade [2] and Max Frenzel [2]

1 German Mineral Resources Agency (DERA), Federal Institute for Geosciences and Natural Resources (BGR), 13593 Berlin, Germany; britta.bookhagen@bgr.de

2 Helmholtz-Zentrum Dresden-Rossendorf, Helmholtz-Institute Freiberg for Resource Technology, 09599 Freiberg, Germany; l.buarque-andrade@hzdr.de (L.B.A.); m.frenzel@hzdr.de (M.F.)

* Correspondence: carolin.kresse@bgr.de

Abstract

The recycling of lithium-ion batteries is picking up rather slowly, although recent rapid growth in consumption and increasing prevalence of battery electric vehicles have increased the quantity of recoverable material from past years of production. Yet, the diversity of different product types i.e., chemistries and product life spans complicates the recovery of raw materials. At present, large-scale industrial recycling of lithium-ion batteries employs (1) pyrometallurgy, with downstream hydrometallurgy for recovery of refined metals/salts; and (2) hydrometallurgy, requiring upstream mechanical shredding of cells and/or modules. Regulatory requirements, especially in Europe, and the high industry concentration along the lithium-ion battery value chain drive recycling efforts forward. The present study aims to quantify the potential contribution of 2nd lithium from recycling to battery production on a global and European scale up to 2050. The overall recycling output of lithium in any given year depends on the interactions between several different factors, including past production, battery lifetime distributions, and recovery rates, all of which are uncertain. The simplest way to propagate input uncertainties to the final results is to use Monte Carlo-type simulations. Calculations were done separately for EVs and portable batteries. The overall supply of lithium from recycling is the sum of the contributions from EVs and portable electronics from both the EU and the RoW in each battery production scenario. Results show a total global supply of recycled lithium below 20% in each scenario until 2050. On the EU level, the contribution of recycled lithium may reach up to 50% due to the high collection and recovery rate targets.

Keywords: lithium; lithium-ion batteries; recycling; Monte Carlo simulation; policy

1. Introduction

Lithium-ion batteries (LIBs) are currently the most important commercially available battery type for applications such as consumer electronics, e-mobility, and energy storage [1,2]. According to the EU's Green Deal, LIB production will increase and needs to be optimised in terms of production scrap rates, materials used, and design for recycling [3]. However, most battery raw materials such as lithium, cobalt, nickel, graphite, and manganese currently come from primary resources characterised by a considerable CO_2 footprint [4]. An increase in the recycled content of the raw materials used for battery production could improve the CO_2 footprint of the batteries [5]. However, recycling LIBs not only faces issues regarding collection and return rates of End-of-Life (EoL) material but is also a technically complex process combined with safety issues [6]. The ramp-up of e-mobility and the resulting increase in demand for battery raw materials have made the recycling of LIBs not only ecologically necessary but also increasingly economically attractive in recent years [5].

According to Circular Energy Storage [7], current global production figures for battery cells are approximately 800 GWh. European production accounts for approximately a quarter, with 200 GWh for 2023. Forecasts for global battery cell production put the figure up to 20,000 GWh in 2050 to achieve the Net Zero Pathway [8]. In Europe, LIB cell production capacity could increase to around 1000 GWh by 2030. At the same time, global new scrap volumes (production scrap) of about 141 GWh are expected by 2030, including about 45 GWh in Europe (Figure 1) [7]. Leaps and bounds in recycling capacities in China, the USA and Europe show the growing trend in this sector.

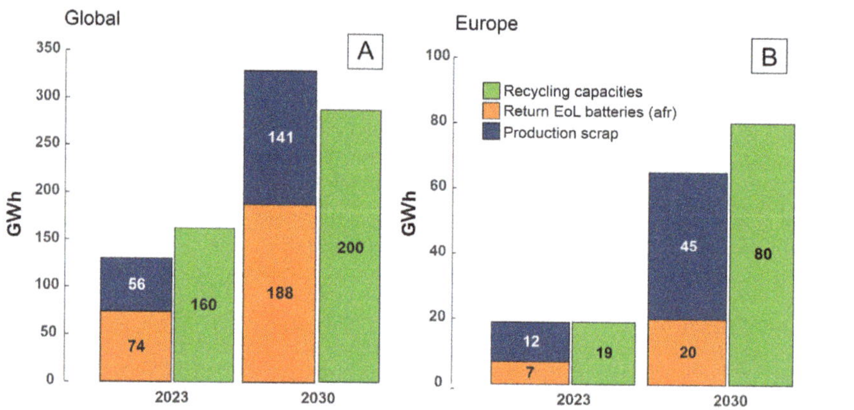

Figure 1. Status quo of the estimated global (**A**) and European (**B**) recycling capacities, return volumes of EoL batteries available for recycling (afr) and production scrap in 2023 and estimates for 2030 in GWh (Data source and forecasts [7].

Despite the rapid growth in consumption and increasing market penetration rates of battery electric vehicles, there are tight limits to recycling, as the volume of material available for recycling does not depend on the current annual battery production, but on past production years, product types, i.e., chemistries, and product lifetime.

Current global return volumes of LIBs available for recycling are about 74 GWh (Figure 1A), and approx. seven GWh at the European level (Figure 1B). Expected volumes in 2030 are about 188 GWh global, or 20 GWh in Europe, which is a considerable increase (Figure 1A,B) [7]. As the number of battery cell production facilities is growing rapidly, production scrap will account for the larger share of materials to be recycled in terms of volume compared to EoL material until 2030 (in Europe). From a global perspective, the EoL volume will already account for the largest share in 2030 due to the Chinese EoL volumes and due to the fact that production scrap, especially in China is already zeroing out. The increased EoL battery volumes are also expected from 2030 onwards in Europe [7]. Considering the volumes of EoL batteries and production scrap, there is currently an overcapacity of recycling facilities on a global scale (combined pre- and post-treatment). However, increased battery cell production and projected recycling capacities point to a large gap in 2030 globally, which implies not enough capacities to process the material available for recycling. In Europe, we might see the opposite in terms of overcapacities and less material availability because of unregulated exports of EoL material as well as recycled spent batteries in the form of black mass. Numerous announcements for new recycling facilities are being published on a weekly basis, which makes it difficult to keep track of recycling capacities, globally and in Europe, in the mid- to long-term. In contrast, however, there are also many cancellations and postponements for recycling plants now.

The aim of the present study is to quantify the potential contribution of secondary lithium from lithium-ion battery recycling to battery production on a global scale up to 2050. Available estimates for the supply of secondary lithium from current studies consider either regional estimations on a country level [9–12] or scenarios based on the recycling potential from EoL-EV batteries only [13,14]. In addition, those studies do not consider uncertainties on all the input parameters relevant to the outputs.

To further refine those estimates, the authors of this study gathered detailed global data on lithium, from EVs and portable electronics as well as production scrap, to enable better estimates of the future production of cathode material for battery cells, based on secondary lithium supply. A key novelty of this paper is the detailed incorporation of uncertainties into the estimation procedure to account for major current knowledge gaps. This requires consideration of the main factors impacting the availability of secondary resources: (1) LIB types and chemical material compositions, (2) recycling technologies and new developments in this research area, (3) battery lifetime and usage profiles, (4) collection and recovery rates, and (5) governmental/regulatory factors (registration, evaluation, authorisation and restriction of chemicals). Furthermore, additional factors such as new cathode and anode material developments, the development of new recycling technologies, as well as the extension of recycling capacities worldwide will also affect secondary supply in the long term and will be discussed.

2. Current Recycling Routes

Recycling routes for lithium-ion batteries have been summarised and reviewed frequently by many authors and are briefly addressed below [15–21]. After the collection and registration of EoL LIBs from either consumer electronics or EVs, the general process for the recycling of LIBs comprises the following steps (Figure 2):

1. Preparation: sorting, disassembly, discharging (optional)
2. Thermal and/or mechanical pre-treatment (optional)
3. Main processes: pyro- and/or hydrometallurgy; direct recycling

After preparation, the aluminium casing, copper cables, and plastic are recycled and returned to the general production cycle. Physical processes typically involve pre-treatment including disassembly, crushing, screening, magnetic separation, washing, and heat treatment. Chemical processes are grouped into pyro- and hydrometallurgical processes, which usually require leaching, separation, extraction and chemical/electrochemical precipitation [22].

Pyrometallurgy uses high-temperature processes usually above 1400 °C [23] with the addition of slag formers to convert waste battery materials (i.e., entire battery systems; dismantling at the cell level is not necessary) into metal alloy containing cobalt, nickel, and some of the original manganese and copper. This metal alloy can then be processed by hydrometallurgy, whereby cobalt, nickel, and manganese sulphates are obtained and subsequently used for new batteries or other products. During pyrometallurgical processing, lithium, manganese, aluminium, and if present oxidized iron, are transferred to the slag, whereas graphite serves as a reducing agent and burns off. Manganese and lithium can be recovered by hydrometallurgy from the slag. Aluminium and iron can currently not be recovered economically. The lithium content of the slag is generally comparable to the lithium content of spodumene concentrates with approx. 2.79% lithium [15].

The purely hydrometallurgical route uses an initial thermal and mechanical pre-treatment of the LIBs. Thermal pre-treatment serves to remove bonding material, which typically consists of polyvinylidene fluoride (PVDF) or polytetrafluoroethylene (PTFE). In the next step, the battery modules are mechanically crushed (shredded) under a protective atmosphere and turned into a wet mass by the released liquid electrolyte. Further fractions with high product contents, which can be recovered mechanically as well, are copper and aluminium concentrate (from the anode and cathode foils of the LIB cells) and the steel/casing fractions. Yield and purity of individual recycled products can be increased by suitable measures (e.g., avoidance of overgrinding). Through drying, sieving, and magnetic separation, the shredded material becomes finer. Finally, the so-called "black mass" is produced, which consists mostly of fine particles of mixed cathode and anode materials, thus containing lithium, manganese, cobalt, nickel, and graphite. Contents of valuable materials, particularly cobalt, nickel, and graphite are highly variable, as the composition of LIBs differs significantly depending on the manufacturer. Black mass is then separated into its components using various dissolution processes (including solvent extraction). Theoretically, graphite could be recovered, but this is currently not done on an industrial scale and in a reusable form [24,25].

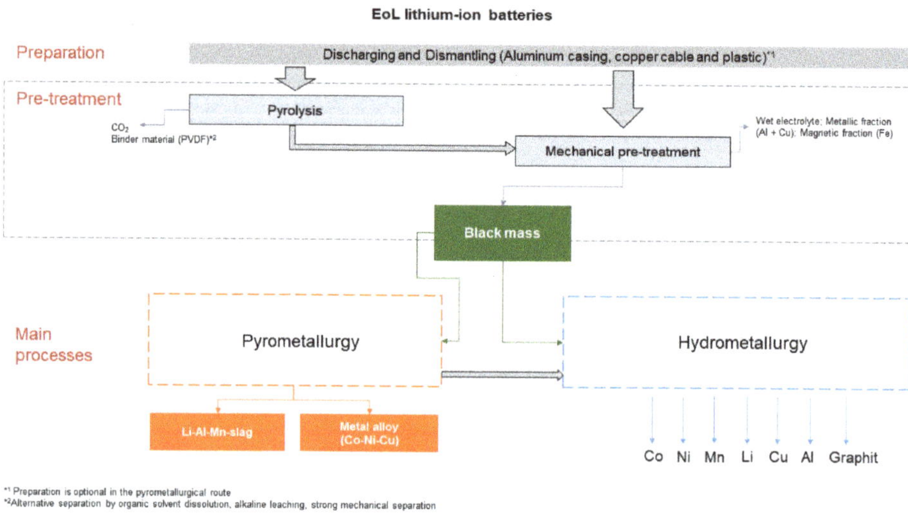

Figure 2. Overview of commercial processes for lithium-ion battery recycling. Modified after [11,21].

3. Materials and Methods

The following subsections initially describe the general estimation approach. They then deal with the input data, their sources and uncertainties, and any assumptions that were made in order to compile the final estimates.

3.1. General Approach

Since the overall recycling output of lithium, in any given year, depends on the interactions between a number of different factors, including past production, battery lifetime distributions, collection rates, and recovery rates, the simplest way to propagate input uncertainties to the final results is to use Monte Carlo-type simulations. Monte Carlo simulations serve as the methodological basis for the solution approach pursued in the present study. These simulations are a widely used tool in the natural sciences, as they allow the modelling of the dynamics of complex systems by means of the composition of relatively simple individual components. One of the most important applications in current research is the quantitative analysis of probabilistic systems, which are relevant in a wide range of areas from solid-state physics to climate research. Monte Carlo simulations are an extremely versatile tool and, due to their great flexibility, are particularly suitable for the task to be solved in the present study. Two specific components are required as input here: firstly, a concrete simulation structure and secondly, probability density distributions for the input variables to be considered.

Monte Carlo simulation allows for the specification of all input data in terms of known or assumed probability distributions reflecting their inherent uncertainties. The outputs are then numerically generated probability distributions of the desired quantities. Similar strategies were previously used to estimate the likely availabilities of by-products such as Ga, Ge, In, and Te from global raw material streams [26,27].

To show the effect of future market growth on the likely supply of secondary lithium in a more simplified way, three separate estimates corresponding to low, medium, and high future production scenarios were made, instead of simulating a continuous distribution of production scenarios. For each of the three scenarios, 1000 simulations were run, using input parameters drawn randomly from distributions described in the following subsections.

Simulated input parameters were then fed through a calculation procedure to quantify lithium recycling flows. Figure 3 schematically shows the overall structure of this procedure. The total amount of EoL batteries available each year (in GWh/yr) is estimated from past production figures and a specified battery lifetime distribution as:

$$B_{EoL}^{T} = \sum_{i=1}^{N} P^{T-i} * \left(F_t^{T-i}(t=i) - F_t^{T-i}(t=i-1) \right) \tag{1}$$

where T is the year, for which the amount of EoL batteries is calculated, P^{T-i} is the battery production in year $T-i$, $F_t^{T-i}(t=i)$ is the cumulative failure rate of the batteries produced in year $T-i$ at i years after their production (from lifetime distribution), and N is the number of years into the past, to which the sum is evaluated. In the present case, $N = 20$ years was used, since it is expected that most batteries will have failed after this time given current and likely future lifetime distributions (see below).

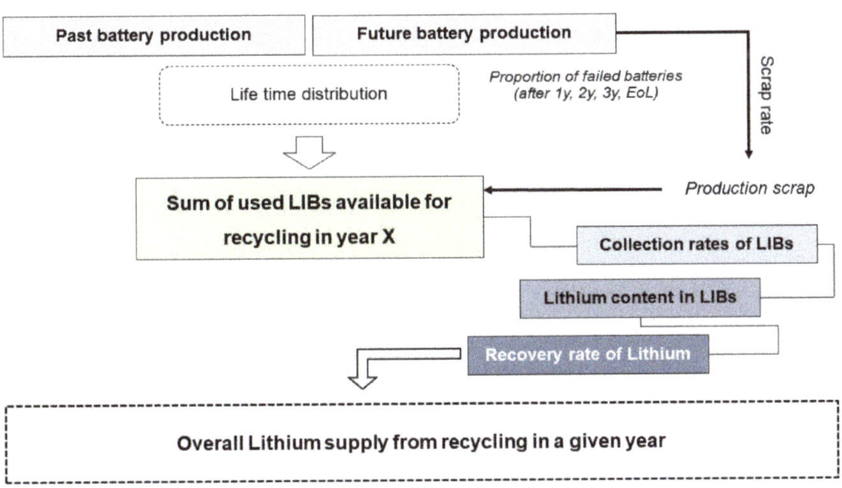

Figure 3. Simulation structure of input parameters to quantify lithium recycling flows.

In addition, production scraps were considered as a fixed percentage of the production, denoted by the scrap rate SR, in each year:

$$Scrap^T = P^T * SR^T \tag{2}$$

The quantity B^T_{EoL} is then multiplied with the corresponding collection rate, $Coll^T$, added to $Scrap$, and the total multiplied by the lithium content in the batteries, C_{Li}, and the recovery rate of lithium for year T, Rec^T, to calculate overall lithium supply from recycling in a given year in each simulation:

$$Li^T_{recycling} = \left(B^T_{EoL} * Coll^T + Scrap^T\right) * C_{Li} * Rec^T \tag{3}$$

The overall recycling contribution of lithium to battery production is then equal to:

$$Li^T_{recycl-pptn} = \frac{Li^T_{recycling}}{(P^T * C_{Li} * (1 + SR^T))} \tag{4}$$

These calculations were done separately for EV and portable batteries, and also for the EU as well as the rest of the world. These separations were made, since lifetime distributions for EV and portable batteries are expected to be different, and also because EU collection and recovery rates were used as upper limits in the models, with the rest of the world either having lower or similar rates. The underlying assumptions are described further below. Finally, the overall supply of lithium from recycling is the sum of the contributions from EVs and portable electronics from both the EU and the rest of the world.

3.2. Past and Future Battery Production

Past production numbers for both EV and portable batteries are based on reports by Circular Energy Storage [7] for Europe and the rest of the world, and go back until the year 2000. The three scenarios that were used for future production are based on an evaluation of the literature studies presented in Table 1 and are included in Supplementary Materials, Table S1.

For portable batteries, only one forecast of future production numbers was used, assuming moderate growth rates of ~1.5% per year until 2050. This is loosely based on the forecasts by Circular Energy Storage [7] up to 2030, and the assumption that the portable electronics market is a relatively mature market already. Since the expected contributions of portable batteries to the future market growth are so low, uncertainties on these forecasts are negligible for the present study.

The main differences between the three production scenarios arise from the forecasts for EV battery production. In the low-production scenario, a combination of the scenarios put forward by Öko Institut [28] and Avicenne Energy [29] was used, as well as the low-demand scenario of Moores [30], with global production at ~1500 GWh/a in 2030, ~4000 GWh/a in 2040, and ~5000 GWh/a in 2050. For the high-production scenario, reference points of ~6000 GWh/a in 2030, ~10,000 GWh/a in 2040, and ~20,000 GWh/a in 2050 were used, based mostly on the high-production scenario of Benchmark Minerals [8] Finally, 50% of the high-production scenario values were used for the medium-production case. In order to interpolate smoothly between current production and these future figures, third-order polynomial fits as shown in Figure 4 were used.

For the proportions of EV and portable batteries consumed in Europe, estimates by Circular Energy Storage [7] were used for past production. For future years, the assumption was made that the proportions of EV and portable batteries used in Europe as a fraction of global demand would be constant with time and fall somewhere between 20 and 30% of the world market. A uniform distribution between 20 and 30% was used to include this uncertainty in the simulations.

Table 1. Summary table of the literature studies showing annual lithium-ion battery production capacities until 2050 in GWh globally.

Data Source	2016	2020	2022	2023	2025	2027	2029	2030	2035	2040	2050
Rho Motion (EVs) [31]	-	154	473	698	1132	1601	2236	2651	4568	6786	-
S. Moores [30]	-	501	-	-	2492	-	-	3010	-	3900	5000
Benchmark Minerals [8]	-	501	-	-	2492	-	-	6700 (NET Zero Pathway)		11,800 (NET Zero Pathway)	20,000 (Net Zero Pathway)
S&P Global Mobility (EVs) [32,33]	-	455	966	1246	2800	-	-	5900	-	-	-
Avicenne Energy [29]	94	232	350	-	609	-	-	1300	-	-	-
IEA, Global EV Outlook (EVs) [34]	-	-	-	1500	-	-	-	5500	-	6500	-
Öko Institut (B2DS Szenario-"Unter 2 Grad Szenario") (EVs) [28]	>100	-	-	-	-	-	-	1500	-	-	6600
Circular Energy Storage [7]	120	265	588	721	1059	1558	2333	2900	-	-	-

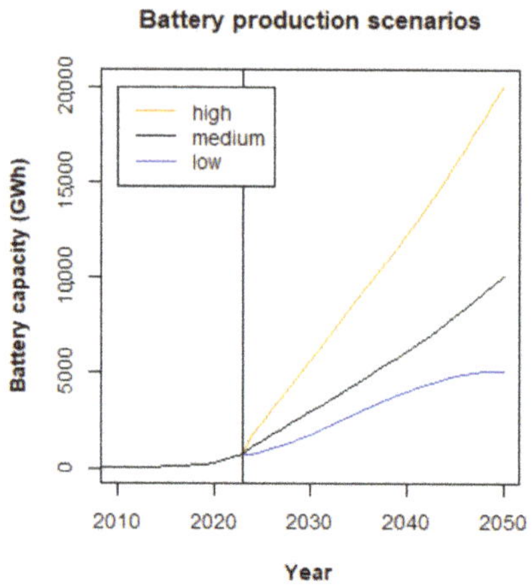

Figure 4. Scenarios for future Li-ion battery production used in our analysis. Low production means the lowest assumption of projected production from studies in Table 1. High production means the upper end of projected production from studies in Table 1. The vertical line marks the year 2023.

3.3. *Lithium Contents in Batteries*

Lithium contents in different battery types were compiled from the literature and are summarized in Table 2. Most of the values range between 0.08 and 0.12 kg/kWh. To convert estimates of recycled LIBs capacity to lithium contents, an average value of 0.10 kg/kWh was applied. Uncertainties for lithium contents in our estimates were not explicitly included. Rather, all calculations were done in terms of battery capacities (GWh) and results were subsequently converted to million tons of lithium carbonate equivalent (LCE).

Table 2. Specific lithium requirements for various cathode materials [kg/kWh]. NMC–Nickel-Manganese-Cobalt, NCA–Nickel-Cobalt-Aluminium, LFP–Lithium-Iron-Phosphate, LMO–Lithium-Manganese-Oxide, LMNO–Lithium-Manganese-Nickel-Oxide, LMFP–Lithium-Manganese-Iron-Phosphate [35].

Cathode Material	Lithium Content
NMC	0.096–0.123
NCA	0.095–0.098
LFP	0.084
LMO	0.080
LMNO	0.065
LMFP	0.082

Note: NMC cathode material includes NMC 111, 532, 622, and 811.

This approach is equivalent to assuming that lithium contents in batteries are constant over time. In this case, the overall lithium contents do not matter, since the proportions of recycled versus consumed lithium in any given year will remain the same as recycled vs. consumed battery capacities in GWh. Only if lithium contents increased or decreased systematically over time, differences in the results should be expected. While this may happen in the future, it is hard to predict in which direction the variation will tend. If more LFP batteries are used, then somewhat lower future lithium contents would be expected. If solid-state batteries gain importance, lithium contents in future batteries might be higher, between 0.26 and 0.52 kg/kWh [36]. There are no data available for the lithium content in Li-sulfur batteries, but [36] assume a lithium content of 1.2 kg/kWh, based on data for high cathode metal demand values for LIBs [37].

Decreasing future lithium contents in batteries would mean a faster growth of potential recycling contributions to overall supply than in our simulations while increasing lithium contents would mean slower growth of the recycling contributions. However, since these effects are expected to be small as long as LIBs are the main battery type ($\leq$10% relative variation), this was not included in our estimation. Other sources of uncertainty, such as battery lifetimes and recycling and collection rates are much more important.

3.4. Battery Lifetimes

Distributions of battery lifetimes are one of the most critical inputs into the simulations since they are an important determinant for the amount of EoL batteries available each year for recycling (cf. Equation (1)). Unfortunately, very little primary information is available on the lifetime distributions for batteries actually used in vehicles and portable electronics. This is probably because lifetime distributions are classified as confidential information by individual companies.

Due to the lack of information, available data points and a set of assumptions were used to estimate the likely shape of battery lifetime distributions for EVs and portable electronics. Available data points are warranty periods offered by manufacturers for their products and a selection is listed in Table 3. Assumptions concern the likely properties of the engineered lifetime distributions compatible with these warranty periods.

Table 3. Summary table of warranty periods offered by manufacturers for their products.

Manufacture	Warranty Period	Reference
BYD	8 years	[38]
Hyundai	8 years	[39]
KIA	8 years	[40]
Tesla	8 years	[41]
Volkswagen	8 years	[42]

For EV batteries, most manufacturers currently offer eight-year warranty periods (Table 3). Offering such a long warranty period can only be economical if the vast majority of batteries are expected to survive for at least this amount of time. Therefore, it was assumed that the likely survival rate of current EV battery packs ranges between 80–95% for the first eight years. On the other hand, manufacturers cannot be assumed to have an interest in producing battery packs that last substantially longer than the warranty period. This is because battery exchanges are expected to be a major source of revenue. Therefore, we assumed that median lifetimes (50% cumulative failure) are around two years longer than the warranty period, i.e., 10 years at present.

Using these constraints and assuming a simple two-parameter probability distribution model, it is possible to estimate the corresponding sets of distribution parameters: one for an 80% survival rate at eight years, the other for a 95% survival rate at eight years; and both with a 50% survival rate at 10 years. A two-parameter Weibull distribution was used here with estimated α (=shape) and β (=scale) parameters for both cases. The results are shown graphically in Figure 5A,B. Random linear mixing of the two-parameter sets according to a uniform distribution was then used to generate the specific parameters for the distributions used in each simulation.

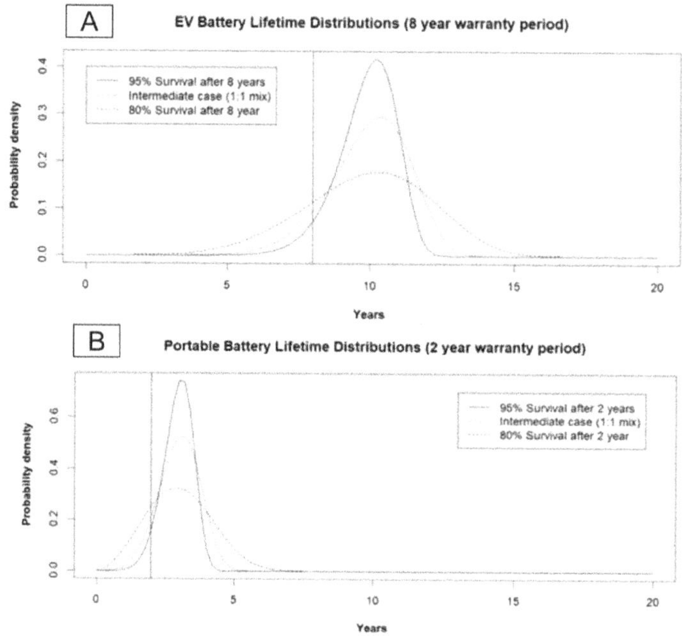

Figure 5. (**A**) Reconstructed EV battery lifetime distributions (Weibull) for the 8-year warranty period, as described in main text. The vertical line shows the end of the 8-year period. (**B**) Reconstructed portable battery lifetime distributions (Weibull) for the 2-year warranty period, as described in main text. The vertical line shows the end of the 2-year period.

Uncertainties with respect to the present shape of battery lifetime distributions are not the only relevant uncertainties for the present paper. Uncertainties also exist for the past and future evolution of these distributions. For the past, a slow increase in a lifetime from a warranty-relevant period of five years in 2007 to eight years in 2020 (at an 80–95% survival rate) was assumed. Ranges for the distribution parameters α and β for the batteries produced in these years were estimated under similar assumptions as above, assuming failure rates of 80–95% at the end of the warranty-relevant period, and 50% two years after. For the future development of EV batteries, two end-member cases were assumed: (a) that EV battery lifetimes would not improve further, i.e., that the current distributions apply in the future, and (b) that battery lifetimes keep increasing until 2030, reaching a value of 15 years for the warranty-relevant period in 2030, and then remain constant. Distribution parameters for the future were estimated as linear mixtures between these two endmember cases, using a uniform distribution (0–1) to simulate the proportion of mixing. Therefore, average battery lifetimes in the simulations increase over time, reaching about ~11 years in 2030 (at the 80–95% failure rate).

For portable electronics, the derivation of lifetime distribution parameters and their uncertainties followed a similar logic as for the EV batteries above. However, warranty periods of two years [43] were used (at 80–95% failure rate), assuming 50% of failures to occur after three years. This was assumed to improve slightly to three years for the warranty-relevant period by 2035, and 4.5 years for the 50% cumulative failure rate. Table 1 in the electronic supplement summarizes the parameter sets used to describe the past and future trends in EVs and portable batteries.

3.5. Production Scrap Rate

As already noted in Equations (2)–(4) the potential contribution of production scraps to total lithium production was also considered. By definition, these are faulty products (e.g., battery packs, cell banks, individual cells), which never enter the market, rather than wastes of cathode or anode materials. These products must be treated in the same way as EoL batteries to recover the contained materials since they are already assembled. It is therefore essential to consider these wastes together with EoL batteries as a source of secondary lithium supply. In addition, scrap rates also affect lithium demand since their production uses up lithium that should normally have gone into finished products (Equation (4)). Therefore, the actual use of lithium for batteries must be higher than the demand for finished batteries by the scrap rate, and thus lithium production was adjusted accordingly in each simulation.

Since current rates of scrap production are unknown, a uniform distribution between two end-member scenarios was used and applied to all battery types: (1) a current scrap rate of 10%, decreasing to 5% in 2035, and (2) a current scrap rate of 5% now, decreasing to 3% in 2035. Values were assumed to be comparatively low, i.e., generally <10% because there are strong economic incentives to minimize them. The assumed decreases over time are intended to reflect the expected maturation of the production technology with increases in volume, and associated reductions in scrap rates. After 2035, production scrap rates are assumed to remain constant.

3.6. Collection Rates

Once batteries reach the end of their service life, they must be collected in order to become available for recycling. This process is generally not perfectly efficient, i.e., collection rates are typically lower than 100%. It is assumed in this paper that collection rates differ by battery type, with EV batteries being more likely to be collected.

In the simulations, current EU collection rates for EV batteries are assumed to be at 90% [44] with a gradual increase to 100% by 2050 in all cases. This is an optimistic scenario, but very high collection rates of >90% are probable since manufacturers are likely to develop their own collection systems. Besides the batteries and their material contents represent a major contribution to the material costs of electric vehicles, which is likely to continue into the future.

For portable batteries, collection rates in the EU are assumed to rise rather linearly from 47.4% in 2018 [45] to 63% in 2027 and then 73% in 2030, following EU targets. After this, portable battery collection rates in the EU are considered to remain approximately constant at the proposed 2030 level.

EU collection rates constitute a best-case scenario since the EU is currently implementing measures and regulations to support a circular economy. For the rest of the world, this is not generally the case [46], and lower collection rates for both EV and portable batteries are therefore probable. However, the exact future rates are uncertain. Therefore, a uniform probability distribution between a best-case and worst-case scenario is assumed in order to describe uncertainties regarding global collection rates. In the worst-case scenario, average collection rates for EV batteries and portable batteries in non-EU countries are assumed to be 50 and 0%, respectively. In the best-case scenario, collection rates similar to EU rates are assumed.

Since scrap is produced within the supply chain, it is reasonable to assume that this material will enter directly into recycling processes. Collection rates for production scrap can therefore be assumed to be 100%, unlike for EoL batteries.

3.7. Recovery Rate

The current recovery rate of lithium from LIBs is highly uncertain, but probably very low, since recovery is not economic at the moment [47]. Therefore, recyclers will probably follow regulations imposed upon them by governments. In the EU, legally mandated recovery rates for lithium will be 50% in 2028, and 80% in 2032 [3]. For our calculations, it was assumed that EU recovery rates for lithium will show a linear increase from 0% today to the target rate of 50% in 2028, and subsequently 80% in 2032. Thereafter, the lithium recovery rate was assumed to remain constant at the 2032 level. For the rest of the world, two scenarios were assumed, a worst-case scenario with a recovery rate of 0% as well as a best-case scenario with a recovery rate similar to those mandated in the EU, with a uniform probability distribution describing the likelihoods of intermediate cases.

4. Global Results

In the following subsections, we briefly summarise the global results of the statistical analysis for each production scenario. Figure 6 summarises these results for all production scenarios, showing both the median estimates and 95% confidence intervals (C.I.s) through time. While differences exist in the absolute production, scrap, and EoL battery quantities (in terms of Mt/a of LCE) between scenarios, it is noteworthy that they all follow similar trends. The estimates for collected batteries plus scrap as a proportion of total production generally stay below 40%, while estimated recycling contributions are unlikely to reach the 20% mark by 2050 in all of the production scenarios.

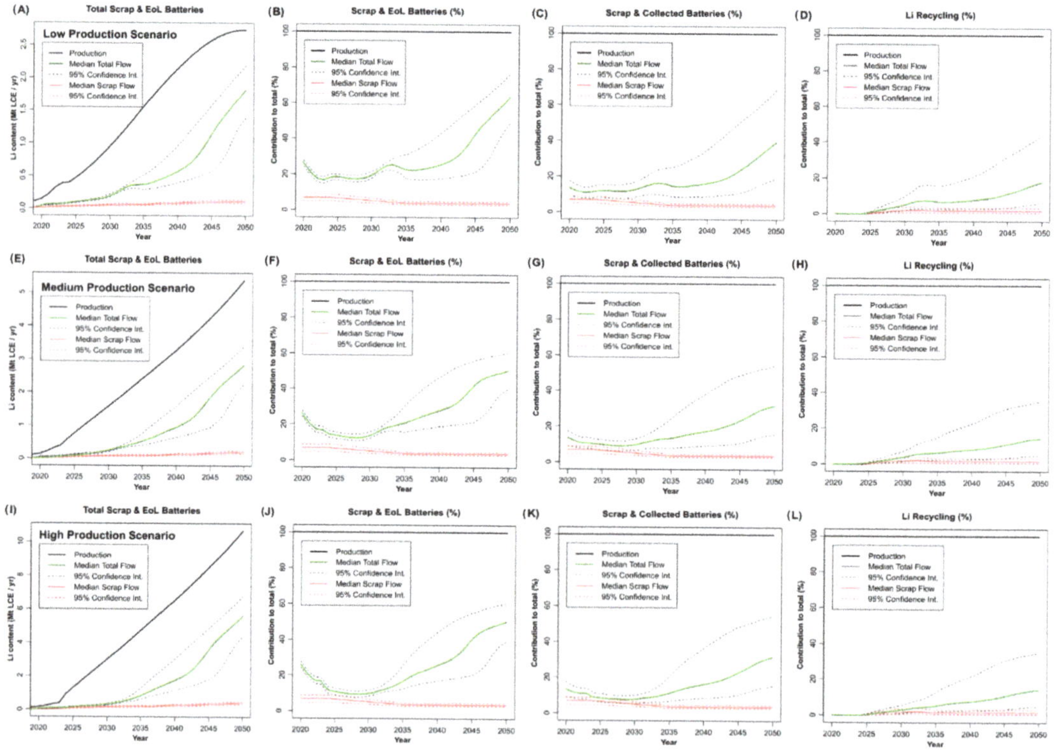

Figure 6. Global results of the statistical analysis. Panel (**A–D**) low production scenario. Panel (**E–H**) medium production scenario. Panel (**I–L**) high production scenario.

4.1. Low Production Scenario

In the low production scenario, the calculated amount of LCE contained in batteries increases from 0.4 Mt/a (million tons/annum) in 2024 up to 2.7 Mt/a in 2050. In the low production scenario (Figure 6A) the median total scrap volumes plus EoL batteries in terms of Mt of lithium carbonate equivalent (LCE) per year increases from less than 0.1 Mt/a in 2024 up to 1.8 Mt/a in 2050. The 95% C.I. of the median total scrap volume and the EoL battery amount ranges between 0.07 and 0.09 Mt/a in the year 2024 and between 1.4 and 2.2 Mt/a in the year 2050. Uncertainties are already substantial for EoL batteries plus scrap rates (Figure 6A). This reflects the uncertainties in the lifetime distribution parameters since there are no other uncertainties that affect the estimated total amounts of EoL batteries in these simulations.

The median of the total return flows is expected to grow from 18% in 2024 (95% C.I.: 16–20%) to 64% in 2050 (95% C.I.: 49–77%), as a fraction of battery production. Based on our model assumptions, new scrap production will contribute around a third of these total return flows (5–9% of production, 95% C.I.) until 2030 (Figure 6B). After 2030, new scrap rates are expected to decrease, down to a contribution to the total return flow of 3 to 5% (95% C.I.) by 2035. These assumptions were made for the global and European markets, resulting in the same values, and will not be mentioned again in the description of the European results.

The total material volume available for recycling (collected EoL batteries and new scrap) is expected to increase from 12% (95% C.I.: 8–15%) in 2024 to 13% (C.I.: 9–17%) in 2030, 14% (C.I.: 8–25%) in 2035, and 39% (19–69%) in 2050 (Figure 6C). The median contribution of lithium recycling to total battery production is expected to be 0.7% (95% C.I.: 0.3–1.3%) in 2025. In 2050, recycling production is expected to grow to a median of 17.7%, with a comparatively wide 95% confidence interval of 6.1–44.0% (Figure 6D).

4.2. Medium Production Scenario

In the medium production scenario, battery production in 2024 is expected to amount to an equivalent of 0.6 Mt/a LCE, going up to 5.4 Mt/a LCE in 2050. Based on this, the median total scrap volumes plus EoL batteries increase from less than 0.1 Mt/a LCE in 2024 up to 2.8 Mt/a LCE in 2050 (Figure 6E), with 95% C.I.s of 0.08–0.11 Mt/a in 2024, and 2.2–3.4 Mt/a in 2050. Uncertainties for these figures are similarly substantial as in the low production scenario, reflecting again the uncertainties in the lifetime distribution parameters.

New scrap production might contribute 0.04 Mt/a LCE in 2024 as part of the total scrap volumes, with a 95% C.I. of 0.03–0.06 Mt/a LCE. In 2050, the new scrap volume might increase to 0.22 Mt/a LCE (95% C.I.: 0.17–0.27 Mt/a LCE) (Figure 6E).

In percentage terms, the median total return flow of EoL batteries and new scrap is expected to grow from 15% of total battery production in 2024 (95% C.I.: 13–16%) to 51% in 2050 (95% C.I.: 40–50%) (Figure 6F). Collected batteries plus new scrap will amount to 10% in 2024, expected to grow to 32% in 2050 (95% C.I.: 16–54%) in 2050 (Figure 6G). Finally, recycling could contribute a median of 14% of total battery production in 2050, with a 95% C.I. ranging from 5 to 35% (Figure 6H).

4.3. High Production Scenario

In the high production scenario, LCE consumption for batteries is projected to grow from 0.9 Mt/a LCE in 2024 to 10.7 Mt/a LCE in 2050 (Figure 6I). In this scenario, the median total return flow of batteries (new scrap and EoL) would amount to 12% of overall battery production in 2024 (with a 95% C.I. of 10–14%), and would grow to 50% in 2050 (95% C.I. of 39–61%) (Figure 6J). Collected EoL batteries and new scrap would be 9% (6–11%) of total battery production in 2024, reaching 32% (15–54%) in 2050 (Figure 6K). Lithium recycling would be expected to contribute 0.5% to total LCE use in batteries in 2025, and 14% in 2050, with 95% C.I.s of 0.2–0.9% and 5–35%, respectively (Figure 6L).

5. EU Results

Similar to the global results, the estimates done at the EU level (Figure 7) show comparable patterns in each scenario for the total percentages of new scrap plus EoL batteries as a fraction of total production/consumption. The median estimates for collected batteries reach between 50–60% by 2050 in all scenarios. A decrease in this proportion from 2020 to 2025 was also noted. This is largely due to the shift from portable batteries to EV batteries as the main constituents of EoL battery flow. In contrast to the global results, overall recycling contributions are expected to be higher, reaching around 50% by 2050 in the low-production scenario. In the medium and high production scenario recycling might contribute ~40% of the lithium demand for battery production by the year 2050. These higher proportions largely reflect the effects of the planned regulatory framework in the EU (cf. methods section).

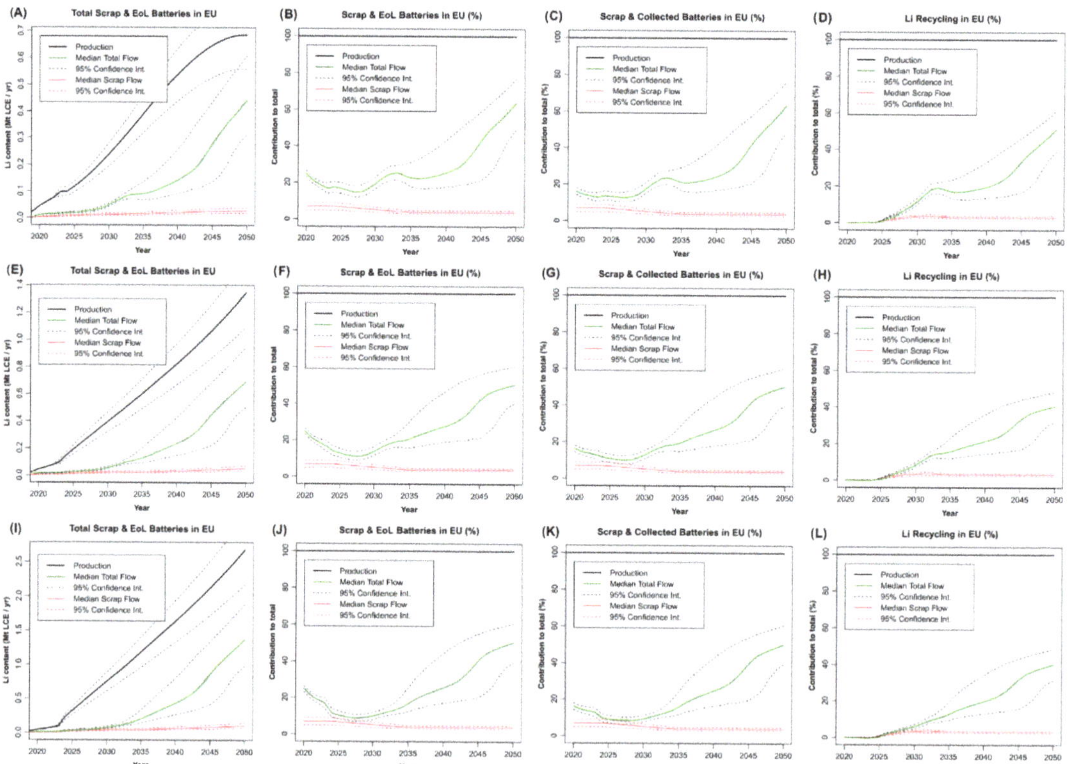

Figure 7. EU results of the statistical analysis. Panel (**A–D**) low production scenario. Panel (**E–H**) medium production scenario. Panel (**I–L**) high production scenario.

5.1. Low Production Scenario

For the EU, the median usage of LCE in batteries is expected to increase from 0.1 Mt/a in 2024 up to 0.7 Mt/a in 2050 (Figure 7A) in the low-production scenario. Thus, the median total scrap volumes plus EoL batteries in terms of contained LCE per year are expected to increase from 0.02 Mt/a (0.016–0.022 Mt/a) in 2024 up to 0.44 Mt/a (0.31–0.61 Mt/a) in 2050. Again, the uncertainties are already substantial for these flows (Figure 7A). The 95% confidence interval of the percentage of collected material and new scrap is 11–16% (median: 14%) in 2024. This range grows wider in the future years to 16–29% in 2035, and 49–76% in 2050 (Figure 7C). The probable recycling contribution of lithium in 2025 amounts to 1.3% (1.1–1.6%) of total lithium use in batteries. In 2050, this could increase to 51% (39–61%) (Figure 7D).

5.2. Medium Production Scenario

In the medium production scenario, the median LCE use in batteries for the EU amounts to 0.15 Mt/a LCE in 2024, increasing to 1.35 Mt/a LCE in 2050 (Figure 7E). The median total return flow would grow from 14% in 2024 (95% C.I.: 12–17%) to 51% in 2050 (95% C.I.: 40–60%) (Figure 7F). The total proportion of collected material plus new scrap would be 12% (95% CI: 9–14%) in 2024 and 50% (95% C.I.: 40–60%) in 2050 (Figure 7G). The contribution of lithium from recycling would then amount to 1.1% of total battery consumption in 2025 (95% C.I.: 0.9–1.3%), reaching 40% (32–48%) in 2050 (Figure 7H).

5.3. High Production Scenario

In the high production scenario, 0.23 Mt/a LCE are consumed by battery production for the EU in 2024 (95% C.I.: 0.19–0.28 Mt/a LCE), increasing to 2.68 (2.17–3.17) Mt/a LCE in 2050 (Figure 7I). The total return flows in this scenario are expected to reach 11% of consumption in 2024 (95% C.I.: 9–14%) and 50% in 2050 (95% C.I.: 39–61%) (Figure 7J). Collected EoL batteries plus new scrap are expected to reach 10% of total consumption in 2024 (95% C.I.: 8–12%), and 50% in 2050 (95% C.I.: 39–60%) (Figure 7K). The recycling contribution of lithium to battery consumption would be expected to be 0.9% (0.7–1.1%) in 2025 and should increase to 40% (31–48%) in 2050 40% (Figure 7L). This is 10% less than in the low-production scenario, but essentially the same as in the medium-production one.

6. Discussion

In the following subsections, we first consider the limitations of our estimates due to unavailable data (producing uncertainties) and factors that we could not consider. The likely effects of these limitations on our results are discussed and summarized. We then consider the policy implications of our results at both the global and EU level.

6.1. Major Sources of Uncertainty

Major sources of uncertainty in the factors we considered in this study include future battery production, battery lifetime distributions and their likely evolution in the future, collection rates, and lithium recovery rates in the recycling processes. These are discussed below.

i. Battery production: Future battery production scenarios are highly uncertain. This is why we used three cases in this study, without specifying the relative probabilities of each of these cases. Globally, battery cell production capacities for LIBs have been between 700 and 800 GWh for the EV sector in 2024 [48]. Current battery cell production for the EV sector in Europe amounts to around 190 GWh/a [48]. Both figures, global and EU, approximately correspond to the production capacities in the low production scenario (Figures 7 and 8). Comparing actual tracked capacities in the literature with simulated production capacities in each scenario globally and on the EU level, the low production scenario is probably most realistic to happen in the short- to mid-term future. Especially the recent slowdown in EV sales and cancellations of battery cell factories in the EU underpin this assumption [49,50]. The uncertainty in the production scenarios naturally increases the further projections are made into the future. However, we note that the final results of expected recycling rates by 2050 do not differ substantially between scenarios and are only slightly higher in the low-production scenario compared to the medium- and high-production scenarios. Therefore, the substantial uncertainty in future production growth does not translate to similar uncertainties for expected recycling rates.

ii. Battery lifetime: Uncertainties in the distributions of battery lifetimes are much more important as a source of uncertainty in our estimates. An increasing lifetime of a maximum 15-year warranty-relevant period in 2030 was assumed, remaining constant afterwards. The current development and rapid dynamics in battery research indicate that the average lifespan of batteries tends to lengthen. However, it is not clear whether this trend will continue indefinitely. It is not clear at what level further improvements will not be economic. Either way, improvements in battery lifetimes beyond those assumed in our simulations would imply that the return flows of EoL batteries would be smaller in the future than we estimated. Furthermore, batteries may remain in use beyond their currently expected lifetime distributions, depending on customer needs or their repurposing for secondary purposes. These factors greatly influence the available EoL material that should ultimately be available for recycling. Assuming lifespans of over 20 years [51] with a subsequent secondary use phase of 5–30 years [52,53], the availability of substantial amounts of EoL material, and consequently its contribution to new battery production as recycled lithium, could be delayed well beyond 2050.

iii. Collection rates: the rates for globally collected batteries plus scrap volumes available for recycling, range approximately between 20% and 60 to 70% of battery production in the medium and high production scenario. The median is below 40% in all the scenarios. The lower limits of these estimates result from the assumed range of collection rates between 0% and EU rates in non-EU countries, where only production scrap is collected and contributes to recycling. The general increasing trend of collected batteries and scrap volumes results mostly from regulated increases in the EU (see methods) and increasing return flows relative to production volumes over time, which are in turn due to the expected slowdown in the relative growth rates of battery production by 2050 (Figure 4). The contribution to the total amounts to maximum values between 60 to 70% in each of the scenarios, which results from the assumed global collection rate of EoL batteries of 100% by 2050, is the ideal case. However, this final rate is highly uncertain. Comparing global and EU results, it is noticeable that the collection rates are higher in the EU due to implemented targets set on a national level and in the EU battery regulation (cf. methods). The significant increase at the end is most probably due to the high volumes of EoL batteries. The little bump between the years 2030 and 2035 (for the proportion of collected EoL batteries plus scrap) is mostly due to the little kink at the beginning of the production time series (2023–2024), where we go from real data to interpolated future production. For collected batteries plus scrap as a proportion of total production, we note an actual decrease in proportion from 2020 to 2022. This is due to the transition from portable to EV batteries being the dominant EoL battery type (longer average battery lifetimes decrease return rates initially). Overall, we believe that our assumptions for global (non-EU) collection rates, which we took to be 45–50% on average, provide a reasonable intermediate scenario. The raw materials contained in LIBs already make their recycling economically attractive, which may incentivize higher rates in the future. However, this may change if battery compositions become dominated by low-value components such as in LFP batteries.

iv. Recovery rates: The contribution of recycled lithium to battery production also depends greatly on the recovery rates of lithium in recycling processes. Effective recovery rates for lithium on the global and European scales are currently unknown. Lithium is currently only recovered by hydrometallurgical processes. In our simulations, the expected contribution of recycled lithium in 2050 ranges between ~40 and 50% in Europe, across all scenarios (Figure 8). These large quantities are a result of the high recovery rates that the EU sets in the EU battery regulation. In case these rates should be revised due to the uneconomic conditions they may induce for the recycling industry, recovery rates could be reduced and recycled lithium would not be available in the estimated amounts. On a global scale, the median recycled lithium contribution to the total production does not exceed 20% by 2050 in any of the scenarios (Figure 8), even though expected total return flows (medians) generally reach around 50–60% by 2050. However, the ranges are quite narrow in the beginning and will get wider in the future years. Besides, the second use of EV LIBs is gaining traction, with numerous real-world projects e.g., Nissan [54], Vattenfall [55] and Mercedes Benz [56], and projects indicating a significant role for repurposed batteries especially in energy storage systems. The parameters of the second life were not taken into account in this study, mainly because there is a lack of reliable and publicly available data. However, it is certain that in case of a large-scale implementation of second-use applications, the return volume of EoL batteries will be delayed by several years and therefore less recycled lithium will be available.

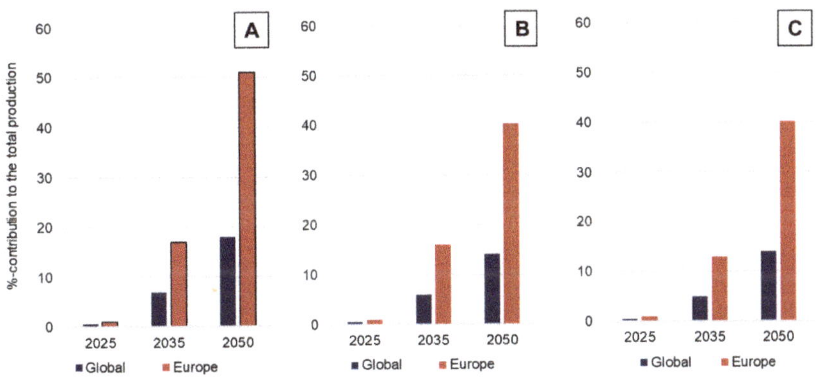

Figure 8. Median recycled lithium volumes as a %-contribution to the total production globally and European level in years 2025, 2035 and 2050. (**A**) Low production scenario; (**B**) Medium production scenario; (**C**) High production scenario.

Global recoveries of lithium from spent batteries are likely to remain smaller as long as recovery is neither economic nor mandatory. The EU regulations are unlikely to have a global impact. In case the recycling capacities will not be installed as announced due to the economic conditions and financial incentives for mining, the global contribution of secondary lithium could only reach about the 10% mark in the year 2050.

In summary, uncertainties in future production scenarios do not appear to have a major impact on estimated future recycling rates, as long as the future follows a more or less steady growth scenario. Major uncertainties are due to battery lifetime distributions, collection rates of batteries, and recycling recovery rates for lithium. While we used a relatively conservative assumption for lifetime distributions, longer lifetimes and second-use of EV batteries in the future may significantly decrease the amounts of EoL batteries available for recycling compared to our simulations.

For recovery and collection rates, the future evolution probably depends on the decisions made in countries outside the EU and technological developments, since lithium is not currently economically recyclable from LIBs. However, we believe that the uncertainties associated with collection and recovery rates are well-captured in our current simulations, and therefore, would not expect reality to diverge substantially from the 95% confidence intervals of our predictions.

6.2. Factors Not Considered in This Study

The quantities of secondary lithium supply simulated in this study represent a case where installed recycling capacities do not constitute a limiting factor for recycling rates. However, detailed market dynamics in the recycling sector do impact installed capacities and thus recycling rates. Furthermore, the sales figures of EVs have impacts on the number of spent LIBs available for recycling in the future, and our study did not consider potential deviations from projected, largely policy-driven scenarios for future EV sales. Finally, global trade in EoL battery materials would likely affect lithium recycling rates, if there is a net outflow (or inflow) of material from the EU, where assumed recycling rates are higher due to regulatory frameworks.

Our battery production scenarios do not take into account actual declining EV sales in Europe but these will have a significant impact on the new scrap volumes and return quantities of EoL material. As of 2024, the market is in a difficult situation. Many companies have become cautious about investments in the areas of e-mobility, battery and cathode production, as well as recycling. Lower capacity utilization in battery cell production has a direct impact on the availability of production scrap, and also EoL batteries in the future.

There are increasing reports of cancellations, delays and postponements in the construction of recycling plants. In case recycling capacities will not be installed in the mid- to long-term future, the material cannot be processed. A subsidy landscape that is currently attracting investors to the US is making it even more difficult for the European players to be competitive.

According to Rho Motion [57], 15.2 million EVs have been sold globally so far in 2024, growing by 25% year-to-date. The EV market of the region of the EU & EFTA and the UK remains down, total EV sales dropped by 3% in November, compared to the same period. The region had 280,000 units sold in November, thus reaching 2.7 million units sold year-to-date. All these factors will ultimately have an impact on the secondary amount of lithium available, as all of them play a crucial part in the value chain. Declining EV sales lead to lower demand for battery cells produced and deployed, which leads to less new scrap volumes, and subsequently less EoL material available for recycling. This represents a clear investment uncertainty, as constant input for recycling plants is necessary for them to operate profitably.

The price instability of primary raw materials and battery cells naturally has an effect on the availability of recycled lithium. Lithium prices have fallen sharply since January 2023. The actual price for battery-grade lithium carbonate amounts to approximately 10,650 US$/t (58). Prices for nickel and cobalt are also at low levels. Since July 2022, the LME prices for cobalt metal declined to 25,000 US$/t (August 2024), and the LME nickel metal price is currently at 16,246 US$/t (August 2024) [58]. The cost of production of lithium-ion batteries has also fallen drastically and will most likely continue to fall in the next few years. According to Wood et al. [59], the estimated cost for NMC111 cells produced in the United States in the year 2015 was around \$271 kWh^{-1}, due to the high costs of electrode materials, current collectors, separator and electrode processing. Recent cost models put cell production costs at \$106 kWh^{-1} (NMC622) and \$98 kWh^{-1} (NMC811) in Europe and the United States [60,61]. Due to these declines, the recycling of LIBs can only progress slowly, as it is not economically viable (at least for lithium) for the industry yet, lowering the amount of recycled lithium available in the end. Recycling costs play a major role. There are no minimum price figures per raw material for when recycling is worthwhile, only prices for production costs of primary cells are described. It would be an important addition if recycling prices of the raw materials were available. However, lithium recycling can be made mandatory by regulation.

In addition to these considerations, the collection infrastructure for EoL batteries presents significant challenges due to the lack of a uniform, globally standardized process. The absence of consistent guidelines complicates the safe and efficient collection and transportation of EoL batteries. Addressing this issue requires international agreements to establish an extensive collection infrastructure, a costly and complex process.

The international trade in lithium-ion batteries and recycled raw materials, especially black mass, is becoming increasingly important. Black mass is a valuable globally traded raw material. It is possible to store black mass in order to secure raw materials, but only under strict conditions in accordance with waste legislation. For European companies, however, the storage of black mass does not currently represent a profitable business model. For this reason, it is usually sold off in order to capitalize on the contained values immediately. Copper, nickel, cobalt and lithium are the most profitable raw materials here. From a global perspective, exports and imports have no influence on the available quantity of recycled lithium. At the European level, however, when the valuable black mass is exported to Asia or the US, it is no longer available for the production of new battery-grade material in Europe. Unfortunately, the trade in black mass cannot be tracked at the moment. Black mass is a non-standardized "waste" material, which complicates the definition in HS codes. However, the classification of black mass as hazardous waste is currently being discussed by the European Commission as part of an amendment to the European list of waste to address waste batteries and wastes from treating them [62,63]. If significant amounts of black mass leave the EU and are processed in countries where lithium recycling is not legally mandated, then the contained lithium would be lost, and global recycling rates may drop below those projected in our simulations.

Research on new cathode and anode material developments and the development of new recycling technologies will also affect the secondary supply in the long term. The development of new systems such as other metal anodes, e.g., sodium-ion or vanadium-flow batteries, and gaseous or liquid cathode systems have the potential to displace at least some of the existing LIB types in certain applications. Those battery systems based on other ions will actually reduce the overall demand for lithium by displacement, even if only a small proportion. If the demand for lithium shrinks in the long-term future, recycling of LIBs potentially could cover a small part of secondary material contents as it depends on previous LIBs production. However, a commercial market penetration of alternative battery systems and thus the potential replacement of lithium is not expected in the next 10 to 15 years. Recently, solid-state batteries have attracted interest as energy stores for electric vehicles. This type requires higher lithium contents, assumingly between 0.26 and 0.52 kg/kWh as mentioned earlier, which would mean a huge increase in demand for lithium in LIBs. Based on electric vehicle manufacturers' technology roadmaps and technological advances, possible commercial viability of solid-state batteries must be awaited within the next 5 years [64,65]. In the case of commercial deployment, it would also imply a high recycling potential due to the high lithium content. It remains to be seen whether the recycling processes will be efficient enough to recover the material in solid-state batteries.

LFP and LMFP battery types increase their market shares on a global scale [19,34]. Both do not require any cobalt and nickel, which is advantageous in terms of a resilient supply chain, but recycling does not deliver a value-creating black mass. Instead, the main valuable materials would be lithium and graphite. Whether the recycling of these two types is a business case remains to be seen, as there is an increasing interest in direct recycling of the cathode and anode materials. At the moment, this process route is still in the early stages without any deployment at an industrial scale but could be promising in terms of efficiency, especially for LFP/LMFP. In case these two battery types increase their market shares drastically, but recycling would not be economic, especially in Europe, as it would impact the supply of secondary lithium.

6.3. Policy Implications

The EU Battery Regulation has been in effect for Europe since August 2023. The regulation specifies labelling rules, information obligations, e.g., CO_2-footprint, and supply-chain due diligence standards. It also mandates metal-specific recycling recovery rates, and the use of recycled content in batteries with a capacity above 2 kWh, most of which are used in EVs. The amended proposal of the battery regulation from June 2023 states that LIBs have to contain a minimum percentage share of 6% recycled lithium from 08/2031, recovered from battery manufacturing waste or postconsumer waste, and 12% recycled lithium from 08/2036.

Our simulations suggest that the EU goals of recovery rates and minimum recycled lithium content should technically be achievable, yet the economic feasibility of the required recycling plants is questionable. Nevertheless, the results of the present study indicate that sufficient recycled lithium in each of the production scenarios (on a European scale), would be available in the years 2031 and 2036 to meet the targets.

Taking the current market developments into consideration, recycling plants will have to process accumulated new scrap volumes ranging between 5 and 9% until the larger return flow of EoL material arrives. These production scraps are highly sought after as material available for recycling will subsequently impact recycling plants. Without enough feedstock, they cannot operate at full capacity. The international trade in LIBs and recycled raw materials, especially black mass, is becoming increasingly important. Black mass is a valuable, globally traded raw material. It is possible to store black mass in order to secure raw materials, but only under strict conditions in accordance with waste legislation. For European companies, the storage of black mass is currently not a profitable business model. For this reason and due to the fact that there is no comprehensive large-scale processing of black mass in Europe, European recycling companies generally export black mass in order to monetise the resources immediately. Black mass is exported as a product or hazardous waste to Asia. It is becoming clear that a regulated and harmonised trading system for this kind of material is missing. Export controls could counteract here but would require the processing capacities for black mass in Europe as well as downstream buyers of recycled materials, ideally pCAM or CAM producers. In addition, lithium recycling is not economical, but battery recycling, in general, is very economical though, otherwise, there would not be international competition for the black mass. It would need an enormous political effort to work against the principles of companies in a market economy.

6.4. Future Work

The results of this study highlight the multiple influences on current and future lithium recycling flows on a global scale, as well as the uncertainties associated with them. The study also addresses general issues with respect to forecasts of future battery production and return quantities of EoL material available for recycling. The distribution of battery lifetimes is one of the most critical inputs in the simulations since it is an important determinant of the amount of EoL batteries available each year for recycling. Battery lifespans not only become longer by optimising the battery itself but also through the increasing secondary use of batteries in other application sectors. The highly dynamic and immature market of battery cell production and battery recycling is driven by political, regulatory and economic factors. As long as lithium recycling from LIBs is not economically viable, mining will be the first choice to meet demand.

To address further gaps in the methodology, additional factors could be considered, e.g., (1) lithium contents in other newly developed battery types and the probable deployment of these types, (2) the probability of installed recycling plants and their actual capacities to recycle LIBs and refine the valuable black mass, (3) marginal costs of adding lithium extraction to existing recycling facilities would need to be balanced against lithium prices to determine whether extraction is likely.

Supplementary Materials: The following supporting information can be downloaded at: https://www.mdpi.com/article/10.3390/recycling10040122/s1.

Author Contributions: Conceptualization: C.K., M.F. and B.B.; Methodology: C.K., M.F. and B.B.; Investigation: C.K. and B.B.; Formal analysis: C.K., M.F. and B.B.; Visualization: C.K., M.F. and L.B.A.; Supervision: M.F. and B.B.; Project Administration: M.F. and B.B.; Resources: C.K., M.F., B.B. and L.B.A.; Funding acquisition: M.F. and B.B.; Writing—original draft: C.K., M.F. and B.B.; Writing—review and editing: all authors. All authors have read and agreed to the published version of the manuscript.

Funding: This research was funded by the Federal Ministry of Education and Research under grant number 03XP0335B. The APC was funded by the Federal Institute for Geosciences and Natural Resources.

Data Availability Statement: The data supporting this article have been included as part of the Supplementary Materials.

Acknowledgments: The project "BatMix", on which this publication is based, was funded by the Federal Ministry of Education and Research. The responsibility for the content of this publication lies with the authors.

Conflicts of Interest: The authors declare no conflict of interest.

References

1. Buchert, M.; Dolega, P.; Degreif, S. *Gigafactories für Lithium-Ionen-Zellen–Rohstoffbedarfe für Die globale Elektromobilität bis 2050*; Oeko-Inst. eV: Darmstadt, Germany, 2019.
2. Clean Energy Institute. Lithium-Ion Battery. University of Washington. 2020. Available online: https://www.cei.washington.edu/research/energy-storage/lithium-ion-battery/ (accessed on 14 December 2023).
3. European Commission. Regulation of the European Parliament and of the Council Concerning Batteries and Waste Batteries, Amending Directive 2008/98/EC and Regulation (EU) 2019/1020 and repealing Directive 2006/66/EC 2023. 2023. Available online: https://eur-lex.europa.eu/legal-content/EN/TXT/PDF/?uri=CELEX:32023R1542 (accessed on 16 June 2025).
4. Agora Verkehrswende.Klimabilanz von Elektroautos. Einflussfaktoren und Verbesserungspotenzial. 2019. Available online: https://www.agora-verkehrswende.de/fileadmin/Projekte/2018/Klimabilanz_von_Elektroautos/Agora-Verkehrswende_22_Klimabilanz-von-Elektroautos_WEB.pdf (accessed on 16 June 2025).
5. World Economic Forum. A Vision for a Sustainable Battery Value Chain in 2030: Unlocking the Full Potential to Power Sustainable Development and Climate Change Mitigation. 2019. Available online: https://www3.weforum.org/docs/WEF_A_Vision_for_a_Sustainable_Battery_Value_Chain_in_2030_Report.pdf (accessed on 16 June 2025).
6. PEM Chair at RWTH and Roland Berger GmbH. Battery Monitor 2023. The Value Chain Between Economy and Ecology. 2023. Available online: https://content.rolandberger.com/hubfs/07_presse/Batteriemonitor_2023_digital_final.pdf (accessed on 16 June 2025).
7. Circular Energy Storage. CES Online Database. 2023. Available online: https://www.circularenergystorage-online.com/ (accessed on 16 June 2025).

8. Benchmark Minerals. Lithium Has to Scale Twenty Times by 2050 as Automakers Face Generational Challenge. 2022. Available online: https://source.benchmarkminerals.com/article/lithium-has-to-scale-twenty-times-by-2050-as-automakers-face-generational-challenge (accessed on 3 July 2024).
9. Castro, F.D.; Cutaia, L.; Vaccari, M. End-of-life automotive lithium-ion batteries (LIBs) in Brazil: Prediction of flows and revenues by 2030. *Resour. Conserv. Recycl.* **2021**, *169*, 105522. [CrossRef]
10. Gaines, L.; Zhang, J.; He, X.; Bouchard, J.; Melin, H.E. Tracking Flows of End-of-Life Battery Materials and Manufacturing Scrap. *Batteries* **2023**, *9*, 360. [CrossRef]
11. Sattar, A.; Greenwood, D.; Dowson, M.; Unadkat, P. Automotive Lithium-ion Battery Recycling in the UK. Warwick University (The British Advance). 2020. Available online: https://hvm.catapult.org.uk/wp-content/uploads/2022/06/22350m-WMG-Battery-Recycling-report-v7.pdf (accessed on 16 June 2025).
12. Weyhe, R.; Yang, X. Investigation about Lithium-Ion Battery Market Evolution and future Potential of Secondary Raw Material from Recycling. 2019. Available online: https://accurec.de/wp-content/uploads/2018/04/0-2Market-Research_YXF_3.0.pdf (accessed on 16 June 2025).
13. Kastanaki, E.; Giannis, A. Dynamic estimation of end-of-life electric vehicle batteries in the EU-27 considering reuse, remanufacturing and recycling options. *J. Clean. Prod.* **2023**, *393*, 136349. [CrossRef]
14. Shafique, M.; Rafiq, M.; Azam, A.; Luo, X. Material flow analysis for end-of-life lithium-ion batteries from battery electric vehicles in the USA and China. *Resour. Conserv. Recycl.* **2022**, *178*, 106061. [CrossRef]
15. Brückner, L.; Frank, J.; Elwert, T. Industrial Recycling of Lithium-Ion Batteries—A Critical Review of Metallurgical Process Routes. *Metals* **2020**, *10*, 1107. [CrossRef]
16. Ciez, R.E.; Whitacre, J.F. Examining different recycling processes for lithium-ion batteries. *Nat. Sustain. Nat.* **2019**, *2*, 148–156. [CrossRef]
17. Hanisch, C.; Diekmann, J.; Stieger, A.; Haselrieder, W.; Kwade, A. Recycling of Lithium-Ion Batteries. In *Handbook of Clean Energy Systems*; Yan, J., Ed.; John Wiley & Sons, Ltd.: Chichester, UK, 2015; pp. 1–24.
18. Harper, G.; Sommerville, R.; Kendrick, E.; Driscoll, L.; Slater, P.; Stolkin, R. Recycling lithium-ion batteries from electric vehicles. *Nature* **2019**, *575*, 75–86. [CrossRef]
19. IEA Recycling of Critical Minerals-Strategies to Scale Up Recycling and Urban Mining. 2024. Available online: https://iea.blob.core.windows.net/assets/3af7fda6-8fd9-46b7-bede-395f7f8f9943/RecyclingofCriticalMinerals.pdf (accessed on 16 June 2025).
20. Mayyas, A.; Steward, D.; Mann, M. The case for recycling: Overview and challenges in the material supply chain for automotive li-ion batteries. *Sustain. Mater. Technol.* **2019**, *19*, e00087. [CrossRef]
21. Sojka, R.; Pan, Q.; Billmann, L. Comparative Study of LIB Recycling Processes. 2020. Available online: https://accurec.de/wp-content/uploads/2021/04/Accurec-Comparative-study.pdf (accessed on 16 June 2025).
22. Zhou, L.F.; Yang, D.; Du, T.; Gong, H.; Luo, W.B. The Current Process for the Recycling of Spent Lithium Ion Batteries. *Front. Chem.* **2020**, *8*, 578044. [CrossRef]
23. Windisch-Kern, S.; Holzer, A.; Ponak, C.; Raupenstrauch, H. Pyrometallurgical Lithium-Ion-Battery Recycling: Approach to Limiting Lithium Slagging with the InduRed Reactor Concept. *Processes* **2021**, *9*, 84. [CrossRef]
24. Olutogun, M.; Vanderbruggen, A.; Frey, C.; Rudolph, M.; Bresser, D.; Passerini, S. Recycled graphite for more sustainable lithium-ion batteries. *Carbon Energy* **2024**, *6*, e483. [CrossRef]
25. Vanderbruggen, A.; Hayagan, N.; Bachmann, K.; Ferreira, A.; Werner, D.; Horn, D.; Peuker, U.; Serna-Guerrero, R.; Rudolph, M. Lithium-Ion Battery Recycling–Influence of Recycling Processes on Component Liberation and Flotation Separation Efficiency. *ACS EST Eng.* **2022**, *2*, 2130–2141. [CrossRef]
26. Frenzel, M.; Mikolajczak, C.; Gutzmer, J.; Reuter, M.A. Quantifying the relative availability of high-tech by-product metals–The cases of gallium, germanium and indium. *Resour. Policy* **2017**, *52*, 327–335. [CrossRef]
27. Nassar, N.T.; Kim, H.; Frenzel, M.; Moats, M.S.; Hayes, S.M. Global tellurium supply potential from electrolytic copper refining. *Resour. Conserv. Recycl.* **2022**, *184*, 106434. [CrossRef]
28. Öko Institut. B2DS Szenario-"Unter 2 Grad Szenario". 2019. Available online: https://www.now-gmbh.de/wp-content/uploads/2020/10/now-factsheet_elektrobomilitaet-und-rohstoffe.pdf (accessed on 16 June 2025).
29. Avicenne Energy. Commercial Data. 2021. Available online: https://www.avicenne.com/pdf/Presentation_Avicenne_Energy_June_2022.pdf (accessed on 16 June 2025).
30. Moores, S. The Global Battery Arms Race: Lithium-Ion Battery Gigafactories and Their Supply Chain. Oxford Institute for Energy Studies. 2021. Available online: https://www.oxfordenergy.org/wpcms/wp-content/uploads/2021/02/THE-GLOBAL-BATTERY-ARMS-RACE-LITHIUM-ION-BATTERY-GIGAFACTORIES-AND-THEIR-SUPPLY-CHAIN.pdf (accessed on 16 June 2025).
31. Rho Motion. EV Battery Outlook. 2022. Available online: https://rhomotion.com/research/ev-battery-forecast/ (accessed on 16 June 2025).

32. S&P Global Mobility (EVs). Press Release. 2022. Available online: https://www.spglobal.com/market-intelligence/en/news-insights/research/investment-in-lithium-ion-batteries-could-deliver-5-point-9-twh-capacity-by-2030 (accessed on 16 June 2025).
33. S&P Global Mobility (EVs). Press Release. 2021. Available online: https://www.spglobal.com/market-intelligence/en/news-insights/research/top-electric-vehicle-markets-dominate-lithium-ion-battery-capacity-growth (accessed on 16 June 2025).
34. IEA. Global EV Outlook 2023. 2023. Available online: https://iea.blob.core.windows.net/assets/dacf14d2-eabc-498a-8263-9f97fd5dc327/GEVO2023.pdf (accessed on 16 June 2025).
35. Marscheider-Weidemann, F.; Langkau, S.; Baur, S.J.; Billaud, M.; Deubzer, O.; Eberling, E.; Erdmann, L.; Haendel, M.; Krail, M.; Loibl, A.; et al. Raw materials for emerging technologies 2021. *DERA Rohst.* **2021**, *50*, 348.
36. Anderson, B.A.; Råde, I. Metal resource constraints for electric-vehicle batteries. *Transp. Res. Part D Transp. Environ.* **2001**, *6*, 297–324. [CrossRef]
37. Gerssen-Gondelach, S.; Faaij, A.P.C. Performance of batteries for electric vehicles on short and longer term. *J. Power Sources* **2012**, *212*, 111–129. [CrossRef]
38. BYD. Warranty Policy. 2025. Available online: https://www.byd.com/eu/service-maintenance/warranty-policy (accessed on 16 June 2025).
39. Hyundai. Warranty Policy. 2025. Available online: https://www.hyundai.com/eu/driving-hyundai/owning-a-hyundai/why-hyundai-services/warranty.html (accessed on 16 June 2025).
40. KIA. Warranty Policy. 2025. Available online: https://www.kia.com/uk/about/news/electric-car-battery-warranty/ (accessed on 16 June 2025).
41. Tesla. Warranty Policy. 2025. Available online: https://www.tesla.com/support/vehicle-warranty (accessed on 16 June 2025).
42. Volkswagen. Warranty Policy. 2025. Available online: https://www.volkswagen.co.uk/en/electric-and-hybrid/living-electric/looking-after-your-ev/vehicle-and-battery-warranties.html (accessed on 16 June 2025).
43. EU. Consumer Guarantees. 2024. Available online: https://europa.eu/youreurope/business/dealing-with-customers/consumer-contracts-guarantees/consumer-guarantees/index_en.htm (accessed on 5 August 2024).
44. Gattiglio, F. The Batteries Directive Review. In Proceedings of the 24th International Congress for Battery Recycling ICBR 2019, Lyon, France, 18–20 September 2019. Panel discussion.
45. EPBA The Collection of Waste Portable Batteries in Europe in View of the Achievability of the Collection Targets Set by Batteries Directive 2006/66/EC. Short Update Covering 2018 Data. 2020. Available online: https://www.epbaeurope.net/assets/resources/Report-on-the-portable-battery-collection-rates-Short-Update-Mar-20-final-1.1.pdf (accessed on 16 June 2025).
46. DOE Energy Department Announces Battery Recycling Prize and Battery Recycling R&D Center. 2019. Available online: https://www.energy.gov/articles/energy-department-announces-battery-recycling-prize-and-battery-recycling-rd-center#:~:text=Currently,%20lithium-ion%20batteries%20are,ion%20batteries%20for%20eventual%20recycling.%20 (accessed on 5 August 2024).
47. Ziemann, S.; Müller, D.B.; Schebek, L.; Weil, M. Modeling the potential impact of lithium recycling from EV batteries on lithium demand: A dynamic MFA approach. *Resour. Conserv. Recycl.* **2018**, *133*, 76–85. [CrossRef]
48. VDI/VDE Innovation + Technik GmbH. Battery Cell Production in Europe: Status Quo and Outlook, Market Analysis Update Q2 2024. 2024. Available online: https://www.ipcei-batteries.eu/accompanying-research/market-updates (accessed on 16 June 2025).
49. EY. How to Retake the Momentum in the EV Transition. Press Release. 2024. Available online: https://www.ey.com/en_gl/insights/automotive/how-to-retake-the-momentum-in-ev-transition#:~:text=Consequently,%20overall%20EV%20market%20growth,2025%20up%20to%20approximately%2023%25 (accessed on 8 October 2024).
50. Transport & Environment. How not to Lose It All: Two-Thirds of Europe's Battery Gigafactories at Risk Without Further Action. 2023. Available online: https://www.transportenvironment.org/uploads/files/TE-Battery-risk-report.pdf (accessed on 16 June 2025).
51. Fastmarkets. EV Battery Lifespans: Unlocking the Secrets to Battery Lifetimes and Battery Recycling. Press Release. 2024. Available online: https://www.fastmarkets.com/insights/ev-battery-lifespans-unlocking-the-secrets-to-battery-lifetimes-and-battery-recycling/#:~:text=The%20longevity%20of%20lithium-ion,to%20extend%20beyond%2015%20years (accessed on 18 December 2024).
52. Bobba, S.; Mathieux, F.; Blengini, G.A. How will second-use of batteries affect stocks and flows in the EU? A model for traction Li-ion batteries. *Resour. Conserv. Recycl.* **2019**, *145*, 279–291. [CrossRef]
53. Casals, L.C.; García, B.A.; Canal, C. Second life batteries lifespan: Rest of useful life and environmental analysis. *J. Environ. Manag.* **2019**, *232*, 354–363. [CrossRef]
54. Nissan. Nissan and Ecobat to Give Used EV Batteries a Second Life Beyond the Car. 2024. Available online: https://uk.nissannews.com/en-GB/releases/nissan-and-ecobat-to-give-used-ev-batteries-a-second-life-beyond-the-car (accessed on 13 January 2025).

55. Vattenfall. Autobatterien Entsorgen und Recyceln. 2023. Available online: https://www.vattenfall.de/infowelt-energie/e-mobility/elektroauto-batterie-recycling (accessed on 13 January 2025).
56. Mercedes-Benz Energy Storage. 2025. Available online: https://www.mercedes-benz.com/en/mercedes-benz-energy/ (accessed on 13 January 2025).
57. Rho Motion. Press Release "Record Number EVs Sold in November" 2024. Available online: https://rhomotion.com/news/record-number-evs-sold-in-november/ (accessed on 16 December 2024).
58. DERA. Preistrendmonitor. 2025. Available online: https://www.deutsche-rohstoffagentur.de/DERA/DE/Aktuelles/Monitore/2025/01-25/2025-01-preistrendmonitor.pdf?__blob=publicationFile&v=2 (accessed on 16 June 2025).
59. Wood, D.L.; Li, J.; Daniel, C. Prospects for reducing the processing cost of lithium ion batteries. *Power Sources* **2015**, *275*, 234–242. [CrossRef]
60. Duffner, F.; Mauler, L.; Wentker, M.; Leker, J.; Winter, M. Large-scale automotive battery cell manufacturing: Analyzing strategic and operational effects on manufacturing costs. *J. Clean. Prod.* **2021**, *232*, 107982. [CrossRef]
61. Mauler, L.; Duffner, F.; Leker, J. Economies of scale in battery cell manufacturing: The impact of material and process innovations. *Appl. Energy* **2021**, *286*, 116499. [CrossRef]
62. European Commission. Waste Treatment–Amendment to the European List of Waste to Address Waste Batteries and Wastes from Treating Them. Published Initiative. 2024. Available online: https://ec.europa.eu/info/law/better-regulation/have-your-say/initiatives/14016-Waste-treatment-Amendment-to-the-European-List-of-Waste-to-address-waste-batteries-and-wastes-from-treating-them_en (accessed on 19 December 2024).
63. European Commission. Critical Raw Materials Act. 2024. Available online: https://single-market-economy.ec.europa.eu/sectors/raw-materials/areas-specific-interest/critical-raw-materials/critical-raw-materials-act_en (accessed on 19 December 2024).
64. Yole Développement. Solid-State Battery 2021, Market & Technology Report. 2021. Available online: https://medias.yolegroup.com/uploads/2021/04/YINTR21187_Solid-State_Battery_2021_Flyer.pdf (accessed on 16 June 2025).
65. Fraunhofer Institute for Systems and Innovation Research ISI. Solid-State Batteries for Electric Vehicles: Still in R&D or on the Verge of Commercialization? Blog Battery Update. 2024. Available online: https://www.isi.fraunhofer.de/en/blog/themen/batterie-update/feststoffbatterien-elektro-autos-kommerzialisierung-stand-forschung-entwicklung.html (accessed on 19 December 2024).

recycling

MDPI

Article

Product and Process Data Structure for Automated Battery Disassembly

Domenic Klohs *, Moritz Frieges, Jonas Gorsch, Philip Ellmann, Heiner Hans Heimes and Achim Kampker

Production Engineering of E-Mobility Components, RWTH Aachen University, 52072 Aachen, Germany; m.frieges@pem.rwth-aachen.de (M.F.); j.gorsch@pem.rwth-aachen.de (J.G.); p.ellmann@pem.rwth-aachen.de (P.E.); h.heimes@pem.rwth-aachen.de (H.H.H.); a.kampker@pem.rwth-aachen.de (A.K.)

* Correspondence: d.klohs@pem.rwth-aachen.de

Abstract: Battery disassembly forms a central jumping-off point for recycling in the context of a sustainable closure of the battery loop. The main objective for economic realization in line with European recycling regulations is therefore a transformation of the battery disassembly from a manual to an automated process. Product-related influences such as design variations and process-side constraints including the selection of disassembly technologies require large amounts of data for implementation in an automated system. This article examines accessible data sources in the literature and the upcoming battery passport to build a basis for a multi-layered methodical analysis of the data required for the automation of battery disassembly. For this purpose, the disassembly sequence and depth of an Audi e-tron battery pack are first identified using a priority matrix and converted into a product and process structure. Definitions for product- and process-related elements are established, and a generalized process model is developed, which is finally converted into a data structure model approach. The result shows that much of the required data to automate the disassembly of used batteries are currently not yet available. Further efforts must be made to establish data structures and standards regarding product- and process-related disassembly data.

Keywords: lithium-ion battery; electric vehicles; automated battery disassembly; dismantling challenges; battery recycling; battery product structure; battery process structure; battery passport

1. Introduction

Electromobility has achieved a considerable market share in the passenger vehicle sector in Europe over the last decade. This development has not yet reached its peak, resulting in further growth in electrically powered vehicles in the coming years. An important challenge associated with such a trend is not only the further development of existing battery technologies but also the creation of a sustainable and economical use of the required raw materials [1]. Closing the battery loop therefore requires a seamless process chain, especially after the battery's first phase of use in the vehicle [2].

During the charging and discharging cycles of traction batteries, various chemical and environmental effects cause the amount of stored energy to slowly decrease over the period of use [3]. This ratio of initial nominal capacity to current nominal capacity is defined as the state of health (SoH). Currently traction batteries in electrically powered vehicles will be retired from their first life in EV applications after they fall below an SoH of 70–80% [4].

 https://doi.org/10.3390/recycling10010025

The average statistics for reaching this state vary, but current estimations range around 12–15 years [5,6].

Combined, the rapidly growing market share of electrically powered vehicles and the estimated average life expectancy of a traction battery of more than ten years result in a significant increase in the number of battery returns over the next years from the current very moderate levels and will pose challenges for existing structures in the areas of battery logistics, second life, disassembly and recycling [7]. New scalable process chains must be developed for the further use and recovery of battery materials, which means large-scale automation of the process chains in order to ensure cost-effectiveness, especially for high-wage regions such as Europe [8,9]. In order to achieve this, framework conditions and standardized data transfers are necessary, which are currently not available or available only to a limited extent and should be addressed more extensively in the future [10,11].

2. State-of-the-Art Battery Disassembly

When looking at the battery life cycle, disassembly plays a central supporting role in the second phase during and after the use of the battery in the vehicle. Dismantling batteries is a fundamental step in repair processes in which the battery pack is opened and modules or peripherals are replaced. Disassembly can also contribute to the conversion of battery packs or the removal of modules for stationary applications as part of a second life approach [12]. For battery recycling, disassembly as an upstream or associated step of mechanical pre-processing forms a central jumping-off point and makes a significant contribution to achieving the quality targets of the black mass and the output material streams by pre-separating components into defined factions [13].

Due to the previously mentioned moderate return quantities today, battery disassembly is currently still largely carried out manually [14]. This means that, typically, at least two high-voltage (HV) trained workers separate the individual components of the battery pack from each other and divide them into predefined factions, which go into different recycling streams depending on the underlying material compositions [15]. At present, the batteries are generally dismantled down to the module level. These battery modules then move on to mechanical pre-processing, where they are fed into a shredding process depending on the process route and sometimes with thermal pre-treatment [16]. Due to the described process approach, this paper focuses on the area of pack-to-module disassembly and its data, as this must be seen as a first step towards automation.

The manual dismantling process also poses safety risks for workers. Battery systems in passenger vehicles generally operate at voltage levels of 400 to 800 volts [17]. Depending on the condition of the battery pack, there may also be an increased hazard from leaking toxic electrolytes [18]. In addition, there is always a risk of thermal runaway, particularly in the case of damaged batteries [19,20].

The major challenge for future automation in the field of battery dismantling is the complex product structure of battery packs along with the large number of variants on the market and the corresponding process structure for the disassembly. In general, battery systems combine functionally equivalent component groups within their underlying sub-levels [21]. For example, battery systems usually have interconnected stacks of cells. In the case of non-cell-to-pack approaches, these are interconnected to form modules, which are brought together in a housing for functional or safety reasons and are monitored with corresponding sensors and controllers [22]. A similar principle is also followed at the pack level, where the previously interconnected modules are in turn combined via busbars or high-voltage (HV) cables to be able to meet the respective capacity and performance requirements of the vehicle. In addition, there are battery management controllers, structural components and corresponding peripherals for the cooling circuit [23]. With a housing cover and corresponding interfaces to the vehicle, the battery is integrated into the vehicle design. Figure 1 illustrates the common component groups mentioned and the conditional states for battery pack disassembly.

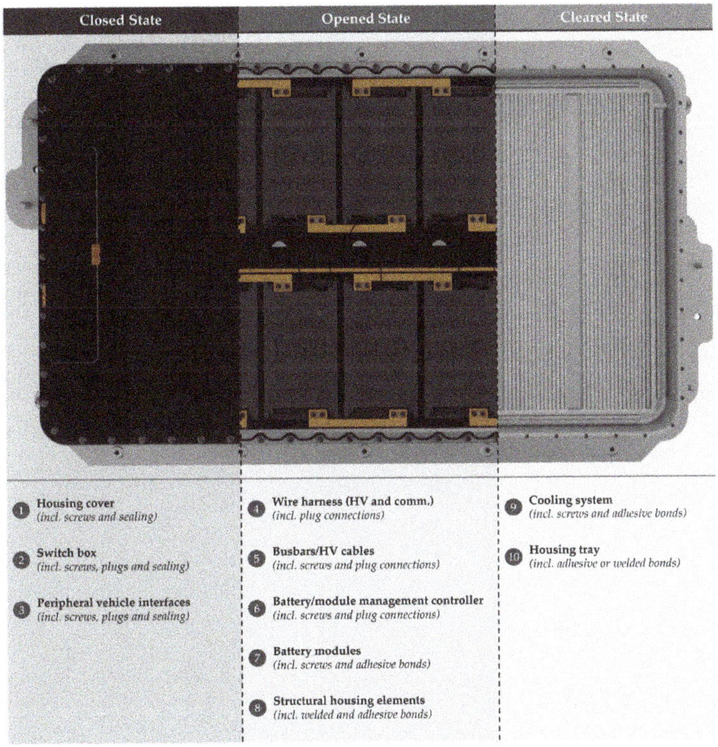

Figure 1. Overview of different states of a battery pack's disassembly depth as well as the relevant components and joint connections.

Despite the functionally similar structures, there are major differences in product topologies. These are based partly on the different shapes, sizes and design characteristics of individual components and secondly on the joints used and the resulting connections between different components as well as the associated disassembly sequences derived from accessibility [24]. Finally, the condition of the components and the joint connections also play a role in the disassembly strategy. There are, for example, different types of joints such as welded, adhesive, plug and screw connections, which need to be handled adequately [25].

Altogether, this means that a huge number of decision criteria and large amounts of data have to be processed for the disassembly of battery packs in general but especially for automation approaches, which in turn require a flexible system with an adaptable yet robustly functioning process chain. However, the problem does not start with the consideration of numerous parameters for a disassembly operation, but rather, with the provision of such information and the standardization of basic battery parameters and data for disassembly [26].

An examination of the existing literature on product- and process-relevant information for battery disassembly in Table 1 reveals that on both the product and process side, the specification of certain types of data is very common, while very important sub-aspects are only considered in a very fundamental way.

Table 1. Review of existing dismantling-specific literature with regard to product- and process-related information specifications.

General Information				*Product-Related Information*					*Process-Related Information*					
Authors	**Ref.**	**Year**	**Specified Battery**	**Pack Detail Level**	**Module Detail Level**	**Cell Detail Level**	**Joint Connections**	**Position/ Quantity**	**Disassembly Sequence**	**Process Step**	**Activities**	**Operations**	**Tools/ Specifications**	**Access Angle/ Vectors**
Wegener et al.	[27]	2014	Audi Q5 hybrid	Types (14)	No	No	Fundamental	No/Yes	Priority matrix	Types (24)	Fundamental	No	Partly detailed/No	No
Cerdas et al.	[28]	2018	Audi Q5 hybrid	Types (16)	Types (2)	No	Fundamental	No/ Partly detailed	Priority matrix	Types (24)	Fundamental	No	Partly detailed/No	No
Gumanová et al.	[29]	2019	VW Jetta hybrid	Types (14)	No	No	Fundamental	No/Yes	Priority matrix	Types (24)	Fundamental	No	Partly detailed/No	No
Alfaro-Algaba et al.	[30]	2020	Audi A3 hybrid	Parts (84)	Parts (96)	No	Fundamental	No/Yes	Disassembly graph	Types (7)	188 activities	No	No/No	No
Gentilini et al.	[31]	2020	Toyota Prius	Parts (33)	No	No	No	No/ Fundamental	Decisional tree	-	Fundamental	No	No/No	No
Rallo et al.	[32]	2020	Smart ForFour	Types (9)	Types (4)	No	Fundamental	No/No	-	Types (24)	Partly detailed	No	No/No	No
Ke et al.	[33]	2020	Unknown pack	Parts (23)	No	No	No	No/No	Priority matrix	Parts (23)	Fundamental	No	No/No	No
Tan et al.	[24]	2021	Audi Q5 hybrid	Types (11)	Types (8)	Types	Fundamental	No/No	Disassembly task list	Types (19)	Partly detailed	Detailed	Fundamental/No	No
Hellmuth et al.	[23]	2021	Chevrolet Bolt	Parts (76)	No	No	Detailed	No/Yes	Disassembly graph	Types (19)	Detailed	Fundamental	Detailed/No	Fundamental
Baazouzi et al.	[34]	2021	Audi A3 hybrid	Parts (32)	No	No	Fundamental	No/Yes	3-step chromosome approach	Parts (77)	166 activities	Fundamental	Fundamental/No	No
Xiao et al.	[35]	2022	Unknown module	-	Parts (23)	No	No	No/Yes	Priority matrix	Types (10)	Fundamental	No	No/No	No
Kong et al.	[36]	2022	Chery EV	Parts (156)	No	No	Fundamental	No/Yes	Task priority diagram	Types (30)	180 activities	No	No/No	No
Wu et al.	[37]	2022	Tesla Model S	-	Types (12)	No	No	No/ Partly detailed	Task priority diagram	Parts (481)	Fundamental	No	No/No	No
Rosenberg et al.	[38]	2022	Unknown hybrid module	Types/ Parts (11)	No	No	Detailed	No/Yes	Disassembly graph	Types/ Parts (10)	57 activities	No	Fundamental/No	No
Zhan et al.	[39]	2023	Tesla battery	Types (22)	No	No	Fundamental	No/No	Improved northern goshawk optimization	Types (22)	Partly detailed	Fundamental	Fundamental/No	Fundamental
Cong et al.	[40]	2023	Unknown module	Parts (61)	No	No	Fundamental	No/No	Precedence graph	Parts (24)	Partly detailed	No	No/No	No

Table 1. *Cont.*

General Information				*Product-Related Information*					*Process-Related Information*					
Authors	**Ref.**	**Year**	**Specified Battery**	**Pack Detail Level**	**Module Detail Level**	**Cell Detail Level**	**Joint Connections**	**Position/ Quantity**	**Disassembly Sequence**	**Process Step**	**Activities**	**Operations**	**Tools/ Specifications**	**Access Angle/ Vectors**
Villagrossi et al.	[41]	2023	Fiat 500e	Parts (18)	Parts (17/29)	No	Detailed	No/Detailed	Disassembly graph	Types (24)	Detailed	No	Detailed/Detailed	Fundamental
Qu et al.	[42]	2024	Unknown hybrid pack	Parts (9)	No	No	Partly detailed	No/ Partly detailed	Decisional tree	Types (4)	115 activities	No	Partly detailed/ Detailed	No
Chu et al.	[43]	2024	Tesla Model S	-	Parts (44)	No	No	No/No	Priority diagram	Parts (44)	Fundamental	No	No/No	No
Hathaway et al.	[44]	2024	Mitsubishi Outlander hybrid	Types (11)	Types (2)	No	Fundamental	No/Detailed	Disassembly task list	Types (13)	68 activities	Yes	Yes/Fundamental	Partly detailed
Jiao et al.	[45]	2024	MB EQS580	Types (13)	No	No	No	No/No	Disassembly task list	Types (13)	Fundamental	No	No/No	Detailed
Rastegarpanah et al.	[46]	2024	Nissan Leaf	Fundamental	No	No	No	No/No	Flowchart tasks	-	Partly detailed	Partly detailed	No/No	No
Rettenmeier et al.	[47]	2024	-	Fundamental	-	-	-	-	-	-	Detailed	Detailed	Detailed/ Fundamental	Fundamental
Rettenmeier et al.	[48]	2024	-	-	-	-	-	-	Technology radar	Types (12)	Detailed	Fundamental	No/Fundamental	No
Wu et al.	[49]	2024	Unknown module	Types (9)	No	No	Fundamental	No/No	Knowledge graph	Types (39)	Fundamental	No	No/No	No
Yang et al.	[50]	2024	Unknown module	-	Types (25)	Yes	Fundamental	No/ Fundamental	Priority diagram	Types (24)	Fundamental	No	No/No	No

Fundamental: brief notation, which is not further specified; Partly detailed: certain individual aspects are described in detail and quantified; Detailed: the majority of all aspects targeted were described and quantified.

Although the literature frequently addresses specific approaches and concepts in detail and in some cases also illustrates these using specific battery systems, the general product- and process-related data are very diverse and only available in a great variety of detail. At the battery pack level, information on the components and their quantity in the battery pack is available in most of the battery disassembly-related publications analyzed, although the joint connections are only considered and quantified in detail in very few cases. Position data usually cannot be derived in detail from this information either. When considering the module and cell levels, the information provided in relation to the components is increasingly reduced. Another important point is the differentiation between component types and parts in the context of the disassembly view. While in some cases, one category of components is mentioned, in other cases, the specific number of parts is mentioned.

Similar findings were obtained for the process-related information. A wide variety of approaches were developed and applied to derive the disassembly sequence. While the level of detail for process steps in relation to components was considered extensively in detail or type-based, the level of information in relation to activities for loosening and separating joints and rudimentary operations such as the use and switching of tools varied greatly. The same applies to the identification of tools and their specifications. Information on access directions, angles or vectors, which are essential for automated execution, is only fundamentally illuminated in a few publications.

In conclusion, the scope and level of detail of the information available in the literature is not sufficient and is too diverse to be fully utilized as a basis for automated process design in battery disassembly. These findings support the underlying thesis of this article: generally valid and standardized definitions are required for the product- and process-related information in battery disassembly in order to build a database for automation.

EU Battery Directive and Battery Passport

The need for more battery-related information and the importance of realizing those data along the entire value chain for a more sustainable circular battery life cycle have also been recognized by legislators. The current proposal of the Battery Directive of the European Union (EU) implements the mandatory introduction of a battery passport for batteries from electric vehicles, light means of transport (LMT) such as electric scooters and industrial batteries over 2 kWh starting at the beginning of 2027 [51]. The fundamental objective of this new battery regulation is to encourage sustainability in the design and production of batteries and to reduce their environmental impact throughout their life cycle. Part of this includes promoting the circular economy of batteries by providing data that enable second life approaches and improves the quality and quantity of recycling [52]. Building on this legislative initiative, a research and industry collaboration is developing framework conditions and a concept for the implementation of such a battery passport, as specified by the EU for 2027 [53]. At present, there are minimum requirements for battery disassembly, which are being supplemented by extensions and will certainly be expanded in the future. These requirements include the part numbers of the components as well as exploded diagrams of the battery system/pack showing the location of the battery cells, disassembly sequences and type and number of fastening techniques. In addition, the battery passport should specify the tools required for disassembly and safety warnings in the event of the risk of damage due to process operation and the quantity and layout of the cells [51,53]. All in all, the accessibility of the different levels of information will vary depending on the type of data, ranging from generally accessible manufacturer data to specific information such as the exact material composition that can only be accessed by persons with a legitimate interest or authorities [54].

The current specifications of the data basis for the disassembly process within the battery passport are a first step in the right direction but will not be sufficient for comprehensive automation of the disassembly process. The extent to which the battery passport facilitates disassembly will depend largely on the quality and detail of the data provided. This includes product-related data on all relevant subcomponents and joints, as well as process-related information such as the disassembly sequence. These data must be precise in order to accurately schedule and execute individual operations on a granular level. In the following methodological approach, an initial proposal regarding possible relevant data and parameters for automated battery disassembly will be developed. This approach will be generated by analyzing the disassembly process of a relevant reference battery pack.

3. Methodological Approach and Results

To implement standardized parameter definitions, it is not only necessary to work out all the relevant data points, but it is also essential to classify these data in a structured form that can be applied to a wide variety of battery systems in the same manner. This objective results in a four-stage methodical approach, which is subdivided into the areas of disassembly depth evaluation, product and process data structure analysis and the set-up of an integrated data model approach. The disassembly depth is based on the disassembly sequence and differs from it in the sense that it can be used to identify components that have no dependencies on each other, which in turn opens parallelization potential for the execution of the disassembly. It also forms one of the links between the downstream product and process structure. As part of the design of the product and process structure, we first introduce general applicable definitions that will be used to derive the layouts of the structures. The layouts resulting from this step are then examined to derive all relevant disassembly data definitions. In the final stage, these data are combined in a model approach and prepared with regard to a possible integration into the battery passport or a local collection in the decentralized disassembly system. The individual stages are applied using an Audi e-tron 50 battery system (71 kWh).

3.1. Disassembly Priority Matrix and Depth Derivation

To derive the disassembly depth of the individual components of the battery system, the disassembly priority matrix approach introduced by Wegener et al. [27] was initially applied, with which the sum of the downstream components for each individual component was initially obtained using an evaluation scheme consisting of -1, 0 and 1. A classification of 0 indicates no dependencies between the two components. A value of -1 indicates that the compared component needs to be removed before the considered component. The sum of positive entries in a row results in a value that is in turn classified as a rank in the disassembly sequence. The complete disassembly priority matrix is shown in Figure 2. There are 17 different component types for the Audi e-tron battery system, whereby the busbars were divided into six separate types as they were present in different shapes. This in turn must be considered in any sensor-based vision approach in process automation. The combination of the module management controller (MMC) and holder was also found in two different configurations.

The existing approach has then been extended by incorporating the influence of independent components (marked with 0) on the sequence, whereby individual components have been moved to one level depending on the rank and independence. In addition, for the special case of the MMC together with its holders, it was considered that these can be removed together in one step. Once sequence-adjusted, this results in the disassembly depth of the individual components, which was then transferred to the product and process data structures.

		1	2	3	4	5	6	7	8	9	10	11	12	13	14	15	16	17	Total	Rank	Up	Depth
1	Battery Junction Box		0	1	1	1	1	1	1	1	1	1	1	1	1	1	1	1	15	1	1	1
2	Communication Interface Cover	0		1	1	1	1	1	1	1	1	1	1	1	1	1	1	1	15	1	1	1
3	Housing Cover	−1	−1		1	1	1	1	1	1	1	1	1	1	1	1	1	1	14	3	3	2
4	Wire Harness	−1	−1	−1		−1	0	−1	0	0	0	1	1	1	1	1	1	1	7	4	4	3
5	Module Management Controller A (MMC)	−1	−1	−1	1		0	1	0	0	0	0	1	1	1	0	1	1	7	4	4	3
6	Module Management Controller B (MMC)	−1	−1	−1	0	0		0	1	0	0	0	0	0	0	0	0	1	2	7	4[2]	3
7	MMC Holder A	−1	−1	−1	1	−1	0		0	0	0	0	1	1	1	0	1	1	6	6	4[1]	3
8	MMC Holder B	−1	−1	−1	0	0	−1	0		0	0	0	0	0	0	0	0	1	1	15	4[1,2]	3
9	HV Connector Cable	−1	−1	−1	0	0	0	0	0		0	0	0	0	0	0	1	1	2	7	9	4
10	Busbar (Type A)	−1	−1	−1	0	0	0	0	0	0		0	0	0	0	0	1	1	2	7	9	4
11	Busbar (Type B)	−1	−1	−1	−1	0	0	0	0	0	0		0	0	0	0	1	1	2	7	9	4
12	Busbar (Type C)	−1	−1	−1	−1	−1	0	−1	0	0	0	0		0	0	0	1	1	2	7	9	4
13	Busbar (Type D)	−1	−1	−1	−1	−1	0	−1	0	0	0	0	0		0	0	1	1	2	7	9	4
14	Busbar (Type E)	−1	−1	−1	−1	−1	0	−1	0	0	0	0	0	0		0	1	1	2	7	9	4
15	Busbar (Type F)	−1	−1	−1	−1	0	0	0	0	0	0	0	0	0	0		1	1	2	7	9	4
16	Battery Module	−1	−1	−1	−1	−1	−1	−1	−1	−1	−1	−1	−1	−1	−1	−1		1	1	15	16	5
17	Housing Tray	−1	−1	−1	−1	−1	−1	−1	−1	−1	−1	−1	−1	−1	−1	−1	−1		0	17	17	6

1: Can be merged and removed as one component (e.g. MMC and holder)
2: Rank upgrade possible, as there is no interference with preceding components of lower rank (indicated by the evaluation a0 for the respective component interface)

Figure 2. Disassembly priority matrix for the battery system of an Audi e-tron 50 (71 kWh) with extended disassembly depth derivation.

3.2. Disassembly-Related Product Definitions and Data Structure

To ensure a uniform understanding and a clear categorization within the overall assessment, the elements relevant to the product-related data structure, such as the battery system, its components and joints, are first defined with reference to dismantling and then transferred to an illustration of the product data structure in Figure 3.

A product system (PS) in the context of disassembly describes the accumulation of individual components that are arranged via the connection of their joints and thus form an overall system. For battery disassembly, the battery pack forms such an overall system.

A product component (PC), or simply component in the context of dismantling battery systems, is a clearly identifiable and functionally definable part within the overall system. It represents an independent unit that fulfills specific functions and can be assembled and disassembled as a whole. A component is connected to other components of the overall system by one or more joint connections. The battery module or the busbar used to connect the battery modules can be mentioned as an example of a component.

A joint connection (JC) is a physical or chemical connection that permanently or reversibly detaches two or more components of a system. In the context of disassembly, the joint refers to the various mechanisms used to hold the individual components of an overall system in place. Examples of joints include screws, plugs, rivets, adhesive or welded joints.

Following these definitions, the product data structure for the components connected to each other via joint connections can now be considered along the previously derived disassembly depth for the special case of the Audi e-tron 50 battery pack. The component links shown in Figure 3 illustrate the complex and multi-layered structure of a battery system at the pack level. The specific objective of disassembly is to remove the product components from the product system efficiently, which in turn involves separating the joint connections and subsequently removing the component. This procedure is now followed along the dismantling depths, whereby the potential for parallelization can already be identified from the illustration. The type and quantity of components and joints processed in the product data structure for the Audi e-tron are summarized in the subsequent Table 2 extended by weight shares and main materials. While the material can be a decision criterion for the faction to be assigned, the weight is also relevant for the end effector and robot selection in automated disassembly.

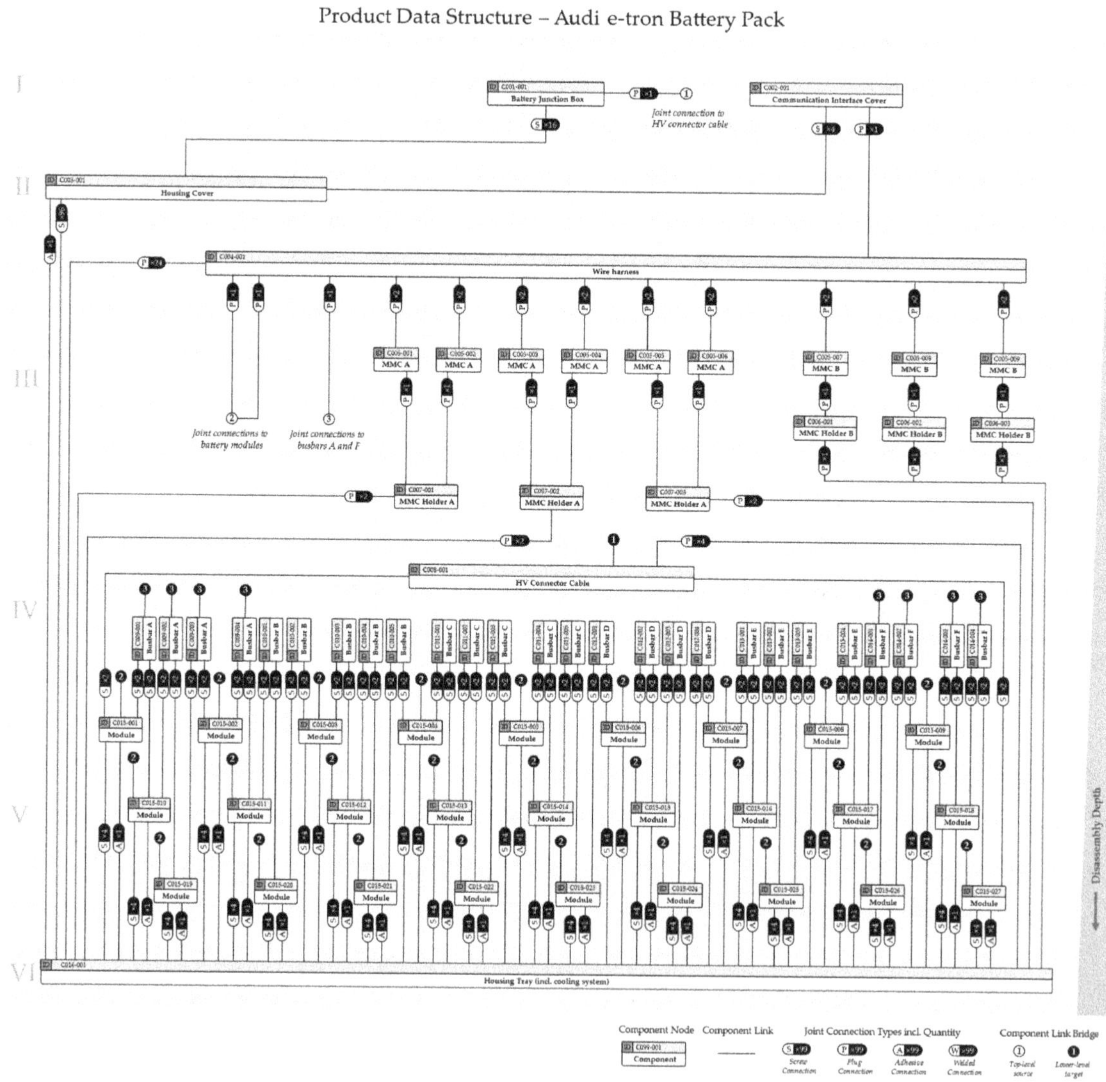

Figure 3. Product data structure of the Audi e-tron 50 battery system consisting of the components and their connections characterized by joints, arranged along the disassembly depth levels.

Table 2. Overview of the components and joints and their quantitative links at the product level for the Audi e-tron battery pack.

Component	QTY	Joint Connection	Joint Quantity [1]	Share in kg	Materials (Main Share)	Linked Subsequent Components [2]
Battery Junction Box	1	Screw connection Plug connection	16 1	12.9	Aluminum/ Others	Housing Cover HV Connector Cable
Communication Interface Cover	1	Screw connection Plug connection	4 1	0.2	Plastics	Housing Cover Wire Harness
Housing Cover	1	Screw connection Adhesive connection	96 1	14.3	Aluminum	Housing Tray
Wire Harness	1	Plug connection	24	0.7		MMC A/B Battery Module Housing Tray
Module Management Controller A (MMC)	6	Plug connection	1	0.6	Plastics/ Others	MMC Holder A
Module Management Controller B (MMC)	3	Plug connection	1	0.3	Plastics/ Others	MMC Holder B
MMC Holder A	3	Plug connection	2	0.5	Plastics	Housing Tray
MMC Holder B	3	Plug connection	1	0.2	Plastics	Housing Tray
HV Connector Cable	1	Screw connection Plug connection	4 4	1.5	Copper/ Others	Battery Module Housing Tray
Busbar (Types A to F)	26	Screw connection	4	3.8	Copper/ Plastics	Battery Module
Battery Module	27	Screw connection Adhesive connection	4 1	364.5		Housing Tray
Housing Tray [3]	1	-	-	~130.2	Aluminum/Metal	-

[1] Quantity of joint connections per component; [2] Subsequent in relation to the disassembly depth/sequence of disassembly; [3] Scope up to the tray (subsequent cooling system and housing components were not considered).

In the case of the Audi e-tron, the majority of the joint connections are screw and plug connections, while adhesive connections in the form of the thermal interface material and the sealing are also present, particularly in the context of module removal and when removing the housing cover. To remove the cover of the pack, it is necessary to first remove the battery junction box and the communication interface cover. This is necessary because the cable connections would otherwise prevent the cover from being removed. The cover itself is attached to the housing with 96 screws and has a sealing that adheres quite firmly. Once the screw and adhesive connections have been separated, the component housing cover can be removed.

In addition to the type and quantity, the position of the components and joints must also be known for the actual execution. Furthermore, tool information and tool specifications are required to separate the joints, which then can be used in combination with access vectors for the specific process sequence in the case of an automated procedure.

3.3. Disassembly-Related Process Definitions and Data Structure

Similar to the product analysis, the elements of the process were also defined in a uniform manner and linked to the respective product data. A distinction was made between process step, disassembly activity and disassembly operation.

A process step in the context of disassembly (DPS) refers to the detachment and removal of a specific component to be separated and removed from the overall product system. Each process step therefore represents a clearly defined, self-contained sequence that comprises one or more disassembly activities.

A disassembly activity (DA) refers to the separation of one or more specific joint connections of the component focused on the respective process step or the removal of the component itself. The separation can be either destructive and therefore irreversible or non-destructive.

The disassembly operation (DO) forms the third level in the process structure and describes primitive actions, all of which add up to a process activity. Examples of this type of primitive action can be physical movements (e.g., moving the robot from A to B), the use of tools (e.g., gripping, unscrewing) and the opening and closing of a quick-change system for changing tools.

Based on these definitions, and in conjunction with the existing product data, an overall process sequence is achieved along the disassembly depth as shown in Figure 4, whereby process steps on one level can also be swapped in their sequence. Level III forms an exception due to the multi-sided links.

For the removal of the housing cover, this means that the process consists of the sum of the disassembly activities of loosening the screws, which is repeated 96 times, the separation of the connection between the housing and cover in the form of the sealing compound and the removal of the cover itself. Using the same analogy, a disassembly activity could be represented as the sum of the disassembly operations. For the screw connection, in an automated approach, this would be, for example, (1) the movement from the initial position to the end effector exchange station (which in turn can be expressed as a chain of movements); (2) the application of the screwdriving tool; (3) the movement to the screw position (whereby a delta position must be approached here depending on the tool); (4) the execution of the unscrewing operation; (5) the removal or picking up of the screw; (6) the movement to a target position within a faction; and (7) the depositing of the screw within the faction.

Similar to the product data structure, dependencies, data and information can also be derived for the process view, which are required for automated execution, particularly at the level of a disassembly operation. In addition to the positions and access vectors already mentioned in the product structure for components and some types of joints, a target position, e.g., derivable via an assigned fraction or the material of the component/joint, is also required to remove it cleanly from the product system and, based on the overall goal of disassembly, to provide homogeneous material flows for recycling. In addition, each component and joint must be assigned a tool and the specifications required for the utilization of the tool.

Using the information and data structures derived from the example of the Audi e-tron battery pack, the subsequent objective is to establish a generalized data flow within the process representation for automated disassembly. Due to the fact that an interaction of hardware and software systems takes place within the automated disassembly, and it can therefore be referred to as a cyber-physical system (CPS), the hardware-side resources or mechanisms as well as any control instances and data flows must also be taken into account alongside the input and output data flows in the modeling. These inputs, outputs, controls and mechanisms (ICOMs) are therefore transferred into a modeling approach in Figure 5 using the Integration Definition for Process Modeling (IDEF) approach for the general process consideration in the automated disassembly of battery systems.

Part A of Figure 5 references the process step level. On the left-hand side are the input data derived from the components, which are necessary to execute the process step. In addition to the position data already discussed (component position and target/deposit position) and a suitable access vector, the joint connections also flow into the process together with their respective referenced data points and the current progress state. While the joint data are used in the detailed view B of Figure 5 for the different activities as an input, the progress state is required for every step to ensure the correct sequence according to the sequence plan (controls). Depending on the system design and procedure, reference data also flow into the process, which are used as a control instance in the context of, e.g., sensor recordings (e.g., 2D or 3D recordings of a vision sensor). The type and scope of these data vary, which is why it is generally described here as reference data. Various types of mechanisms are required to ensure the process execution of the CPS. In addition to any sensors, one or more tools and their process-related specifications are used. Depending on the automation approach, robots can be used for execution, for example. These and other approaches are generally summarized under the term automation system. The output of the process step is the respective component and, depending on the type, the individual joint connections and the new progress state.

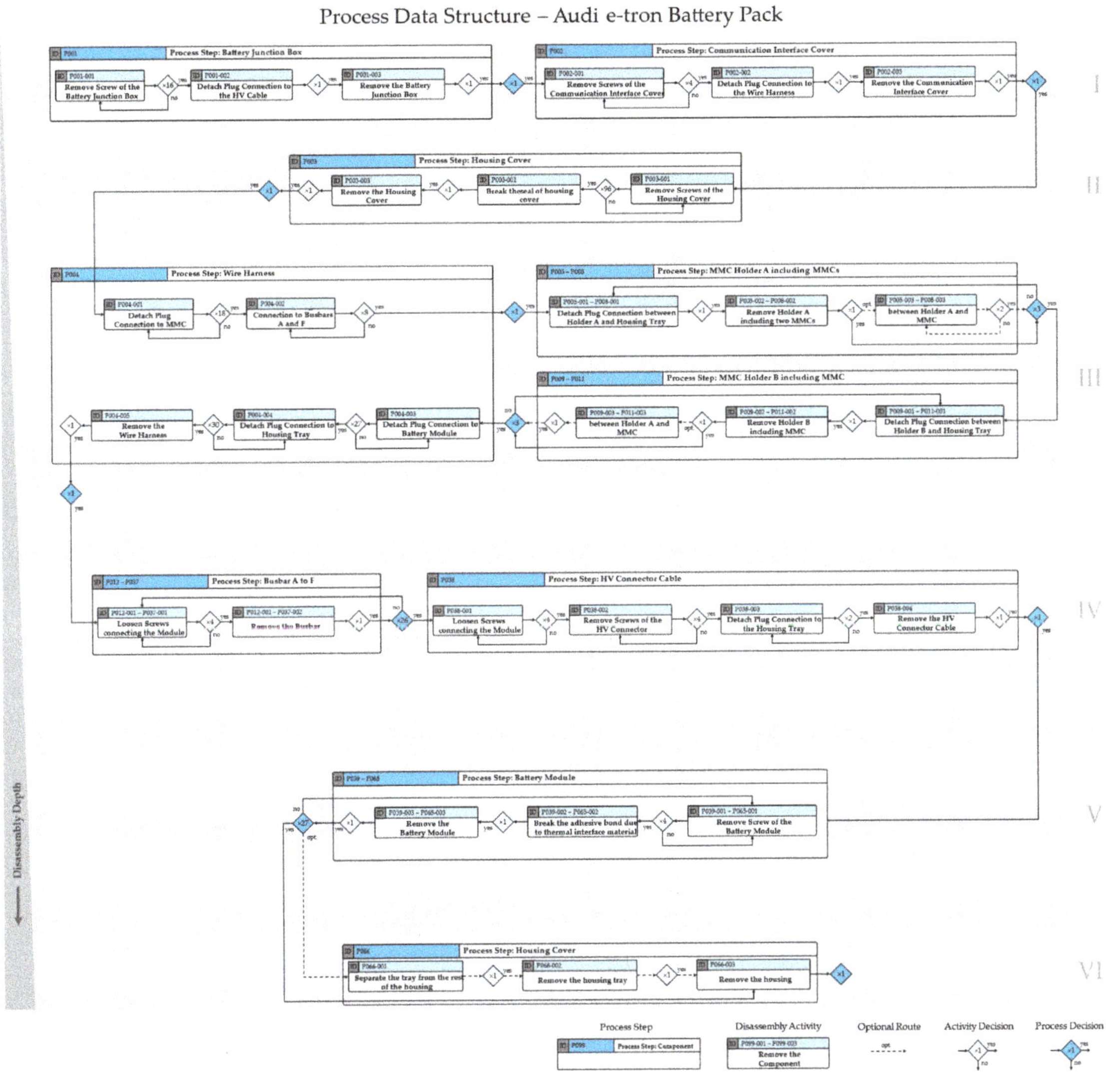

Figure 4. Process data structure for the disassembly of the Audi e-tron 50 battery system consisting of the sequence of process steps and their respective assigned process activities, arranged along the disassembly depth levels.

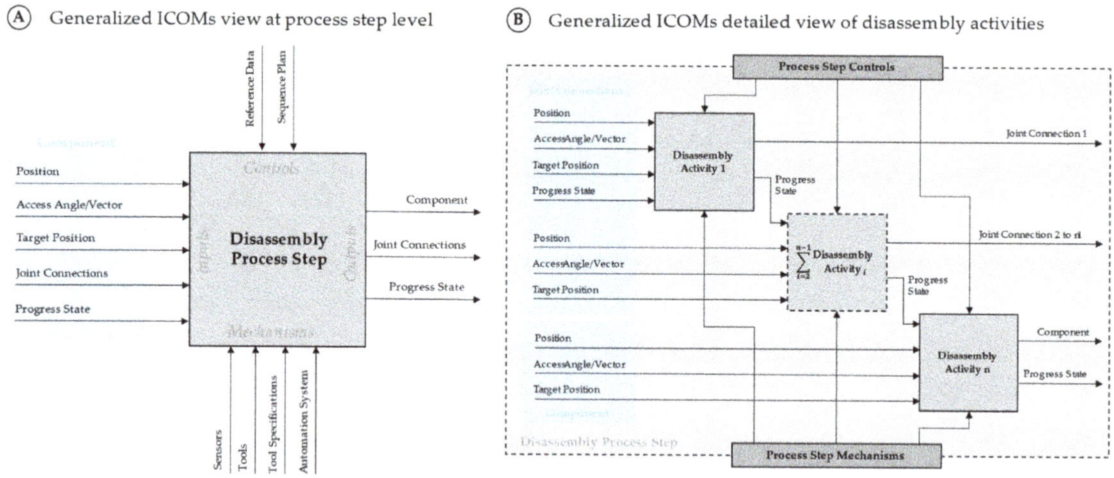

Figure 5. Generalized process flow diagram according to IDEF0 with all basic inputs, controls, mechanisms and outputs (ICOMs) that are required to carry out an automated battery disassembly step derived from the product and process data structure.

Part B of Figure 5 is structured in the same way, whereby the respective disassembly activities are executed here, each of which references the data of the joints as inputs before the component itself is removed from the product system in a final step. The progress state depends on the previous activity. The type of joint determines whether or not it will be carried out as a physical output from the activity step. Adhesive and welded joints are usually separated and then remain as part of the respective component for removal from the product system. This can also apply in part to plug connections, depending on the disassembly technology selected. Depending on the hardware and software structures within the disassembly system, the ICOMs must be expanded to include additional elements. The approach therefore serves as a generalized basis. For combined tools that process several joint connections simultaneously, the sequencing would have to be parallelized.

3.4. *Product and Process Data Model Approach*

In conclusion, a data model approach can be developed from the cooperations of the product and process data structure, the respective definitions and the general process approach derived from the specific example of the Audi e-tron. On the product side, this is made up of the JC, PC and the battery pack as a PS and is supplemented on the process side by the DO, DA and DPS. The two interlinked data hierarchies can be brought together in a higher-level battery model, which describes, for the example of the Audi pack, the e-tron battery model as it exists for a vehicle generation or time period, whereas a specific installed Audi e-tron battery would represent the battery pack.

The structure described is visualized in Figure 6 as a data structure model approach and contains the data properties identified in the previous stages.

The data type designations of the respective data properties required for a software-side implementation are also part of the structure as well as a categorization of possible data sources for the implementation based on the initial consideration of the battery passport. Based on the current state of discussion for the battery passport, the data that will be provided in the prototype application of the passport are highlighted (green). There are also data that are partially available or where the level of detail could not yet be fully derived (blue) and where there are no verified data sources present (white) [53]. Some data properties can also be generated via the disassembly plant (DP) and its structure (e.g., during the process via sensors). These are mostly located at the lower hierarchical levels. At present, this means that the upcoming battery passport will not provide sufficient data for the automation of battery disassembly. The more data points listed that are integrated into the battery passport in the future, the greater the potential is for implementing automated battery disassembly for a variety of different battery pack designs. For independent use without external sources such as the battery passport, there is also the possibility of an integrated teach-in strategy for new battery models, which would require systematic recording of the data during a manual disassembly procedure.

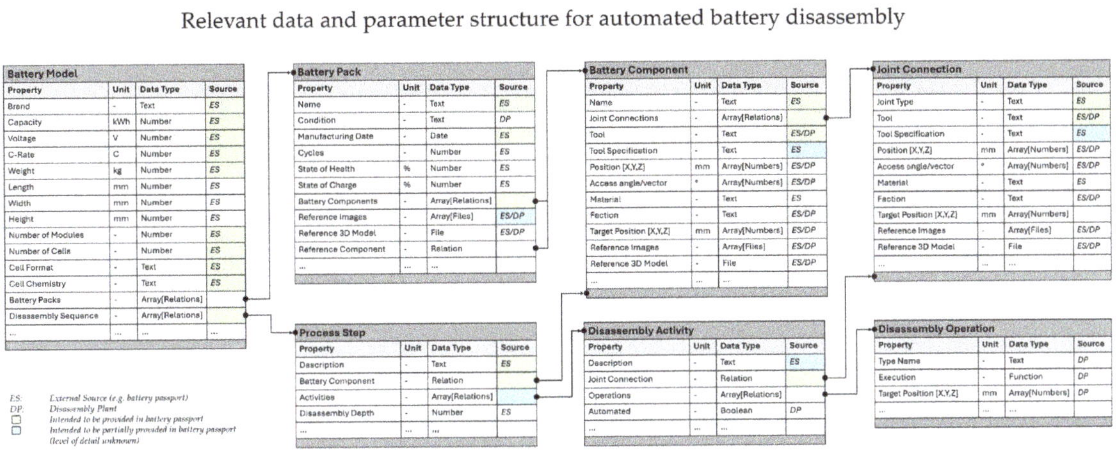

Figure 6. Derived data structure model approach for automated battery disassembly based on the integration of product and process structure with the boundary conditions of automation and the battery passport as a partial data source.

The data model approach shown includes further data properties at the battery model level and in parts for the subsequent levels, which are relevant for the selection of dismantling technologies and the related interface processes. Capacity, c-rates, state of charge (SoC) and voltage level, for example, are important parameters for executing the discharging of the battery pack. The cell chemistry, the charge and discharge cycles already performed, the state of health (SoH) and the condition of the battery pack (e.g., mechanical deformation in the case of crash batteries) can be used to derive decisions regarding reuse strategies and subsequently the use of destructive or non-destructive disassembly methods, for example. The condition also influences the disassembly sequence, the tools required and the fundamental assessment of the successful automated execution of a disassembly step. Depending on the respective type, the disassembly operation either requires a target position in addition to the current position (e.g., movement operations) or a type of execution (e.g., unscrewing, gripping, end effector release, etc.). Depending on the information required, associated classes may be needed as inputs. Extensions to this structure are conceivable for both the product-side and the process-side class structure.

4. Conclusions

This article focused on the identification and accessibility of product- and process-related data in the context of the automated disassembly of battery systems. For this purpose, the current status and scope of battery disassembly-specific data in the literature were first analyzed and discussed. The result showed a very diverse level of information depth, which also varied in the type and scope of its structures depending on the respective disassembly topic being discussed. This led to the conclusion that such data were not widely applicable for a general utilization in connection with automated disassembly processes. The battery passport, which was also considered, is also not yet a sufficient data source at the current planning stage for the specific implementation of automated disassembly steps.

Based on these findings, a methodical approach was first applied to establish a product and process data structure by deriving the disassembly sequence and depth for the specific case of an Audi e-tron battery pack. Product-related elements and relationships were identified and defined along the disassembly depth. Based on this, the process-side level was also analyzed according to the same principle. With the identified data, structures, relationships and definitions, a generalized process model was designed and then transferred into a holistic data structure model approach.

On the product side, this approach comprises the joints, components and the battery pack as a product system and is extended by the process step associated with the component, the disassembly activity referencing the joint and the level of rudimentary disassembly operations. This hierarchical structure is complemented at the top by the battery model. Data properties that are relevant for the implementation in the context of automated disassembly and any interfaces have been assigned to all the elements mentioned.

While this model approach forms a basis for the planning of an automated disassembly system and has largely been applied within a research project for individual automated disassembly steps, an overall validation and the further development of the approach into a fully comprehensive CPS are still required. Furthermore, the battery passport should be further developed as a source of information and data, especially in the context of the challenges of automated battery disassembly as an important step towards establishing homogeneous material flows for recycling. The interaction of a detached teach-in strategy for recording the relevant data together with the current structural design of the model approach must also be further investigated to ensure that the automation of battery disassembly required in the future can be used economically and comprehensively. For the future further development of the approach presented here, special attention should be paid to data deviations in the case of damaged batteries, deformations, external influences or changes due to repairs, as these circumstances have so far only been included in the data via the items' condition and possible reference images and data. On the process side, there need to be process alternatives or fallback approaches for these cases.

Author Contributions: Conceptualization, D.K.; methodology, D.K.; validation, M.F., J.G. and A.K.; formal analysis, D.K. and M.F.; investigation, D.K. and P.E.; data curation, D.K. and P.E.; writing—original draft preparation, D.K.; writing—review and editing, M.F., J.G., H.H.H. and A.K.; visualization, D.K.; supervision, H.H.H. and A.K.; project administration, A.K.; funding acquisition, H.H.H. and A.K. All authors have read and agreed to the published version of the manuscript.

Funding: The projects within which this research was conducted are funded by the German Federal Ministry of Education and Research under the funding codes 03XP0314C (DemoSens) and 03XP0552C (DemoRec). The responsibility for the content of this publication lies with the author.

Data Availability Statement: The data presented in this study are available on request from the corresponding author. The data are not publicly available due to sensitive manufacturer and company information.

Acknowledgments: A special thank you goes to all current and past project partners in both projects, the student research assistants and the chair management for their support in realizing this research.

Conflicts of Interest: The authors declare no conflicts of interest.

References

1. Baars, J.; Domenech, T.; Bleischwitz, R.; Melin, H.E.; Heidrich, O. Circular economy strategies for electric vehicle batteries reduce reliance on raw materials. *Nat. Sustain.* **2021**, *4*, 71–79. [CrossRef]
2. Sommerville, R.; Zhu, P.; Rajaeifar, M.A.; Heidrich, O.; Goodship, V.; Kendrick, E. A qualitative assessment of lithium ion battery recycling processes. *Resour. Conserv. Recycl.* **2021**, *165*, 105219. [CrossRef]
3. Edge, J.S.; O'Kane, S.; Prosser, R.; Kirkaldy, N.D.; Patel, A.N.; Hales, A.; Ghosh, A.; Ai, W.; Chen, J.; Yang, J.; et al. Lithium ion battery degradation: What you need to know. *Phys. Chem. Chem. Phys.* **2021**, *23*, 8200–8221. [CrossRef]
4. Teng, J.-H.; Chen, R.-J.; Lee, P.-T.; Hsu, C.-W. Accurate and Efficient SOH Estimation for Retired Batteries. *Energies* **2023**, *16*, 1240. [CrossRef]
5. Kampker, A.; Heimes, H.H.; Offermanns, C.; Frank, M.; Klohs, D.; Nguyen, K. Prediction of Battery Return Volumes for 3R: Remanufacturing, Reuse, and Recycling. *Energies* **2023**, *16*, 6873. [CrossRef]
6. Hu, X.; Xu, L.; Lin, X.; Pecht, M. Battery Lifetime Prognostics. *Joule* **2020**, *4*, 310–346. [CrossRef]
7. Zhao, Y.; Pohl, O.; Bhatt, A.I.; Collis, G.E.; Mahon, P.J.; Rüther, T.; Hollenkamp, A.F. A Review on Battery Market Trends, Second-Life Reuse, and Recycling. *Sustain. Chem.* **2021**, *2*, 167–205. [CrossRef]
8. Gerbers, R.; Wegener, K.; Dietrich, F.; Dröder, K. Safe, Flexible and Productive Human-Robot-Collaboration for Disassembly of Lithium-Ion Batteries. *Recycl. Lithium-Ion Batter.* **2018**, 99–126. [CrossRef]
9. Lander, L.; Tagnon, C.; Nguyen-Tien, V.; Kendrick, E.; Elliott, R.J.; Abbott, A.P.; Edge, J.S.; Offer, G.J. Breaking it down: A techno-economic assessment of the impact of battery pack design on disassembly costs. *Appl. Energy* **2023**, *331*, 120437. [CrossRef]
10. Meng, K.; Xu, G.; Peng, X.; Youcef-Toumi, K.; Li, J. Intelligent disassembly of electric-vehicle batteries: A forward-looking overview. *Resour. Conserv. Recycl.* **2022**, *182*, 106207. [CrossRef]
11. Li, W.; Peng, Y.; Zhu, Y.; Pham, D.T.; Nee, A.; Ong, S.K. End-of-life electric vehicle battery disassembly enabled by intelligent and human-robot collaboration technologies: A review. *Robot. Comput. Integr. Manuf.* **2024**, *89*, 102758. [CrossRef]
12. Glöser-Chahoud, S.; Huster, S.; Rosenberg, S.; Baazouzi, S.; Kiemel, S.; Singh, S.; Schneider, C.; Weeber, M.; Miehe, R.; Schultmann, F. Industrial disassembling as a key enabler of circular economy solutions for obsolete electric vehicle battery systems. *Resour. Conserv. Recycl.* **2021**, *174*, 105735. [CrossRef]
13. Sommerville, R.; Shaw-Stewart, J.; Goodship, V.; Rowson, N.; Kendrick, E. A review of physical processes used in the safe recycling of lithium ion batteries. *Sustain. Mater. Technol.* **2020**, *25*, e00197. [CrossRef]
14. Choux, M.; Marti Bigorra, E.; Tyapin, I. Task Planner for Robotic Disassembly of Electric Vehicle Battery Pack. *Metals* **2021**, *11*, 387. [CrossRef]
15. Zorn, M.; Ionescu, C.; Klohs, D.; Zähl, K.; Kisseler, N.; Daldrup, A.; Hams, S.; Zheng, Y.; Offermanns, C.; Flamme, S.; et al. An Approach for Automated Disassembly of Lithium-Ion Battery Packs and High-Quality Recycling Using Computer Vision, Labeling, and Material Characterization. *Recycling* **2022**, *7*, 48. [CrossRef]
16. Choux, M.; Pripp, S.W.; Kvalnes, F.; Hellström, M. To shred or to disassemble—A techno-economic assessment of automated disassembly vs. shredding in lithium-ion battery module recycling. *Resour. Conserv. Recycl.* **2024**, *203*, 107430. [CrossRef]
17. Aghabali, I.; Bauman, J.; Kollmeyer, P.J.; Wang, Y.; Bilgin, B.; Emadi, A. 800-V Electric Vehicle Powertrains: Review and Analysis of Benefits, Challenges, and Future Trends. *IEEE Trans. Transp. Electrific.* **2021**, *7*, 927–948. [CrossRef]
18. Beghi, M.; Braghin, F.; Roveda, L. Enhancing Disassembly Practices for Electric Vehicle Battery Packs: A Narrative Comprehensive Review. *Designs* **2023**, *7*, 109. [CrossRef]
19. Gerlitz, E.; Greifenstein, M.; Kaiser, J.-P.; Mayer, D.; Lanza, G.; Fleischer, J. Systematic Identification of Hazardous States and Approach for Condition Monitoring in the Context of Li-ion Battery Disassembly. *Procedia CIRP* **2022**, *107*, 308–313. [CrossRef]
20. Jia, Y.; Liu, B.; Hong, Z.; Yin, S.; Finegan, D.P.; Xu, J. Safety issues of defective lithium-ion batteries: Identification and risk evaluation. *J. Mater. Chem. A* **2020**, *8*, 12472–12484. [CrossRef]
21. Kampker, A.; Heimes, H.H.; Offermanns, C.; Sasse, K.; Frieges, M.H.; Spath, B. Domain based product architecture approach for innovative battery system design. In Proceedings of the 2022 International Symposium on Electromobility (ISEM) IEEE, Puebla, Mexico, 7–19 October 2022; pp. 1–6.
22. Naguib, M.; Kollmeyer, P.; Emadi, A. Lithium-Ion Battery Pack Robust State of Charge Estimation, Cell Inconsistency, and Balancing: Review. *IEEE Access* **2021**, *9*, 50570–50582. [CrossRef]
23. Hellmuth, J.F.; DiFilippo, N.M.; Jouaneh, M.K. Assessment of the automation potential of electric vehicle battery disassembly. *J. Manuf. Syst.* **2021**, *59*, 398–412. [CrossRef]
24. Tan, W.J.; Chin, C.M.M.; Garg, A.; Gao, L. A hybrid disassembly framework for disassembly of electric vehicle batteries. *Int. J. Energy Res.* **2021**, *45*, 8073–8082. [CrossRef]
25. Scott, S.; Islam, Z.; Allen, J.; Yingnakorn, T.; Alflakian, A.; Hathaway, J.; Rastegarpanah, A.; Harper, G.D.; Kendrick, E.; Anderson, P.A.; et al. Designing lithium-ion batteries for recycle: The role of adhesives. *Next Energy* **2023**, *1*, 100023. [CrossRef]
26. Klohs, D.; Offermanns, C.; Heimes, H.; Kampker, A. Automated Battery Disassembly—Examination of the Product- and Process-Related Challenges for Automotive Traction Batteries. *Recycling* **2023**, *8*, 89. [CrossRef]

27. Wegener, K.; Andrew, S.; Raatz, A.; Dröder, K.; Herrmann, C. Disassembly of Electric Vehicle Batteries Using the Example of the Audi Q5 Hybrid System. *Procedia CIRP* **2014**, *23*, 155–160. [CrossRef]
28. Cerdas, F.; Gerbers, R.; Andrew, S.; Schmitt, J.; Dietrich, F.; Thiede, S.; Dröder, K.; Herrmann, C. Disassembly Planning and Assessment of Automation Potentials for Lithium-Ion Batteries. *Recycl. Lithium-Ion Batter.* **2018**, 83–97. [CrossRef]
29. Gumanova, V.; Sobotova, L. Proposal for Disassembly of Electric Vehicle Batteries used in the Volkswagen Jetta Hybrid System. In Proceedings of the 2019 International Council on Technologies of Environmental Protection (ICTEP), Starý Smokovec, Slovakia, 23–25 October 2019; pp. 100–106, ISBN 978-1-7281-4925-7.
30. Alfaro-Algaba, M.; Ramirez, F.J. Techno-economic and environmental disassembly planning of lithium-ion electric vehicle battery packs for remanufacturing. *Resour. Conserv. Recycl.* **2020**, *154*, 104461. [CrossRef]
31. Gentilini, L.; Mossali, E.; Angius, A.; Colledani, M. A safety oriented decision support tool for the remanufacturing and recycling of post-use H&EVs Lithium-Ion batteries. *Procedia CIRP* **2020**, *90*, 73–78. [CrossRef]
32. Rallo, H.; Benveniste, G.; Gestoso, I.; Amante, B. Economic analysis of the disassembling activities to the reuse of electric vehicles Li-ion batteries. *Resour. Conserv. Recycl.* **2020**, *159*, 104785. [CrossRef]
33. Ke, Q.; Zhang, P.; Zhang, L.; Song, S. Electric Vehicle Battery Disassembly Sequence Planning Based on Frame-Subgroup Structure Combined with Genetic Algorithm. *Front. Mech. Eng.* **2020**, *6*, 104461. [CrossRef]
34. Baazouzi, S.; Rist, F.P.; Weeber, M.; Birke, K.P. Optimization of Disassembly Strategies for Electric Vehicle Batteries. *Batteries* **2021**, *7*, 74. [CrossRef]
35. Xiao, J.; Anwer, N.; Li, W.; Eynard, B.; Zheng, C. Dynamic Bayesian network-based disassembly sequencing optimization for electric vehicle battery. *CIRP J. Manuf. Sci. Technol.* **2022**, *38*, 824–835. [CrossRef]
36. Kong, S.; Zhang, Y.; Liu, W. Parallel Disassembly Sequence Planning of Retired Lithium-ion-battery Pack based on Heuristic Algorithm. *J. Phys. Conf. Ser.* **2022**, *2254*, 12010. [CrossRef]
37. Wu, T.; Zhang, Z.; Yin, T.; Zhang, Y. Multi-objective optimisation for cell-level disassembly of waste power battery modules in human-machine hybrid mode. *Waste Manag.* **2022**, *144*, 513–526. [CrossRef] [PubMed]
38. Rosenberg, S.; Huster, S.; Baazouzi, S.; Glöser-Chahoud, S.; Al Assadi, A.; Schultmann, F. Field Study and Multimethod Analysis of an EV Battery System Disassembly. *Energies* **2022**, *15*, 5324. [CrossRef]
39. Zhan, C.; Zhang, X.; Tian, G.; Pham, D.T.; Ivanov, M.; Aleksandrov, A.; Fu, C.; Zhang, J.; Wu, Z. Environment-oriented disassembly planning for end-of-life vehicle batteries based on an improved northern goshawk optimisation algorithm. *Environ. Sci. Pollut. Res. Int.* **2023**, *30*, 47956–47971. [CrossRef] [PubMed]
40. Cong, L.; Zhou, K.; Liu, W.; Li, R. Retired Lithium-Ion Battery Pack Disassembly Line Balancing Based on Precedence Graph Using a Hybrid Genetic-Firework Algorithm for Remanufacturing. *J. Manuf. Sci. Eng.* **2023**, *145*, 051007. [CrossRef]
41. Villagrossi, E.; Dinon, T. Robotics for electric vehicles battery packs disassembly towards sustainable remanufacturing. *J. Remanuf.* **2023**, *13*, 355–379. [CrossRef]
42. Qu, M.; Pham, D.T.; Altumi, F.; Gbadebo, A.; Hartono, N.; Jiang, K.; Kerin, M.; Lan, F.; Micheli, M.; Xu, S.; et al. Robotic Disassembly Platform for Disassembly of a Plug-In Hybrid Electric Vehicle Battery: A Case Study. *Automation* **2024**, *5*, 50–67. [CrossRef]
43. Chu, M.; Chen, W. Multi-manned disassembly line balancing problems for retired power batteries based on hyper-heuristic reinforcement. *Comput. Ind. Eng.* **2024**, *194*, 110400. [CrossRef]
44. Hathaway, J.; Contreras, C.A.; Stolkin, R.; Asif, M.E.; Rastegarpanah, A. *Technoeconomic Assessment of Electric Vehicle Battery Disassembly—Challenges and Opportunities from a Robotics Perspective*; Elsevier: Amsterdam, The Netherlands, 2024.
45. Jiao, J.; Feng, G.; Yuan, G. Research on the Human–Robot Collaborative Disassembly Line Balancing of Spent Lithium Batteries with a Human Factor Load. *Batteries* **2024**, *10*, 196. [CrossRef]
46. Rastegarpanah, A.; Mineo, C.; Contreras, C.A.; Aflakian, A.; Paragliola, G.; Stolkin, R. Electric Vehicle Battery Disassembly Using Interfacing Toolbox for Robotic Arms. *Batteries* **2024**, *10*, 147. [CrossRef]
47. Rettenmeier, M.; Sauer, A.; Möller, M. Laser-based battery pack disassembly: A compact benchmark analysis for separation technologies. *Procedia CIRP* **2024**, *122*, 181–186. [CrossRef]
48. Rettenmeier, M.; Möller, M.; Sauer, A. Disassembly technologies of end-of-life automotive battery packs as the cornerstone for a circular battery value chain: A process-oriented analysis. *Resour. Conserv. Recycl.* **2024**, *209*, 107786. [CrossRef]
49. Wu, H.; Jiang, Z.; Zhu, S.; Zhang, H. A Knowledge Graph Based Disassembly Sequence Planning for End-of-Life Power Battery. *Int. J. Precis. Eng. Manuf. Green Technol.* **2024**, *11*, 849–861. [CrossRef]
50. Yang, S.; Zhuo, X.; Ning, W.; Xia, X.; Huang, Y. Integrated Risk-Aware Smart Disassembly Planning for Scrap Electric Vehicle Batteries. *Energies* **2024**, *17*, 2946. [CrossRef]
51. The European Parliament and the Council. *Regulation (EU) 2023/1542 of the European Parliament and of the Council of 12 July 2023 Concerning Batteries and Waste Batteries, Amending Directive 2008/98/EC and Regulation (EU) 2019/1020 and Repealing Directive 2006/66/EC*; European Parliament and the Council: Luxembourg, 2024.

52. Gianvincenzi, M.; Marconi, M.; Mosconi, E.M.; Favi, C.; Tola, F. Systematic Review of Battery Life Cycle Management: A Framework for European Regulation Compliance. *Sustainability* **2024**, *16*, 10026. [CrossRef]
53. Global Battery Alliance. Circular Design Rulebook. Available online: https://www.globalbattery.org/ (accessed on 20 November 2024).
54. Butera, A.; Gatteschi, V. EU Battery Regulation and the Role of Blockchain in Creating Digital Battery Passports. In Proceedings of the 2024 6th Conference on Blockchain Research & Applications for Innovative Networks and Services (BRAINS), Berlin, Germany, 8–11 October 2024; pp. 1–4. [CrossRef]

recycling

MDPI

Review

Pathways to Circular Economy for Electric Vehicle Batteries

Subin Antony Jose, Lyndsey Dworkin, Saihan Montano, William Charles Noack, Nick Rusche, Daniel Williams and Pradeep L. Menezes *

Department of Mechanical Engineering, University of Nevada-Reno, Reno, NV 89557, USA; subinj@unr.edu (S.A.J.); lyndseydworkin@gmail.com (L.D.); saihanm@unr.edu (S.M.); willnoack22@gmail.com (W.C.N.); nick.rusche123@gmail.com (N.R.); danielwilliams@unr.edu (D.W.)

* Correspondence: pmenezes@unr.edu

Abstract: The global shift towards sustainability is driving the electrification of transportation and the adoption of clean energy storage solutions, moving away from internal combustion engines. This transition significantly impacts lithium-ion battery production in the electric vehicle (EV) market. This paper summarizes specialized topics to highlight regional differences and specific challenges related to electric batteries, focusing on how pollution from gas consumption, distribution, usage, and lithium production affects society. EV batteries offer promising opportunities for a sustainable future, considering their economic and environmental impacts and the importance of understanding their lifecycle. This analysis delves into the recovery of materials and various methods for extracting lithium and manufacturing EV batteries. Efficient lithium recovery is crucial and globally significant, with liquid extraction presenting a more environmentally friendly option. By addressing these challenges, this paper provides an overview of the rationale behind supporting the future of EVs.

Keywords: circular economy; Li-ion battery; recovery; recycling; reuse

1. Introduction

1.1. Overview of Electric Vehicles and Battery Technology

Electric vehicle (EV) batteries are rapidly becoming one of the most significant products of our time. As the climate crisis intensifies, it is imperative to find solutions that reduce greenhouse gas emissions. Traditional vehicles with internal combustion engines are major contributors to these emissions. In contrast, EVs are seen as a viable alternative since they emit no greenhouse gases during operation. The key to enabling EVs to function without internal combustion engines lies in the battery. The lithium-ion battery is the most popular choice for EVs due to its compact size, lightweight nature, and extremely high energy density [1]. Optimizing lithium-ion battery technology is crucial for enhancing the efficiency and performance of EVs, paving the way for a sustainable future.

1.1.1. Growth of the Electric Vehicle Market

The demand for EVs is rapidly increasing and is projected to soar in the coming decades, as illustrated in Figure 1. The global electric vehicle market is expected to grow significantly over the next decade. According to the International Energy Agency (IEA), the number of electric vehicles on the road is projected to reach 145 million by 2030 under current policy scenarios, with the potential to reach 230 million in an accelerated scenario. This growth is driven by government policies, declining battery costs, and increased consumer demand for cleaner transportation options [2]. Several factors contribute to this trend, with financial incentives being a major driving force. Initially, EVs were prohibitively expensive, but as technology has advanced and manufacturing processes have become more efficient, costs have steadily decreased. Additionally, the unreliability and rising costs of gasoline make EVs an increasingly attractive option, despite their higher upfront cost [3,4]. Moreover, societal trends are shifting towards greater environmental consciousness. Many people are becoming more aware of their contributions to climate change and feel good about driving vehicles with a reduced carbon footprint [5]. Many countries are aiming to reduce or eliminate the production of traditional vehicles while encouraging the manufacturing and adoption of EVs. The UK, France, Germany, the Netherlands, and several other countries have committed to the objectives of the Paris Agreement and are developing strategies to phase out the production of internal combustion engine vehicles (ICEVs) by 2040 [6]. Governments are also introducing supportive measures, such as purchase incentives and electricity subsidies, to encourage consumers to choose electric vehicles (EVs) in countries like Japan, Germany, Finland, France, and Austria [7,8]. These factors collectively suggest that the demand for EVs will continue to grow, reinforcing the positive outlook for the EV market [9].

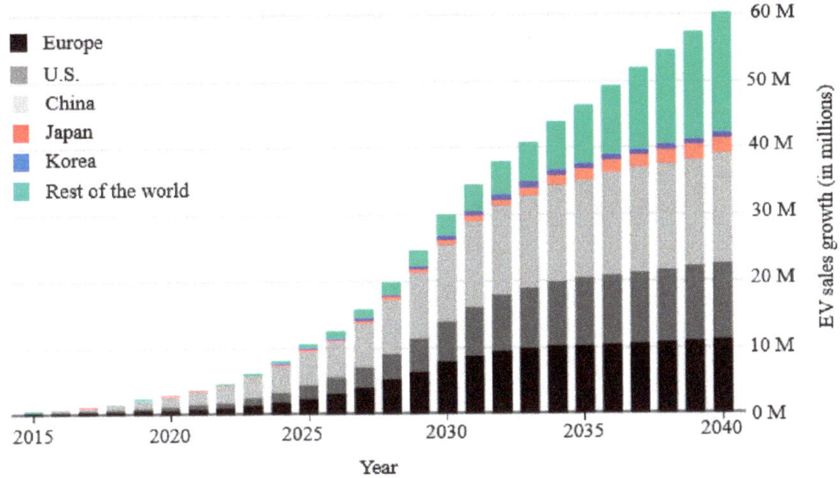

Figure 1. Global sales growth and demand for EVs (reproduced with permission from [10]).

1.1.2. Importance of Batteries in Electric Vehicles

An EV relies entirely on its battery to operate, allowing it to charge and run without an internal combustion engine. While gas-powered vehicles use fuel combustion to generate power, EVs use batteries that convert stored chemical energy into a continuous flow of electrical energy to drive the vehicle [11]. Consequently, the rising demand for EVs directly translates to an increased demand for batteries. The growth in lithium battery demand has been remarkable. In 2010, the demand was approximately 0.5 gigawatt-hours, which surged to 526 gigawatt-hours by 2020 and is projected to reach 9300 gigawatt-hours by 2030 [12]. Lithium batteries are highly appealing due to their ability to store and discharge large amounts of energy over extended periods with relatively low maintenance. These batteries are the cornerstone of EV functionality, necessitating ongoing research to enhance efficiency and optimize their production, usage, and end-of-life management.

1.2. End-of-Life Challenges for Electric Vehicle Batteries

One of the main challenges with large-scale battery use is managing their end-of-life phase. Although this issue has been relatively minor so far owing to the long lifespan of lithium-ion batteries and the recent surge in production over the past decade, it is expected to become significant around 2030 [13]. This is when many of the batteries produced in the last 5–10 years will start to reach the end of their useful life [14]. Fortunately, advances in battery recycling technology are being developed to address this impending challenge. The key will be integrating various recycling systems to ensure a smooth transition through the different stages of a battery's lifecycle, from production to end-of-life management.

1.2.1. Environmental Impact of Disposal

The demand for disposing of EV batteries has surged, attracting significant interest from companies. EV batteries contain rare minerals and materials that are difficult to source, making the recycling process complex. While recycling is essential for nonfunctional or broken batteries, lithium-ion batteries can also be repurposed for other powerful applications, such as solar panels or power generators. These practices are more environmentally friendly than extracting new materials [15]. Although old EV batteries may no longer be suitable for vehicle use, they can still be repurposed in other fields, extending their lifespan and providing additional value.

1.2.2. Resource Recovery Opportunities

The need to conserve and recover valuable minerals from old EV batteries is increasing, helping to minimize environmental impact and promote a more sustainable future. Understanding the lifecycle of EVs and the process of mineral recovery is crucial, as it influences recycling practices and recovery methods. By optimizing these processes, we can reduce waste, minimize resource extraction, and lower carbon emissions [16]. This not only creates economic opportunities but also aligns global interests towards the production of cleaner energy.

1.3. Circular Economy Principles and Their Application

The concept of a circular economy focuses on reusing existing materials and products for as long as possible to reduce the need for raw materials and minimize environmental impact. Figure 2 illustrates the ideal lifecycle of a material within this framework. This approach can be applied across all economic sectors, primarily through recycling or repurposing materials. The benefits are significant: less waste production, reduced raw material extraction, and long-term cost savings. However, the widespread implementation of a circular economy is limited by current technological constraints and the lack of established processes for effectively recycling or repurposing many products. Developing these technologies and systems is crucial for expanding the circular economy across all sectors [17].

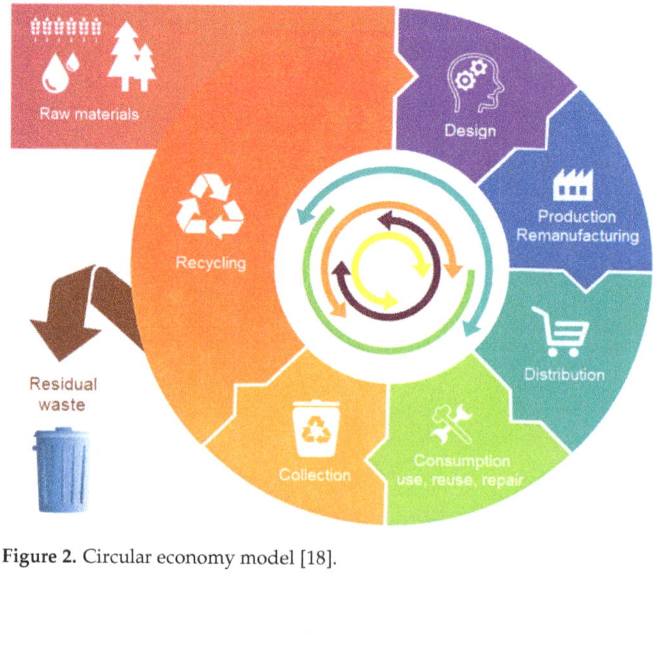

Figure 2. Circular economy model [18].

1.3.1. Circular Economy in the Context of EV Batteries

EV batteries exemplify how products can fit into a circular economy, primarily due to the valuable materials used in their construction. The lifecycle of EV batteries begins with the mining of rare raw materials such as lithium, cobalt, and nickel [19]. These materials are then used in the manufacturing process to create the batteries. With an average lifespan of 10–15 years, the critical question is what happens to these batteries at the end of their life [20]. By applying the principles of a circular economy, the materials from spent batteries can be stripped down and recovered. These reclaimed materials can then be used to manufacture new batteries, reducing the need for fresh raw material extraction. This process not only cuts down costs but also enhances efficiency and sustainability [21].

1.3.2. Objectives and Scope of the Review

The primary objective of this review is to explore pathways to a circular economy for EV batteries. This involves examining the properties and lifecycle of the batteries to understand their societal and environmental impacts and the importance of alternative end-of-life solutions. The review will then analyze how to implement circular economy practices within the lifecycle of EV batteries, including the necessary technologies. Additionally, it will discuss the potential environmental and societal impacts, existing challenges, and barriers that must be addressed. Finally, the review will outline what is needed for this approach to shape the future.

2. Lifecycle of Electric Vehicle Batteries

2.1. Battery Composition and Materials

When discussing EVs, it is important to recognize that various battery types are used, depending on manufacturer preference. The four main battery types used in EVs are lithium-ion, nickel-metal hydride, lead-acid, and ultracapacitor batteries [22]. For this paper, the focus will specifically be on lithium-ion batteries, as they are the most commonly used in the EV industry today. Lithium-ion batteries are preferred in the EV industry due to their long service life, manageable temperature range, and high energy efficiency relative to their physical size [23]. These characteristics make them ideal for a circular economy, as their longevity reduces the need for frequent replacements, thereby decreasing long-term material demand. However, the increasing prevalence of EVs, driven by a global shift towards reducing carbon footprints and enhancing sustainability, has led to a corresponding rise in lithium-ion battery production [24]. This increased production can counteract the reduced material demand benefits typically associated with the batteries' long lifespan [25]. Understanding the detailed properties and lifecycle of lithium-ion batteries is crucial for integrating them into a circular economy model. This knowledge allows for the development of more efficient recycling and repurposing practices, ultimately supporting sustainability goals.

2.1.1. Overview of Battery Components

Lithium-ion batteries consist of four main components: the anode, cathode, electrolyte, and separator. The anode is the negative electrode, and the cathode is the positive electrode. These electrodes work together, with the anode releasing excess electrons into the external circuit while the cathode draws them in. The electrolyte serves as the medium through which electrons are transferred between the anode and cathode. Finally, the separator keeps the anode and cathode apart, preventing short circuits and helping regulate the strength of the electric current [25]. A schematic view of the battery is shown in Figure 3.

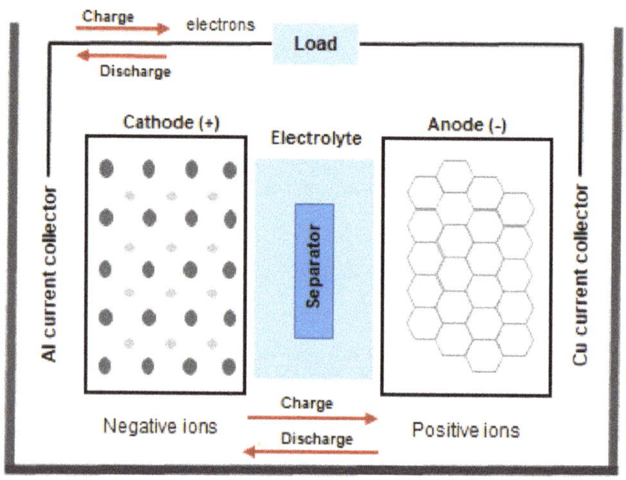

Figure 3. A schematic diagram shows the lithium–ion battery's components and working [26].

2.1.2. Identification of Critical Materials

When discussing the role of EV batteries in a circular economy, the first stage focuses on materials. The raw materials required to make a lithium-ion battery typically include metals such as lithium (with variations in alloys), manganese, copper, graphite, cobalt, nickel, and iron [27]. Figure 4 illustrates the composition of these materials within the overall battery [28]. These metals are chosen for their efficient properties, like conductivity and availability. However, not all materials required for these batteries are economically advantageous to obtain. For instance, cobalt can be expensive because it often needs to be sourced from unstable regions. Despite the cost, cobalt's properties are crucial for better temperature management within the battery, making it difficult to find suitable alternatives [29].

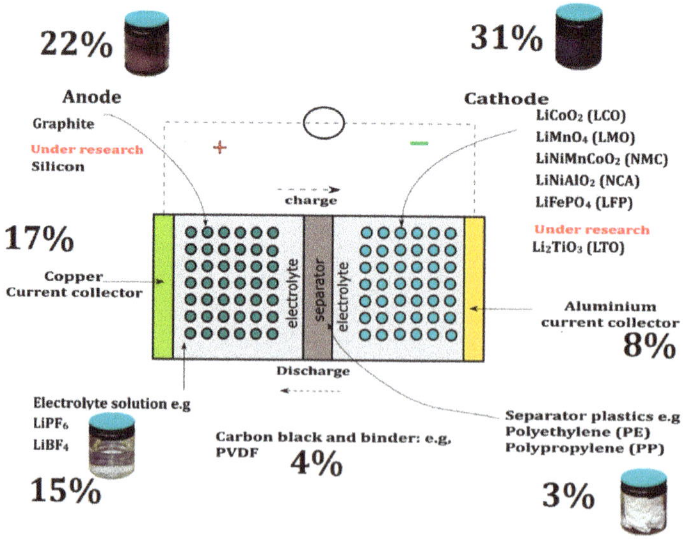

Figure 4. Schematic diagram showing the various components of a lithium-ion battery (reproduced with permission from [30]).

2.2. Manufacturing and Use Phase

The manufacturing process for lithium-ion batteries begins with the selection of their cell type, which can be cylindrical, pouch, or prismatic. In the EV industry, cylindrical cells are typically used. The manufacturing process consists of three main stages: electrode preparation, cell assembly, and battery electrochemistry activation [31]. During the electrode preparation stage, a slurry is created from an active material, a conductive additive, and a binder. The active material, which facilitates chemical reactions that release electrons, can include lithium cobalt for lithium-ion batteries [32]. The conductive additive, often graphite, enhances electron flow between the cathode and anode [33]. The binder, usually a type of polymer, holds the active material and conductive additive together, functioning like glue [34]. The cell assembly stage involves assembling all the battery components into the final cell structure, ensuring it is ready for installation. The final stage, battery electrochemistry activation, involves cycling the battery to establish electrical flow throughout its components, effectively jump-starting its service life [35].

2.2.1. Environmental Impact during Production

The prevalence of EVs has surged in recent years. In an effort to reduce emissions and promote sustainable living practices, transportation has become a key area for implementing these standards, both on individual and governmental levels. However, this growth in EVs brings forth discussions about the environmental impact of their production, particularly the production of the lithium-ion batteries they require. This raises questions about the potential contradictions between the goal of creating a cleaner environment and the environmental footprint of battery production.

An important aspect to consider in the production stages of lithium-ion batteries is the environmental impact of raw material extraction. As previously mentioned, the materials used in lithium-ion batteries include various metals and alloys, which are natural resources. Extracting these materials requires a significant amount of energy, contributing substantially to the environmental impact of lithium-ion battery production [36]. The primary concern with this energy consumption is the associated carbon dioxide emissions. There is a positive correlation between the extraction of natural resources, carbon dioxide emissions, and the ecological footprint. This means that as the mining of natural resources increases, so too do carbon dioxide emissions and the ecological footprint [36]. The ecological footprint measures the amount of land required to sustain the demand for these resources. In summary, increased extraction of natural resources, such as the metals and alloys needed for lithium-ion batteries, leads to greater negative environmental impacts.

Another significant environmental impact of lithium-ion battery production comes from material preparation. After raw materials are extracted, they must undergo various processes before being incorporated into the final battery structure. These processes include sintering, grinding, spraying, coating, drying, and sieving. Each of these steps consumes a substantial amount of energy and resources, such as natural gas, to be carried out effectively [37].

Lastly, the electrode preparation stage in the production of lithium-ion batteries also significantly impacts the environment. During this stage, a slurry is created from active material, a conductive additive, and a binder. This process inevitably releases toxic vapors into the air through chemical reactions. Due to the known toxicity of these vapors, regulations are in place to contain them and prevent their release into the external environment [38].

Overall, the production of lithium-ion batteries for EVs has several negative environmental impacts. These include increased extraction of natural resources, higher carbon dioxide emissions, elevated energy consumption, and the creation of toxic vapors. While EVs are often seen as environmentally beneficial, the environmental effects of battery production can counteract this perception. It is crucial to consider these impacts across the different stages of production to fully understand the environmental footprint of EV batteries.

2.2.2. Considerations during the Operational Life of EV Batteries

Following the production stage, the next phase in the lifecycle of EV lithium-ion batteries is their operational life. This stage involves the actual use of the batteries within EVs. Key considerations during this phase include temperature degradation of the battery, charging practices, and the habits of EV users. From the standpoint of an EV manufacturer, producing a high-quality product is paramount. For EVs, this means ensuring that the lithium-ion battery, the vehicle's power source, has a long service life. Many aspects of the operational phase are directly related to maximizing this lifespan. Therefore, the following considerations will focus on how various factors can impact the longevity of lithium-ion batteries in EVs.

The first consideration is temperature degradation of the battery. EVs are used wherever consumers are, meaning that these batteries can be exposed to a wide range of environments depending on the user's location and the local climate. Seasonal variations pose a challenge to the longevity of lithium-ion batteries, as they may experience overheating in high temperatures or damage from exposure to low temperatures. This cycling of extreme temperatures can significantly reduce the battery's expected service life. Lithium-ion batteries perform best at moderate temperatures, typically when they are in a state of storage. For example, in EVs, the battery is at an ideal temperature when the vehicle is not in operation [39]. However, this is impractical, as the primary purpose of the vehicle is transportation, not storage.

The second consideration is the consumer's charging practices. EVs require regular charging, but the frequency and manner of charging can impact the degradation of lithium-ion batteries. A common charging routine among EV users is to fully charge the battery overnight, drive until the battery is nearly or completely depleted, and then repeat this cycle. This practice is harmful to lithium-ion batteries. Charging the battery for extended periods, such as overnight, can cause overheating, which contributes to battery degradation [40]. As mentioned previously, extreme temperature variations can negatively affect the battery's lifespan. To promote a healthier charging cycle, EV users should aim to maintain a 20% charge at all times and avoid charging beyond an additional 50% at any given time. This balanced approach helps mitigate the thermal stress on the battery and extends its operational life.

The third consideration in the operational life of a lithium-ion battery is the driving habits of the EV user. Just as charging practices can influence battery health, the manner in which the vehicle is driven also plays a significant role. Battery degradation can vary based on the user's routine driving patterns. For example, one user might commute to work five days a week, spending a significant amount of time on the highway at high speeds. In contrast, another user with a similar work schedule might have a shorter commute that only involves driving on smaller streets at lower speeds. These variations in driving conditions between users significantly impact the battery's lifespan [41]. The user with the longer, high-speed commute is likely to experience faster battery degradation, whereas the user with the shorter, lower-speed commute imposes less strain on the battery, resulting in a longer lifespan.

Overall, the operational stage has a significant impact on the service life of EV batteries. Factors such as environmental conditions, charging practices, and driving routines all influence battery longevity. By understanding and optimizing these factors, users can enhance the efficiency and extend the lifespan of their EV batteries.

2.3. *End-of-Life Phase*

Following the operational stage, the end-of-life phase is concerned with the final disposition of the EV's lithium-ion battery. Several options are available for battery disposal, including complete disposal, material recovery and reuse, or various recycling methods, which will be explored in detail later. However, each option comes with its own set of challenges, including potential inefficiencies and costs, which may impact the sustainability of the chosen method for the battery's end-of-life phase.

2.3.1. Challenges in Battery Disposal

Before selecting a disposal method, it is crucial to consider the inherent challenges associated with battery disposal. These challenges include additional costs, limited opportunities for secondary applications, and the competitive market for new batteries.

One significant challenge is the additional expense involved in partial recycling of battery components. Specifically, the valuable metals within batteries, such as lithium and cobalt, are highly sought after. However, the cost of extracting these metals from used batteries is often higher compared to sourcing them through traditional mining. This disparity arises because the process of recovering metals from spent batteries is more complex and resource-intensive than mining new materials. Another challenge is the limited potential for second-life applications of used batteries. Second-life applications involve repurposing a used battery for a different function, such as using an EV's battery to power industrial machinery. However, the chemical composition of lithium-ion batteries changes significantly over time, making it difficult to effectively repurpose them for new uses. The rapid growth in EV production has led to a competitive market for new batteries. As demand for new batteries increases, manufacturers compete by lowering prices to attract customers. This price competition reduces the incentive to seek out and reuse old batteries, making new batteries more attractive despite their lower cost [29].

2.3.2. Opportunities for Circular Practices

Despite the challenges associated with lithium-ion battery disposal, there are significant opportunities to integrate these batteries into a circular economy. The circular economic model envisions a cycle where end-of-life stages connect back to the beginning life stages, promoting sustainability and resource efficiency. While there are hurdles to effective disposal, they do not preclude the potential for reusing and recycling these batteries within a circular framework. Recycling methods, which will be discussed in detail in subsequent sections, offer a pathway to align with circular practices. Additionally, the EV industry is expanding rapidly, driven by increasing efforts to address climate change and enhance sustainability [42]. As the industry grows, so too will the regulatory frameworks designed to mitigate environmental impacts and promote the efficient disposal of EV batteries [43]. Government regulations and industry standards are likely to evolve, reducing the challenges associated with battery disposal and fostering more effective circular economy practices. Current strategies for integrating lithium-ion batteries into a circular economy are already in place and will be explored further in the following sections.

3. Circular Economy Strategies for End-of-Life EV Batteries

3.1. Battery Recycling Technologies

Battery recycling is crucial for sustainability, as it facilitates the recovery of valuable materials and reduces energy consumption. Recycling could potentially lower the primary demand by 25–64% between 2040 and 2050, according to projected demand figures. This suggests that waste streams might significantly contribute to meeting future raw material needs [44]. The battery recycling market is also projected to expand, with estimates suggesting that it could grow from USD 17.2 billion in 2021 to USD 23.2 billion by 2025, at a CAGR of 6.1%. This growth is fueled by the increasing adoption of EVs, stringent environmental regulations, and the rising value of recycled battery materials such as lithium, cobalt, and nickel [45]. In line with government policies that promote green initiatives, efforts such as those in Storey County, Nevada—where American Battery Technology Company (ABTC) has received over USD 70 million in U.S. DOE grants—are advancing the recycling and commercialization of battery metals to create a domestically-sourced circular supply chain [46]. Li-Cycle Corp., Toronto, ON, Canada, is a leading lithium-ion battery resource recovery company that operates several recycling facilities across North America, including Ontario, Canada. The company uses a proprietary hydrometallurgical process to recover up to 95% of all constituent materials found in lithium-ion batteries, including lithium, cobalt, and nickel. Li-Cycle's approach not only helps mitigate the environmental impact associated with battery disposal but also ensures that valuable materials are efficiently reintroduced into the supply chain, supporting sustainability goals [47]. Northvolt, a Swedish battery manufacturer, is developing a comprehensive recycling program called Revolt, which focuses on recovering metals such as lithium, cobalt, and nickel from old EV batteries. Northvolt's recycling facility in Vasteras, Sweden, uses a combination of mechanical and hydrometallurgical processes to achieve high recovery rates for valuable battery materials. The company aims to produce batteries with up to 50% recycled material by 2030, aligning with broader sustainability and circular economy goals [48]. The battery recycling employs different methods such as pyrometallurgical, hydrometallurgical, bioleaching, and direct recycling processes to efficiently extract and repurpose materials for new battery production. A flow chart of the conventional treatment processes of spent Li-batteries (LIBs) is given in Figure 5.

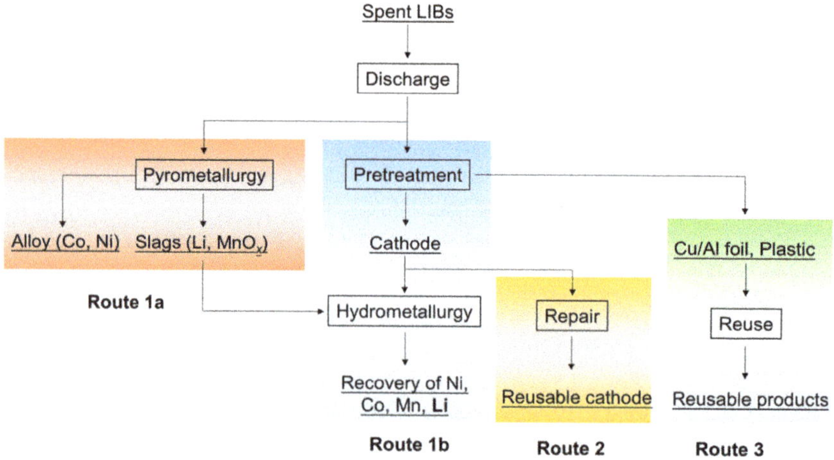

Figure 5. Conventional flow chart of treatment processes of spent LIBs (reproduced with permission from [49]).

Pyrometallurgical processes involve high-temperature smelting to recover valuable metals, while hydrometallurgical methods use aqueous solutions to leach out metals, offering lower energy consumption. Bioleaching utilizes microorganisms to extract metals from spent batteries, presenting a potentially eco-friendly alternative. Direct recycling aims to preserve battery materials' original structure, facilitating their reuse with minimal processing. Each method offers distinct advantages and limitations, and their combined use can optimize resource recovery and reduce environmental impact [50]. Table 1 provides a comparison of these recycling methods. These advanced technologies not only enhance the effectiveness of battery disposal but also contribute to the development of a more robust circular economy. By refining these recycling methods, we can improve resource recovery and reduce environmental impact, paving the way for more sustainable practices in battery management.

Table 1. Comparison of recycling methods [51–54].

Recycle Method	Highlights	Advantages	Disadvantages
Pyrometallurgical	• Involvement of high-temperature for smelting of batteries. • Frequently used for extraction of Co and Ni	• Simple process • High productivity • Industrial scale-capacity	• Requirement of high energy • Emission of hazardous gases • Limited number of materials reclaimed
Hydrometallurgical	• Involves the use of acids and reductants • Most of battery components are recovered as metals (Cu, Al) or salts (Li, Ni, Mn, etc.)	• Low energy consumption • Ease of operation • Low emission of toxic gases • Low cost • Higher energy efficiency	• Complexity in the selection of chemicals • Waste water management problem • Higher consumption of reagents
Bioleaching	• Microorganisms such as bacteria or archaea are used to extract metals.	• Environmentally competitive • Low cost • The addition of a proper catalyst can increase the rate of reaction and leaching yield	• Higher processing cost and environmental impact • Slow process kinetics • Susceptible to contamination
Direct recycling	• Preserves cathode morphology with minimal processing	• Low pollution • Low energy consumption • Less greenhouse gas emission	• Still needs time to develop and mature

3.1.1. Overview of Existing Recycling Methods

EV batteries comprise cells that include a cathode, anode, electrolyte, and separator, arranged in series or parallel configurations. These batteries typically have a lifespan of around 10 years; however, factors such as overcharging and excessive discharging can accelerate their degradation [39]. As EV batteries approach the end of their useful life, they are often discarded or replaced with newer models. Repurposing or reusing these batteries for alternative storage applications presents an opportunity to extract additional value. The growing demand for materials used in EV batteries highlights their significance. This increasing demand underscores the importance of efficient recycling methods to manage the supply and sustainability of these critical materials.

3.1.2. Innovations in Battery Recycling Technologies

The rapid evolution of EV battery technology has significantly impacted both the transportation and production sectors. Advances in charging capabilities and vehicle design have not only improved performance and affordability but have also enhanced the safety and efficiency of discharging and recycling EV batteries. Companies like Li-Cycle, Redwood Materials, and American Battery Technology Company are developing proprietary recycling technologies to maximize material recovery and reduce environmental impact [55]. Innovations in battery recycling technologies are influenced by several factors, including advancements in processing techniques, supply-chain stability, decarbonization efforts, and regulatory pressures and incentives [56]. Research-driven improvements in these areas are helping to ensure the quality and efficiency of battery recycling processes. These developments support the goal of making recycling more effective and economically viable, while also securing local raw materials at competitive prices. Innovations in battery recycling technologies are also addressing the challenge of handling emerging battery chemistries, such as solid-state and lithium–sulfur batteries, which require specialized recycling methods. Efforts are underway to develop scalable processes for these new types of batteries to ensure they can be efficiently recycled as their use becomes more widespread [57,58]. Furthermore, advancements in machine learning and artificial intelligence are being integrated into recycling systems to optimize sorting, identify materials with greater precision, and improve overall recovery rates [59]. These technological advancements not only enhance the effectiveness of recycling but also contribute to the long-term sustainability and economic viability of battery recycling initiatives.

3.2. Second-Life Applications

EV batteries have a finite lifespan, but their repurposing and reuse offer valuable opportunities for creating alternative energy solutions. For instance, these batteries are increasingly being utilized in solar panel systems to store energy for use when sunlight or wind is unavailable [60]. Additionally, they are being employed in electric fast-charging stations that store alternative energy, fostering innovation in transportation infrastructure. Although EV batteries contain harmful minerals, their recycling could act as a catalyst for improving the environmental practices of the mining industry [61]. This not only helps reduce the carbon footprint but also promotes the growth of renewable energy sources. A utilization chain of an EV battery throughout its lifecycle is shown in Figure 6.

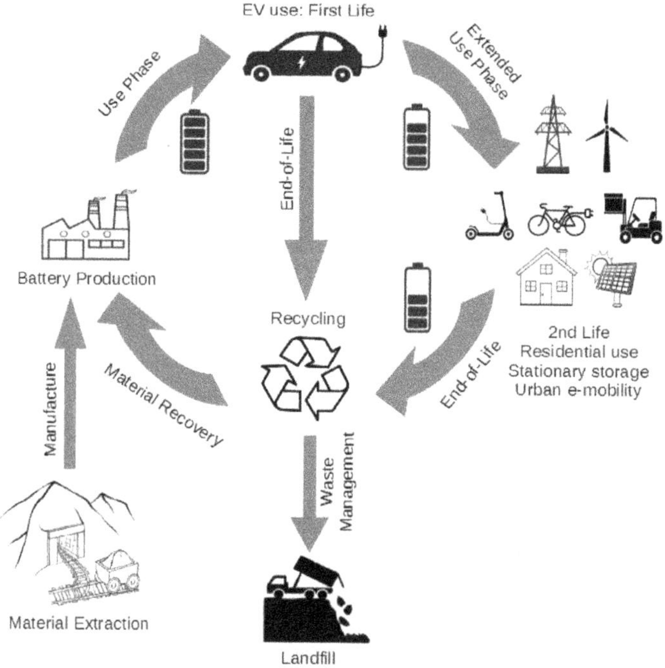

Figure 6. EV battery utilization chain throughout its lifecycle [62].

When an EV battery reaches the end of its life, there are several options for its management. One option is disposal, typically used when the battery is severely damaged or destroyed beyond repair, rendering it unsuitable for reuse. Another option is recycling, particularly valuable when the battery contains metals that can be recovered and repurposed for new applications. Additionally, reusing batteries can offer significant market value. Batteries with reduced capacity or remaining energy can be utilized in secondary applications that require lower energy levels [62]. As EV technology and battery storage continue to advance, the demand for raw materials for these second-life applications is growing, highlighting the importance of effective battery management. With the advancement of EV technology and battery storage solutions, innovations in battery diagnostics, predictive analytics, and automated processing are improving the management of end-of-life batteries. These developments enhance the accuracy of battery health assessments, increase recycling efficiency, and facilitate the adaptation of batteries for new applications. The growing demand for raw materials for second-life applications underscores the importance of effective and sustainable battery management strategies.

3.2.1. Reuse of EV Batteries for Stationary Energy Storage

Reusing and recycling EV batteries offers valuable opportunities to extend the life of newer batteries and support sustainability. Many EV batteries find a second life in stationary energy storage systems, such as power grids or backup power sources. These batteries can be repurposed for applications requiring lower energy levels or integrated into other energy storage systems [63]. Several companies disassemble old EV batteries, recover and recycle valuable components, and supply them to industries in need of these materials. Additionally, many independent companies are investing in recycling facilities, boosting the demand for cathode materials and increasing profitability. While battery recycling is not yet a critical issue, its importance is expected to grow significantly in the future as technology advances and secondary applications expand. For instance, companies like B2U are repurposing old EV batteries to create solar panels, which can maintain grid operations cleanly for over five years before being recycled into new batteries [64]. The recycling process for EV batteries typically involves four main steps. First, the batteries are collected and sorted based on their type and condition. Second, they are disassembled and discharged to safely remove any residual energy. Third, the batteries undergo pretreatment, which involves shredding them into smaller pieces to facilitate further processing. Finally, the materials are recovered using chemical processes [65,66]. To extract and purify the valuable materials, such as nickel, cobalt, and lithium-ion phosphate, different methods like pyrometallurgy, hydrometallurgy, and direct recycling methods are employed. Due to the complex nature of these materials, specialized techniques and equipment are required to effectively dismantle and process them [51].

3.2.2. Repurposing Batteries for Nonvehicular Applications

EV batteries, which convert and store chemical energy as electricity, inevitably experience degradation over time, leading to reduced capacity and performance. Factors such as temperature exposure and usage conditions can accelerate this degradation. When these batteries reach the end of their useful life, proper disposal is crucial to prevent environmental harm while maximizing societal benefits. While recycling is an option, it poses challenges, including potential contamination of wildlife and the high costs associated with extracting valuable materials like lithium, cobalt, and nickel through smelting. Repurposing offers a promising alternative to recycling. By finding new uses for degraded batteries, we can extend their value beyond vehicular applications. For example, Nissan has repurposed batteries to power streetlights in Japan, Renault has used them to back up elevators in Paris, and GM has employed repurposed Chevy Bolt batteries to support its data center in Michigan [67]. The Nissan Leaf Energy Storage Project is a notable example of repurposing end-of-life EV batteries for stationary energy storage. Launched in 2016 in collaboration with Eaton and a local utility, the project involves selecting and testing used batteries from Nissan Leaf vehicles based on their remaining capacity. These batteries are integrated into modular energy storage systems that are connected to the grid. The systems store excess renewable energy and provide backup power during peak demand, thus extending battery life and reducing waste while enhancing grid stability and supporting renewable energy integration [68]. BMW's Second-Life Battery Systems represent another successful approach to repurposing EV batteries. The company utilizes batteries from its i3 and i8 models to create energy storage solutions for commercial and residential use. These repurposed batteries store energy generated from solar panels and release it during high-demand periods, demonstrating how extending battery life can contribute to energy efficiency and support the adoption of renewable energy sources [69]. As EV battery technology evolves and cathode materials become more complex, repurposing and recycling processes may face increased difficulty, particularly with newer, heavier batteries. However, advancements in technology present opportunities to streamline these processes. By focusing on direct recycling and repurposing, costs can be minimized and handling techniques standardized, potentially improving regulatory practices and fostering innovative partnerships within the industry [65]. Despite the challenges, there are significant opportunities for companies to develop technologies and strategies that enhance battery repurposing and recycling in the future.

3.3. Material Recovery and Resource Conservation

The transportation sector is undergoing a rapid transformation as automotive manufacturers shift from gas-powered vehicles to EVs. This shift has intensified the demand for key materials used in EV batteries, such as lithium, cobalt, and nickel, leading to significant advancements in extraction and processing technologies [70]. For instance, improvements in lithium retrieval methods have increased efficiency by over 80%, making the costs comparable to newly produced materials [21]. By the end of 2021, a total of at least nineteen critical materials recovery plants were set up across nine countries, including China, the United States, and Canada, with a combined capacity of around 322,500 tons annually [50]. As EV batteries approach the end of their lifecycle, the critical question arises: Should they be discarded, or can they be recycled? Proper disposal is essential to minimize environmental impact and maximize resource recovery. This is where Total Manufacturing Recovery (TMR) initiatives come into play [71]. A notable example is B2U, a company specializing in repurposing used EV battery components. B2U integrates these components into their solar panel systems, allowing them to store energy when sunlight is insufficient. When solar panels are not producing enough power, the repurposed batteries discharge energy back into the grid, functioning similarly to a generator [72]. This approach not only enhances the efficiency of renewable energy systems but also helps reduce the carbon footprint by providing a sustainable source of electricity for homes and businesses. By focusing on material recovery and resource conservation, companies like B2U are pioneering solutions that address the environmental challenges associated with EV battery disposal and contribute to a more sustainable energy future.

3.3.1. Recovery of Valuable Materials from Spent Batteries

Recovering valuable materials from spent batteries is crucial for environmental sustainability but presents several challenges. Spent batteries, once used in various settings such as households, offices, and schools, must be sorted and processed to extract reusable materials. The recovery process often begins with acid leaching, a technique that uses acids to dissolve metals and convert them into more soluble forms [73]. While this method is effective in extracting valuable metals, it can produce acid residues that may pose environmental risks. To address these challenges, both physical and chemical methods are employed. Physical methods, such as crushing and heating, facilitate the disassembly of battery components and increase the efficiency of recycling. Chemical methods, including hydrometallurgical and pyrometallurgical processes, are used to recover metals like lithium at high temperatures. These techniques enable the effective extraction of metals while minimizing environmental impact [74]. As demand for raw materials continues to rise, advancements in these recovery processes are essential to ensure sustainable management of battery waste and reduce the ecological footprint of battery disposal.

3.3.2. Strategies for Minimizing Waste and Environmental Impact

Processing plants that handle resources and raw minerals depend heavily on trucking for transportation and waste removal. This reliance results in significant pollution and substantial waste generation [50]. As populations grow and operational costs increase, the distances traveled by trucks lengthen, leading to higher CO_2 emissions and further environmental degradation [75]. Despite discussions about the potential for newer technologies to address these issues, tangible progress in reducing the environmental impact of trucking in the industry has been minimal.

Figure 7 illustrates the challenges associated with Environmental Policy Stringency (EPS) and its impact on adhering to environmental policies. While EPS is designed to address and mitigate pollution in various areas, including air quality and greenhouse gas emissions, it often has limited influence on the mining industry. Instead, these policies are more focused on controlling broader environmental issues rather than directly regulating the specific practices of mining operations [76].

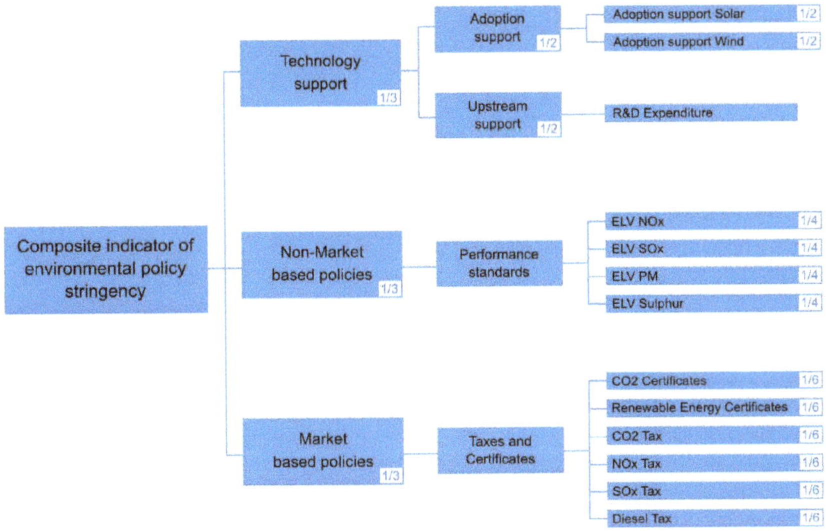

Figure 7. Environmental policy stringency (EPS) index showing advantages of environmental policies [76].

EV batteries have been shown to reduce gas consumption and waste, promoting cleaner and safer transportation. However, focusing solely on enhancing EV battery recycling without considering its environmental impact and energy lifespan could lead to significant consequences [77]. Effective long-distance travel requires a balance between reducing the carbon footprint of transport operations, minimizing fuel consumption, and transitioning to EVs. This shift results in increased distances for transporting raw minerals and materials [76]. While these changes contribute to energy efficiency and improved infrastructure, they also underscore the need for sustainable practices in both battery recycling and mineral transportation to ensure long-term environmental benefits.

4. Environmental and Social Impacts

4.1. Environmental Benefits of Circular EV Batteries

In a circular economy, the aim is not merely to recycle products at the end of their lifecycle but to extend the overall useful life of the products [78]. Although maintaining and repurposing these old products requires fresh raw materials, the environmental benefits of a circular economy are not solely derived from reducing raw material consumption. Instead, the primary advantage lies in decreasing the number of new EV batteries that need to be produced. This reduction results from a slower decline in the number of units in use, as these units are frequently refurbished or repurposed. Consequently, fewer batteries end up in landfills, and the cost of production decreases, leading to significant environmental and economic benefits.

4.1.1. Reduction in Environmental Pollution and Resource Depletion

The lithium industry significantly impacts the environment, from ecosystem destruction during mining to pollution during refinement, production, and disposal. Disposing of lithium batteries in landfills poses unique risks, including heavy metal contamination and the potential for fires [79]. A comprehensive cradle-to-grave lifecycle analysis by Bawankar et al. [80] utilized data from the Ecoinvent and BatPaC databases for NMC811 Lithium-Ion Batteries. They identified 15 areas where environmental impacts could occur, providing insights into the total lifecycle environmental impact, normalized results, weighted results, and the environmental impact during the use stage. Their findings revealed that, while unnormalized data highlighted resource use in energy carriers, land use, and fossil fuels as the top contributors to environmental effects, the normalized and weighted data indicated fossil fuels, energy carriers, and ionizing radiation as the primary contributors. The study emphasized that these findings were heavily influenced by the energy sources of the electrical grid at the production site. Therefore, a significant way to reduce the environmental impact of lithium-ion batteries is to manufacture them using an electrical grid powered by renewable energy. Additionally, the authors noted the substantial water usage required by the lithium industry. This issue has been highlighted by The New York Times, which reported significant water use as a point of contention between the lithium industry, environmental activists, and indigenous groups [81].

4.1.2. Contribution to a Sustainable Materials Lifecycle

To understand how EV batteries can contribute to a sustainable economy, it is essential to first identify their composition. The materials within these batteries can be reused or recycled into various products, depending on their nature. A typical lithium-ion battery comprises four main components: a positive cathode made of lithium and other metals, an anode made of graphite, a barrier that separates the two while allowing electron flow, and an electrolytic solution [82]. However, it is important to note that chemical changes occur in these materials over the battery's lifespan, which contribute to the battery's eventual failure. As electrons travel between the anode and cathode, a solid electrolyte film begins to form on the anode [83]. To make the graphite useful again, it must also go through the recycling process. Understanding these chemical changes is crucial for recovering and repurposing these materials effectively. While knowing how to recover materials is one step, applying them to create new products in a manner that supports a sustainable lifecycle presents a different challenge. Identifying potential new products that can be made from each of the spent components is essential for advancing towards a circular economy. For example, the cathode, due to its valuable metal content, is often directly recycled into new cathodes. This process varies depending on the recycling method. An emerging technique, known as "cathode-to-cathode" recycling, is being employed on a small scale by certain companies. This method involves extracting the cathode first to preserve it before processing the rest of the battery [84]. The preserved cathode can then be incorporated into a new battery. However, the majority of cathode material is typically shredded along with the rest of the battery, mixing with the electrolytic solution and the anode to form a material industrially referred to as "black mass" [84]. The goal during further processing of this black mass is to extract individual metals, as discussed in Section 3.3. In addition to lithium, these metals can include nickel, cobalt, and manganese [85]. Each of these metals has applications beyond being turned back into cathodes. For instance, nickel can be used to plate other metals, making them corrosion-resistant or serving as catalysts for chemical reactions. It is also crucial in desalination plants when combined with copper into an alloy [86]. Similarly, cobalt has various uses outside of batteries, such as in the production of airbags, hard metals, and diamond tools. Like nickel, cobalt can also be used for corrosion resistance and as a catalyst for chemical reactions [87]. The spent graphite anode offers several recycling options. Traditionally, graphite is used in products like writing implements, solid lubricants, and nuclear core stabilizers, or converted into graphene [88]. It can also be recycled into new anodes for lithium-ion batteries. Innovative recycling methods have emerged, allowing spent anodes to be transformed into cathodes for dual-ion batteries [89]. This versatility makes graphite a highly sustainable material, given the various ways it can be processed and repurposed. In fact, the recycling process for graphite is very competitive with the production of new graphite [90]. Additionally, beyond its conventional uses, pencil lead can be synthesized into a graphite–silica composite anode that performs comparably to standard lithium-ion battery anodes [91].

4.2. Social Considerations and Stakeholder Engagement

In society, the pervasive attitude of consumerism needs to be addressed to reduce the demand for raw materials, production, and prevent losses in the recycling cycle. This requires decreasing consumer demand and promoting the longevity of functional technology. For instance, cellphones are typically replaced every 12 to 18 months [61]. Shockingly, as of 2022, less than 5% of lithium-ion batteries are recycled globally, according to a division of the American Chemical Society [92]. This section will explore the current state of community involvement in recycling, societal attitudes, and external factors that both aid and hinder public participation in recycling efforts.

4.2.1. Community Involvement in Circular Battery Initiatives

To effect meaningful change, the environmental impact of lithium-ion batteries requires action on multiple levels, especially in policy, which will become increasingly relevant in the coming years. However, significant steps can also be taken at the local and business levels. Companies like Apple and Samsung offer trade-ins for their devices, providing credit toward new purchases or ensuring responsible recycling of old devices. Encouraging consumers to make fewer frivolous phone purchases could significantly reduce the number of discarded lithium-ion batteries. While trade-in policies are a good start, they do not necessarily mobilize communities toward broader action. It is crucial for governments to disseminate easily accessible information about which products contain lithium-ion batteries and their detrimental environmental effects. Additionally, providing convenient recycling options is essential to encourage responsible disposal and recycling of these batteries.

To create effective recycling policies, the United States could look to successful examples from Europe and other regions. For instance, Germany enforces a deposit scheme for plastic bottles and cans: consumers pay a deposit when purchasing these items and receive a refund upon returning them to the vendor [93]. This system contrasts with the American approach, where financial compensation is provided at the point of recycling, creating an incentive rather than an economic obligation. Germany's approach effectively ensures a higher degree of participation in recycling, reflected in their leading global recycling rate of 69.1% as of 2022 [94]. In the United States, a significant issue is the contamination of recycling streams with nonrecyclable items—one in four items sent for recycling cannot be processed [95]. This discrepancy between public support for recycling and actual recycling behavior highlights the need for better public education and policy enforcement. Germany's deposit system effectively educates consumers about what can be recycled by associating deposits with recyclable items. Even if consumers cannot return their recyclable items, they are more likely to recognize that these items should be recycled and take appropriate action. The United States could also consider policies from other countries, such as South Korea's phase-out of single-use items like disposable cups [93]. This strategy increases recycling rates by reducing the total amount of recyclable material in circulation, thus improving the recycling percentage. These policy frameworks could be adapted to address the recycling of EV batteries, ensuring that valuable materials are recovered efficiently and environmental impacts are minimized.

4.2.2. Social Acceptance and Ethical Considerations in End-of-Life Practices

Current end-of-life practices for lithium-ion batteries are generally not socially accepted, despite strong public support for recycling. A qualitative survey conducted by the World Economic Forum found that 94% of Americans favor recycling, but only 35% actively participate due to a lack of convenient access [96]. This gap highlights the need for more public infrastructure to facilitate the recycling of waste, especially batteries, which require specialized processes. Many states with recycling initiatives have specific rules about what can and cannot be recycled, with processes in place to measure and manage the materials collected. An evaluation matrix by Ball Corporation identified Maine as having the best recycling performance in the United States, based on metrics for recycling cardboard boxes,

PET bottles, various metals, glasses, and general packaging [97]. However, this evaluation did not include electronics recycling, underscoring a critical area that needs attention. Electronics recycling remains largely inaccessible to the public. Given the ubiquity of electronic devices in modern life, it is essential to make electronics recycling equally pervasive. One potential solution is to implement dedicated electronic recycling dumpsters in neighborhoods, allowing residents to dispose of their electronics conveniently. These dumpsters could be serviced on a regular schedule, integrating electronic waste management into the routine waste collection system and significantly improving recycling rates for lithium-ion batteries and other electronic components.

Aside from social considerations, there are significant ethical improvements to the lithium lifecycle as unit recycling becomes more viable. Strip mining in the lithium industry consumes excessive amounts of water and land, often at the expense of native tribes. This has led to protests from environmental and indigenous advocacy groups against further land development projects for lithium mines, including acts of civil disobedience such as setting up encampments on planned development sites [81]. In addition to encroaching on native cultural lands and disturbing local wildlife, lithium mining can have severe impacts on farming and agriculture. There are concerns that the water usage required for lithium mining operations could negatively affect local farmers by diverting water away from their livestock and crops [98]. By increasing the recycling of lithium-ion batteries, the need for new lithium mining could be reduced, thereby lessening the environmental and social impacts associated with mining. Theoretically, if more recycled units are used in manufacturing, lithium-ion battery production could rely less on mining, reducing the overall output demands on new lithium extraction. This shift towards recycling could help mitigate the negative effects on indigenous lands, local ecosystems, and agricultural activities, promoting a more sustainable and ethical lifecycle for lithium batteries.

5. Challenges and Barriers

5.1. Technological Challenges

The concept of a circular economy is still emerging within the EV battery industry. While EVs have gained popularity only in recent years, strategies for managing EV batteries at the end of their lifecycle are even less developed. As the number of batteries approaching the end of their useful life increases, it becomes crucial to address these challenges now. By doing so, we can ensure that valuable materials are not wasted or improperly disposed of. Proactively developing and implementing effective recycling and reuse strategies will be essential for managing the growing volume of end-of-life batteries and advancing the sustainability of the industry.

5.1.1. Limitations in Current Recycling Technologies

As lithium battery technology evolves rapidly, keeping pace with recycling advancements becomes increasingly challenging [51]. The EV industry, still in its formative years, is undergoing frequent trials of new battery designs to determine the most effective solutions. Ongoing modifications aimed at enhancing battery efficiency and vehicle performance continually alter the materials used, complicating the recycling process. Perfecting recycling methods will be difficult until the industry stabilizes and a well-established standard for effective battery design emerges over time.

5.1.2. Technical Obstacles in Repurposing and Second-Life Applications

Repurposing lithium batteries for applications such as home energy storage, grid integration, or low-speed EVs presents significant technical challenges [99]. One major hurdle is assessing the state of health (SoH) of the battery, as different applications have varying regulatory standards and performance requirements. Currently, the technology for accurately evaluating SoH is not advanced enough to facilitate the timely and efficient repurposing of lithium batteries [100]. Advances in research and development are essential to overcoming these obstacles and expanding the viable options for utilizing spent lithium batteries beyond their initial lifecycle.

5.2. Regulatory and Policy Challenges

Even with the ideal technology for recycling and repurposing batteries, its effectiveness would be limited without the appropriate regulations and policies to guide funding and define target applications. Given the rapid advancements in the industry, it is anticipated that growing interest will lead to more opportunities for refining these technologies and establishing the necessary frameworks to support their optimal use [50].

5.2.1. Evaluation of Existing Regulations and Policies

Currently, governmental legislation regarding battery recycling remains underdeveloped, reflecting the industry's relative novelty [56]. Among the major markets, China leads with stringent regulations, and it is projected to have recycled 2.312 million tons of spent lithium batteries by 2026 [101]. In contrast, the U.S. lags behind, with only 1% of spent lithium batteries currently being recycled [102]. Recent advancements, such as the Infrastructure Investment and Jobs Act of 2021, have initiated funding for battery recycling programs [103]. Given the rapid growth of the industry, it is anticipated that new regulations and policies will emerge in the coming years to address technological advancements and ensure effective recycling practices, provided continued investment and development in this area.

5.2.2. Recommendations for Policy Improvements to Support Circular Practices

To establish a circular economy for lithium batteries, increased funding and support for battery recycling technologies are crucial. In the U.S., both national and local efforts must be intensified. Although some legislation related to battery recycling has been enacted, these measures need to be expanded to unlock the full economic potential of this burgeoning industry [104]. Enhanced funding and robust regulations would encourage more companies to innovate and develop safer, more sustainable recycling technologies, driving significant advancements in the sector.

6. Future Directions and Research Needs

6.1. Research Gaps and Opportunities

6.1.1. Areas Requiring Further Investigation and Development

Recycling presents a significant challenge in establishing a circular economy for EV batteries. For a truly circular economy, effective recycling is essential. The concept involves not only reusing old batteries to create new ones but also repurposing them for less demanding applications, such as energy storage in solar power systems, which can extend their operational life [105]. Achieving a nearly perfect cycle of resource use hinges on overcoming current recycling hurdles. One major obstacle in recycling EV batteries is the complexity of their disassembly. EV batteries are highly intricate and packed with a variety of cells arranged differently by vehicle brand and model. This complexity, combined with the diverse materials used in each battery type, makes disassembly and material extraction challenging. The urgent need is for new technology or solutions that address the challenge of separate extraction processes for various elements [106]. Batteries are often assembled with various methods such as adhesive bonding, welding, and other connectors, which complicates the separation of materials. Additionally, battery degradation over time affects recyclability by altering the battery's form, which can pose health risks to those handling the recycling process [107]. This degradation contributes to the uncertainty in the recycling process, making it difficult to develop effective recycling solutions.

6.1.2. Potential for Interdisciplinary Collaboration and Knowledge Transfer

In the business world, battery knowledge is often closely guarded to maintain competitive advantage. While this secrecy serves business interests, it poses challenges for the engineering community, which relies on detailed data to understand battery lifecycles and improve technologies. Access to comprehensive battery information could lead to significant advancements in several areas. For instance, tracking the state of a battery throughout its life could enable the development of preventative maintenance measures. Additionally, understanding the battery's condition at the end of its life would facilitate the creation of more sophisticated disassembly robots. An example is the robot Daisy, which uses −176 °F freezing air to detach batteries from iPhone bodies [107]. To enhance recycling and battery technology, there needs to be greater knowledge sharing among companies. Tesla, for example, has embraced the open-source movement to advance EV technology, making valuable insights publicly available [108]. Embracing similar collaborative approaches, including integrating robotics and AI, could lead to more efficient and effective battery management and recycling solutions.

6.2. Roadmap for Future Circular EV Battery Initiatives

6.2.1. Key Steps toward Widespread Adoption of Circular Practices

The discussion of circular practices for EV batteries encompasses two key aspects: the technology enabling these practices and the businesses or policies that support them. In Europe, for instance, the concept of European Conformity (CE) embodies a comprehensive approach to circularity. CE involves both forward processes—such as production, material use, assembly, distribution, and consumption—and reverse processes—like repair, reconditioning, remanufacturing, recycling, and disposal [44]. This system evaluates products at the end of their lifecycle to determine which circular practices they qualify for, ensuring that materials are effectively managed and reused. However, the effectiveness of these circular practices is also heavily dependent on the technology available. Advanced technologies are crucial for efficient recycling and repurposing, and without them, the potential benefits of circular practices may be limited. Thus, while policies like CE play a vital role in promoting circularity, they must be supported by technological advancements to fully realize their potential.

6.2.2. Collaborative Efforts and Partnerships for Industry Transformation

From a business standpoint, maintaining secrecy and competitiveness is often seen as a strategic advantage. However, this approach can hinder the advancement of a circular economy for EV batteries. Managers and executives within the EV battery sector should consider adopting circular business models (CBMs) as an innovative means to extend product lifecycles and unlock new value propositions. Such models could foster cross-company and sector-wide collaborations, which are crucial for advancing recycling and repurposing technologies [109]. Recycling technology must be highly sophisticated to efficiently extract all valuable materials from EV batteries. Developing unique recycling processes for each individual EV design is impractical. Instead, focusing on creating a few versatile recycling processes for common battery designs would be more efficient and cost-effective. Some organizations are already pioneering this approach. For instance, South Korea has initiated a groundbreaking public-private partnership to establish a sustainable EV battery ecosystem through a memorandum of understanding (MOU) [110]. Such collaborations demonstrate the potential economic benefits of forming a circular economy. Research indicates that partnerships between manufacturers and retailers can achieve higher collection rates and overall welfare benefits [111].

This study highlights that willing collaboration, rather than stringent policies, is more effective in promoting recycling while minimizing adverse effects on welfare.

6.3. Trends towards a Circular Economy

6.3.1. Future Scientific Breakthroughs in EV Battery Recycling

There are numerous untapped breakthroughs in EV battery recycling that remain unexplored. While various methods are being investigated, the effects of thermal treatment in an oxidative atmosphere on the microstructure and composition of cathode and anode materials remain largely unstudied [112]. As noted, degradation of batteries can produce harmful substances, but promising advancements are on the horizon. Recent developments offer hope: new recycling processes are being researched that do not require the use of expensive or harmful chemicals [113]. Previously, battery recycling was considered hazardous due to the use of toxic chemicals, but this perception is shifting with these innovations. As the industry evolves, particularly with the growing prominence of EVs over internal combustion engines, a diverse range of facilities, capabilities, and approaches will emerge as more companies enter the battery recycling space [114]. Given the rapid pace of technological innovation, the transition from ICE to EVs holds significant potential for establishing a sustainable circular economy for EV batteries.

6.3.2. Increased Interest in EV Battery Research

In an era of rapid technological advancement, it is crucial to keep the future in focus. Developing a circular economy for EV batteries involves two main aspects: engineering innovations and incentive structures. Research into EV technology is supported by various sources, including government grants (e.g., National Science Foundation, National Institutes of Health), corporate R&D, and nonprofit foundations [115]. While these stakeholders are invested in research, their motivations can vary significantly. A key challenge to achieving a circular economy is overcoming planned obsolescence—the practice where products are intentionally designed to have a limited lifespan, encouraging repeat purchases. Addressing this issue is crucial for promoting recycling. The EV battery recycling market is projected to reach USD 15.8 billion by 2030, growing at a compound annual growth rate (CAGR) of 32.1% from 2023 to 2030 [116]. This indicates that a circular economy is not only feasible but also economically promising. However, obstacles like corporate greed can impede progress. For instance, a study suggests that if consumer demand were strong enough, EV manufacturers would recycle all batteries without additional incentives [117]. Despite this, EVs face social challenges, including skepticism about their environmental benefits and perceptions of their "coolness". If consumer pressure alone is not enough to drive a circular economy, government intervention becomes essential. Policies must be enacted to encourage collaboration and innovation in recycling. Suggested strategies include increasing funding for both incremental innovations and breakthroughs in recycling technology, supporting pilot projects that foster collaboration across the recycling value chain, and implementing market-pull measures to create a favorable economic and regulatory environment for large-scale EV battery recycling [27]. By aligning private sector interests with regulatory measures, we can foster a thriving, sustainable recycling ecosystem in this rapidly evolving technological landscape.

7. Conclusions

This review has explored various facets of the EV battery lifecycle, with a particular focus on achieving a circular economy. The crucial factor in this effort is managing the product's end-of-life effectively. Implementing processes that enable the recycling or repurposing of all EV batteries once they are no longer usable is essential for closing the loop of a circular economy. Such practices would not only reduce costs and enhance efficiency but also significantly improve environmental and societal outcomes. Despite the current challenges in scaling these processes, it is evident that the future holds promising advancements. As the demand for EV batteries grows, technological innovations will likely address these emerging issues. With ongoing research and development, the solutions needed to establish a robust circular economy for EV batteries are within reach.

The lifecycle of an EV battery remains a significant unknown, which contributes to consumer skepticism about purchasing EVs. One of the primary obstacles facing the EV industry is the limited transparency regarding battery lifespan. Addressing this issue requires both corporate initiatives and government regulations to drive meaningful change. For example, Cox Automotive's ALFRED system and health score aim to enhance transparency and build consumer confidence in EV transactions, akin to the trusted vehicle valuations provided by Kelley Blue Book for nearly a century [118]. While battery health is a critical concern for consumers, particularly those increasingly focused on environmental issues, there is also a demand for broader information beyond just battery life. To address this, the Global Battery Alliance (GBA) has introduced a novel approach to increase transparency. Their "battery passport" functions as a digital twin of the physical battery, capturing and storing vital information—such as provenance, composition, material flows, and manufacturing history—in a QR code [119]. This concept of maintaining a comprehensive record throughout a product's lifecycle could revolutionize how we manage and perceive product histories. Although the initial focus on batteries may seem niche, the principles of this approach have the potential to drive significant changes across various industries and even influence broader economic behaviors. The rise of EVs is undeniable, but achieving a full transition from internal combustion engines (ICEs) to EVs requires substantial effort and ambition. As stated, "Making the 2020s the decade of transition to EVs requires more ambition and action among both market leaders and followers" [120]. While EVs are gaining traction, they are not yet dominating the market. They remain more expensive than ICE vehicles, and their environmental benefits are somewhat tempered by the current state of power grids. This transition is a complex and labor-intensive process. Replacing the established ICE infrastructure with a clean, efficient EV ecosystem demands comprehensive changes, including local community preparations for this shift. For example, in Michigan, state and local leaders are actively preparing for the growth of EV and battery production facilities by creating programs to equip the workforce with necessary skills [121]. Just a few decades ago, the concept of EVs was virtually nonexistent in the minds of most consumers and workers. The rapid development of the EV industry aims to replace the ICE sector in a relatively short time, but the pace of this transition is slower than many would like, given the environmental urgency. President Biden's goal of achieving a 50% market share for EVs in the U.S. by 2030 underscores the high level of commitment needed [122]. While many countries have committed to this transition, the path to a circular economy for EV batteries is still in its early stages. Progress is being made at various levels, and while significant work remains, there is a clear and accelerating momentum towards a more sustainable future.

Funding: This research received no external funding.

Conflicts of Interest: The authors declare no conflict of interest.

References

1. Herrmann, F.; Rothfuss, F. Introduction to Hybrid Electric Vehicles, Battery Electric Vehicles, and off-Road Electric Vehicles. In *Advances in Battery Technologies for Electric Vehicles*; Elsevier: Amsterdam, The Netherlands, 2015; pp. 3–16.
2. Electric Vehicles. Available online: https://www.iea.org/energy-system/transport/electric-vehicles (accessed on 30 August 2024).
3. Trends in the Electric Vehicle Industry—Global EV Outlook 2024—Analysis. Available online: https://www.iea.org/reports/global-ev-outlook-2024/trends-in-the-electric-vehicle-industry (accessed on 6 August 2024).
4. Electric Vehicles—Worldwide | Statista Market Forecast. Available online: https://www.statista.com/outlook/mmo/electric-vehicles/worldwide (accessed on 10 August 2024).
5. What Are the Consumer and Societal Reactions towards Electric Vehicles? Available online: https://www.greenbook.org/insights/what-are-consumer-reactions-to-electric-vehicles (accessed on 6 August 2024).
6. Ibrahim, A.; Jiang, F. The Electric Vehicle Energy Management: An Overview of the Energy System and Related Modeling and Simulation. *Renew. Sustain. Energy Rev.* **2021**, *144*, 111049. [CrossRef]
7. Zhang, X.; Liang, Y.; Yu, E.; Rao, R.; Xie, J. Review of Electric Vehicle Policies in China: Content Summary and Effect Analysis. *Renew. Sustain. Energy Rev.* **2017**, *70*, 698–714. [CrossRef]

8. Mohammadzadeh, N.; Zegordi, S.H.; Kashan, A.H.; Nikbakhsh, E. Optimal Government Policy-Making for the Electric Vehicle Adoption Using the Total Cost of Ownership under the Budget Constraint. *Sustain. Prod. Consum.* **2022**, *33*, 477–507. [CrossRef]
9. Ralls, A.M.; Leong, K.; Clayton, J.; Fuelling, P.; Mercer, C.; Navarro, V.; Menezes, P.L. The Role of Lithium-Ion Batteries in the Growing Trend of Electric Vehicles. *Materials* **2023**, *16*, 6063. [CrossRef]
10. Das, U.K.; Shrivastava, P.; Tey, K.S.; Idris, M.Y.I.B.; Mekhilef, S.; Jamei, E.; Seyedmahmoudian, M.; Stojcevski, A. Advancement of Lithium-Ion Battery Cells Voltage Equalization Techniques: A Review. *Renew. Sustain. Energy Rev.* **2020**, *134*, 110227. [CrossRef]
11. How Do Electric Cars Work? An In-Depth Guide. Available online: https://www.edmunds.com/electric-car/articles/how-do-electric-cars-work.html (accessed on 6 August 2024).
12. Batteries Are a Key Part of the Energy Transition. Here's Why. Available online: https://www.weforum.org/agenda/2021/09/batteries-lithium-ion-energy-storage-circular-economy/ (accessed on 6 August 2024).
13. Morse, I. A Dead Battery Dilemma. Available online: https://www.sciencemagazinedigital.org/sciencemagazine/21_may_2021/MobilePagedArticle.action?articleId=1692939 (accessed on 6 August 2024).
14. Woollacott, E. Electric Cars: What Will Happen to All the Dead Batteries? *BBC News*, 27 April 2021.
15. Inc, C. Cyient | Blogs | Shyam Sundar Pal. Available online: https://www.cyient.com/blog/author/shyam-sundar-pal (accessed on 6 August 2024).
16. EV Manufacturers Looking to Decarbonize—And Profit from It | HERE. Available online: https://www.here.com/learn/blog/ev-battery-sustainability-recycling (accessed on 6 August 2024).
17. What Is the Circular Economy, and Why Does It Matter That It's Shrinking? Available online: https://www.weforum.org/agenda/2022/06/what-is-the-circular-economy/ (accessed on 6 August 2024).
18. Pereira, P.M.; Vieira, C.S. A Literature Review on the Use of Recycled Construction and Demolition Materials in Unbound Pavement Applications. *Sustainability* **2022**, *14*, 13918. [CrossRef]
19. Bhutada, G. The Key Minerals in an EV Battery. Available online: https://elements.visualcapitalist.com/the-key-minerals-in-an-ev-battery/ (accessed on 10 August 2024).
20. Policy Challenges for Accessing Critical Minerals to Electrify Vehicle Transport. Available online: https://www.rff.org/publications/reports/policy-challenges-for-accessing-critical-minerals-for-electric-vehicles/ (accessed on 10 August 2024).
21. Battery Recycling: 10 Breakthrough Technologies. 2023. Available online: https://www.technologyreview.com/2023/01/09/1064886/battery-recycling-10-breakthrough-technologies-2023/ (accessed on 6 August 2024).
22. Alternative Fuels Data Center: Batteries for Electric Vehicles. Available online: https://afdc.energy.gov/vehicles/electric-batteries (accessed on 6 August 2024).
23. Zhang, R.; Fujimori, S. The Role of Transport Electrification in Global Climate Change Mitigation Scenarios. *Environ. Res. Lett.* **2020**, *15*, 034019. [CrossRef]
24. Zubi, G.; Dufo-López, R.; Carvalho, M.; Pasaoglu, G. The Lithium-Ion Battery: State of the Art and Future Perspectives. *Renew. Sustain. Energy Rev.* **2018**, *89*, 292–308. [CrossRef]
25. How Lithium-Ion Batteries Work. Available online: https://www.energy.gov/energysaver/articles/how-lithium-ion-batteries-work (accessed on 6 August 2024).
26. Thi, M.T.; Kim, C.; Kwon, S.; Kang, H.; Ko, J.M.; Kim, J.; Choi, W. Investigation of the Properties of Anode Electrodes for Lithium–Ion Batteries Manufactured Using Cu, and Si-Coated Carbon Nanowall Materials. *Energies* **2023**, *16*, 1935. [CrossRef]
27. Beaudet, A.; Larouche, F.; Amouzegar, K.; Bouchard, P.; Zaghib, K. Key Challenges and Opportunities for Recycling Electric Vehicle Battery Materials. *Sustainability* **2020**, *12*, 5837. [CrossRef]
28. Larouche, F.; Tedjar, F.; Amouzegar, K.; Houlachi, G.; Bouchard, P.; Demopoulos, G.P.; Zaghib, K. Progress and Status of Hydrometallurgical and Direct Recycling of Li-Ion Batteries and Beyond. *Materials* **2020**, *13*, 801. [CrossRef] [PubMed]
29. Deng, J.; Bae, C.; Denlinger, A.; Miller, T. Electric Vehicles Batteries: Requirements and Challenges. *Joule* **2020**, *4*, 511–515. [CrossRef]
30. Yu, D.; Huang, Z.; Makuza, B.; Guo, X.; Tian, Q. Pretreatment Options for the Recycling of Spent Lithium-Ion Batteries: A Comprehensive Review. *Miner. Eng.* **2021**, *173*, 107218. [CrossRef]
31. Liu, Y.; Zhang, R.; Wang, J.; Wang, Y. Current and Future Lithium-Ion Battery Manufacturing. *iScience* **2021**, *24*, 102332. [CrossRef]
32. Konda, K.; Moodakare, S.B.; Kumar, P.L.; Battabyal, M.; Seth, J.R.; Juvekar, V.A.; Gopalan, R. Comprehensive Effort on Electrode Slurry Preparation for Better Electrochemical Performance of $LiFePO_4$ Battery. *J. Power Sources* **2020**, *480*, 228837. [CrossRef]
33. Jiao, X.; Kirianova, A.V.; Xu, X.; Kapitanova, O.O.; Krivchenko, V.A.; Napolskiy, F.S.; Volkov, V.S.; Gallyamov, M.O.; Liu, Y. Conductive Additives for Improving the Rate Capability of Cathode Materials in Secondary Lithium Batteries. *ACS Appl. Energy Mater.* **2023**, *6*, 2855–2862. [CrossRef]
34. Lithium-Ion Batteries and Fluoromaterials | Fluorochemicals | Daikin Global. Available online: https://www.daikinchemicals.com/magazine/column-lithium-ion-battery-vol1.html (accessed on 6 August 2024).
35. Luchkin, S.Y.; Kirsanova, M.A.; Aksyonov, D.A.; Lipovskikh, S.A.; Nikitina, V.A.; Abakumov, A.M.; Stevenson, K.J. Cycling-Driven Electrochemical Activation of Li-Rich NMC Positive Electrodes for Li-Ion Batteries. *ACS Appl. Energy Mater.* **2022**, *5*, 7758–7769. [CrossRef]
36. Lai, X.; Chen, Q.; Tang, X.; Zhou, Y.; Gao, F.; Guo, Y.; Bhagat, R.; Zheng, Y. Critical Review of Life Cycle Assessment of Lithium-Ion Batteries for Electric Vehicles: A Lifespan Perspective. *eTransportation* **2022**, *12*, 100169. [CrossRef]

37. Khan, I.; Hou, F.; Le, H.P. The Impact of Natural Resources, Energy Consumption, and Population Growth on Environmental Quality: Fresh Evidence from the United States of America. *Sci. Total Environ.* **2021**, *754*, 142222. [CrossRef]
38. Yuan, C.; Cao, H.; Shen, K.; Deng, Y.; Zeng, D.; Dong, Y.; Hauschild, M. Water-Based Manufacturing of Lithium Ion Battery for Life Cycle Impact Mitigation. *CIRP Ann.* **2021**, *70*, 25–28. [CrossRef]
39. Han, X.; Lu, L.; Zheng, Y.; Feng, X.; Li, Z.; Li, J.; Ouyang, M. A Review on the Key Issues of the Lithium Ion Battery Degradation among the Whole Life Cycle. *eTransportation* **2019**, *1*, 100005. [CrossRef]
40. Liu, K.; Hu, X.; Yang, Z.; Xie, Y.; Feng, S. Lithium-Ion Battery Charging Management Considering Economic Costs of Electrical Energy Loss and Battery Degradation. *Energy Convers. Manag.* **2019**, *195*, 167–179. [CrossRef]
41. Guo, J.; Li, Y.; Pedersen, K.; Stroe, D.-I. Lithium-Ion Battery Operation, Degradation, and Aging Mechanism in Electric Vehicles: An Overview. *Energies* **2021**, *14*, 5220. [CrossRef]
42. Requia, W.J.; Mohamed, M.; Higgins, C.D.; Arain, A.; Ferguson, M. How Clean Are Electric Vehicles? Evidence-Based Review of the Effects of Electric Mobility on Air Pollutants, Greenhouse Gas Emissions and Human Health. *Atmos. Environ.* **2018**, *185*, 64–77. [CrossRef]
43. Mossali, E.; Picone, N.; Gentilini, L.; Rodrìguez, O.; Pérez, J.M.; Colledani, M. Lithium-Ion Batteries towards Circular Economy: A Literature Review of Opportunities and Issues of Recycling Treatments. *J. Environ. Manage.* **2020**, *264*, 110500. [CrossRef]
44. Nurdiawati, A.; Agrawal, T.K. Creating a Circular EV Battery Value Chain: End-of-Life Strategies and Future Perspective. *Resour. Conserv. Recycl.* **2022**, *185*, 106484. [CrossRef]
45. MarketsandMarkets Battery Recycling Market Worth $23.2 Billion by 2025—Exclusive Report by MarketsandMarketsTM. Available online: https://www.prnewswire.com/news-releases/battery-recycling-market-worth-23-2-billion-by-2025--exclusive-report-by-marketsandmarkets-301051786.html (accessed on 29 August 2024).
46. Company, A.B.T. American Battery Technology Company Receives Contract for $57M Grant from US DOE for Construction of Commercial-Scale Lithium Hydroxide Refinery. Available online: https://www.prnewswire.com/news-releases/american-battery-technology-company-receives-contract-for-57m-grant-from-us-doe-for-construction-of-commercial-scale-lithium-hydroxide-refinery-301919344.html (accessed on 29 August 2024).
47. Krishnasamy, R. Li-Cycle Leads the Recycling of Lithium-Ion Batteries with High Resource Recovery Technology. Available online: https://techblitz.com/li-cycle/ (accessed on 29 August 2024).
48. Revolt: Closing the Loop on Batteries. Available online: https://northvolt.com/articles/revolt/ (accessed on 29 August 2024).
49. Liu, C.; Lin, J.; Cao, H.; Zhang, Y.; Sun, Z. Recycling of Spent Lithium-Ion Batteries in View of Lithium Recovery: A Critical Review. *J. Clean. Prod.* **2019**, *228*, 801–813. [CrossRef]
50. Baum, Z.J.; Bird, R.E.; Yu, X.; Ma, J. Lithium-Ion Battery Recycling—Overview of Techniques and Trends. *ACS Energy Lett.* **2022**, *7*, 712–719. [CrossRef]
51. Harper, G.; Sommerville, R.; Kendrick, E.; Driscoll, L.; Slater, P.; Stolkin, R.; Walton, A.; Christensen, P.; Heidrich, O.; Lambert, S.; et al. Recycling Lithium-Ion Batteries from Electric Vehicles. *Nature* **2019**, *575*, 75–86. [CrossRef]
52. Alipanah, M.; Reed, D.; Thompson, V.; Fujita, Y.; Jin, H. Sustainable Bioleaching of Lithium-Ion Batteries for Critical Materials Recovery. *J. Clean. Prod.* **2023**, *382*, 135274. [CrossRef]
53. Hua, Y.; Zhou, S.; Huang, Y.; Liu, X.; Ling, H.; Zhou, X.; Zhang, C.; Yang, S. Sustainable Value Chain of Retired Lithium-Ion Batteries for Electric Vehicles. *J. Power Sources* **2020**, *478*, 228753. [CrossRef]
54. Tanong, K.; Coudert, L.; Mercier, G.; Blais, J.-F. Recovery of Metals from a Mixture of Various Spent Batteries by a Hydrometallurgical Process. *J. Environ. Manag.* **2016**, *181*, 95–107. [CrossRef] [PubMed]
55. The World Is Powered by Batteries, but Who Is Leading the Charge in Battery Recycling? | Nasdaq. Available online: https://www.nasdaq.com/press-release/the-world-is-powered-by-batteries-but-who-is-leading-the-charge-in-battery-recycling (accessed on 29 August 2024).
56. Velázquez-Martínez, O.; Valio, J.; Santasalo-Aarnio, A.; Reuter, M.; Serna-Guerrero, R. A Critical Review of Lithium-Ion Battery Recycling Processes from a Circular Economy Perspective. *Batteries* **2019**, *5*, 68. [CrossRef]
57. Ahuis, M.; Doose, S.; Vogt, D.; Michalowski, P.; Zellmer, S.; Kwade, A. Recycling of Solid-State Batteries. *Nat. Energy* **2024**, *9*, 373–385. [CrossRef]
58. Wang, H.; Cao, X.; Liu, W.; Sun, X. Research Progress of the Solid State Lithium-Sulfur Batteries. *Front. Energy Res.* **2019**, *7*, 112. [CrossRef]
59. Pregowska, A.; Osial, M.; Urbańska, W. The Application of Artificial Intelligence in the Effective Battery Life Cycle in the Closed Circular Economy Model—A Perspective. *Recycling* **2022**, *7*, 81. [CrossRef]
60. Hua, Y.; Liu, X.; Zhou, S.; Huang, Y.; Ling, H.; Yang, S. Toward Sustainable Reuse of Retired Lithium-Ion Batteries from Electric Vehicles. *Resour. Conserv. Recycl.* **2021**, *168*, 105249. [CrossRef]
61. Costa, C.M.; Barbosa, J.C.; Gonçalves, R.; Castro, H.; Campo, F.J.D.; Lanceros-Méndez, S. Recycling and Environmental Issues of Lithium-Ion Batteries: Advances, Challenges and Opportunities. *Energy Storage Mater.* **2021**, *37*, 433–465. [CrossRef]
62. Kotak, Y.; Marchante Fernández, C.; Canals Casals, L.; Kotak, B.S.; Koch, D.; Geisbauer, C.; Trilla, L.; Gómez-Núñez, A.; Schweiger, H.-G. End of Electric Vehicle Batteries: Reuse vs. Recycle. *Energies* **2021**, *14*, 2217. [CrossRef]
63. Ahmadi, L.; Young, S.B.; Fowler, M.; Fraser, R.A.; Achachlouei, M.A. A Cascaded Life Cycle: Reuse of Electric Vehicle Lithium-Ion Battery Packs in Energy Storage Systems. *Int. J. Life Cycle Assess.* **2017**, *22*, 111–124. [CrossRef]

64. Olsson, L.; Fallahi, S.; Schnurr, M.; Diener, D.; Van Loon, P. Circular Business Models for Extended EV Battery Life. *Batteries* **2018**, *4*, 57. [CrossRef]
65. Striking Gold with EV Battery Recycling. Available online: https://www.bcg.com/publications/2023/striking-gold-with-ev-battery-recycling (accessed on 7 August 2024).
66. Sommerville, R.; Shaw-Stewart, J.; Goodship, V.; Rowson, N.; Kendrick, E. A Review of Physical Processes Used in the Safe Recycling of Lithium Ion Batteries. *Sustain. Mater. Technol.* **2020**, *25*, e00197. [CrossRef]
67. Overview of EV Batteries | GreenCars. Available online: https://www.greencars.com/greencars-101/overview-of-ev-batteries (accessed on 7 August 2024).
68. Nissan Partners with Enel to Launch Innovative "Second Life" Storage System for Used Electric Car Batteries. Available online: http://europe.nissannews.com/en-GB/releases/nissan-partners-with-enel-to-launch-innovative-second-life-storage-system-for-used-electric-car-batteries (accessed on 29 August 2024).
69. BMW Group UK Second-Life Battery Solution in Partnership with off Grid Energy. Available online: https://www.press.bmwgroup.com/united-kingdom/article/detail/T0318650EN_GB/bmw-group-uk-second-life-battery-solution-in-partnership-with-off-grid-energy?language=en_GB (accessed on 29 August 2024).
70. Mining for Electric Car Batteries: What You Need to Know. Available online: https://www.capitalone.com/cars/learn/finding-the-right-car/mining-for-electric-car-batteries-what-you-need-to-know/2395 (accessed on 7 August 2024).
71. Potts, E. Total Material Recovery: State-of-the-Art Battery Recycling for the EV Supply Chain. *Innovation News Network*, 23 November 2023.
72. Your Depleted EV Battery Pack Could Power the Grid after Sunset. Available online: https://www.autoweek.com/news/technology/a43937159/b2u-company-reusing-ev-battery-packs-to-store-solar-energy/ (accessed on 7 August 2024).
73. Tawonezvi, T.; Nomnqa, M.; Petrik, L.; Bladergroen, B.J. Recovery and Recycling of Valuable Metals from Spent Lithium-Ion Batteries: A Comprehensive Review and Analysis. *Energies* **2023**, *16*, 1365. [CrossRef]
74. Yun, L.; Linh, D.; Shui, L.; Peng, X.; Garg, A.; LE, M.L.P.; Asghari, S.; Sandoval, J. Metallurgical and Mechanical Methods for Recycling of Lithium-Ion Battery Pack for Electric Vehicles. *Resour. Conserv. Recycl.* **2018**, *136*, 198–208. [CrossRef]
75. Pulling The Weight of Heavy Truck Decarbonization. Available online: https://rmi.org/insight/pulling-the-weight-of-heavy-truck-decarbonization/ (accessed on 7 August 2024).
76. Mihai, D.M.; Doran, M.D.; Puiu, S.; Doran, N.M.; Jianu, E.; Cojocaru, T.M. Managing Environmental Policy Stringency to Ensure Sustainable Development in OECD Countries. *Sustainability* **2023**, *15*, 15427. [CrossRef]
77. Picatoste, A.; Justel, D.; Mendoza, J.M.F. Circularity and Life Cycle Environmental Impact Assessment of Batteries for Electric Vehicles: Industrial Challenges, Best Practices and Research Guidelines. *Renew. Sustain. Energy Rev.* **2022**, *169*, 112941. [CrossRef]
78. Yang, Y.; Okonkwo, E.G.; Huang, G.; Xu, S.; Sun, W.; He, Y. On the Sustainability of Lithium Ion Battery Industry—A Review and Perspective. *Energy Storage Mater.* **2021**, *36*, 186–212. [CrossRef]
79. Heelan, J.; Gratz, E.; Zheng, Z.; Wang, Q.; Chen, M.; Apelian, D.; Wang, Y. Current and Prospective Li-Ion Battery Recycling and Recovery Processes. *JOM* **2016**, *68*, 2632–2638. [CrossRef]
80. Bawankar, S.; Dwivedi, G.; Nanda, I.; Macedo, V.D.J.; Kesharvani, S.; Meshram, K.; Jain, S.; Mishra, S.; Singh, V.P.; Verma, P. Environmental Impact Assessment of Lithium Ion Battery Employing Cradle to Grave. *Sustain. Energy Technol. Assess.* **2023**, *60*, 103530. [CrossRef]
81. Ethics, Politics and Lithium Mining: In Conversation with Zayn Kaylan of Infinity Stone. Available online: https://mine.nridigital.com/mine_may23/lithium_ethics_infinity_stone (accessed on 10 August 2024).
82. Mekonnen, Y.; Sundararajan, A.; Sarwat, A.I. A Review of Cathode and Anode Materials for Lithium-Ion Batteries. In Proceedings of the SoutheastCon, Norfolk, VA, USA, 30 March–3 April 2016; pp. 1–6.
83. BU-808b: What Causes Li-Ion to Die? Available online: https://batteryuniversity.com/article/bu-808b-what-causes-li-ion-to-die (accessed on 7 August 2024).
84. U.S. Environmental Protection Agency. Lithium-Ion Battery Recycling. Available online: https://www.epa.gov/hw/lithium-ion-battery-recycling (accessed on 7 August 2024).
85. Miao, Y.; Hynan, P.; von Jouanne, A.; Yokochi, A. Current Li-Ion Battery Technologies in Electric Vehicles and Opportunities for Advancements. *Energies* **2019**, *12*, 1074. [CrossRef]
86. Nickel—Element Information, Properties and Uses | Periodic Table. Available online: https://www.rsc.org/periodic-table/element/28/nickel (accessed on 7 August 2024).
87. Cobalt Statistics and Information | U.S. Geological Survey. Available online: https://www.usgs.gov/centers/national-minerals-information-center/cobalt-statistics-and-information (accessed on 7 August 2024).
88. Graphite—Definition, Structures, Applications, Properties, Use with Videos and FAQs of Graphite. Available online: https://byjus.com/chemistry/graphite/ (accessed on 7 August 2024).
89. Yang, J.-L.; Zhao, X.-X.; Li, W.-H.; Liang, H.-J.; Gu, Z.-Y.; Liu, Y.; Du, M.; Wu, X.-L. Advanced Cathode for Dual-Ion Batteries: Waste-to-Wealth Reuse of Spent Graphite from Lithium-Ion Batteries. *eScience* **2022**, *2*, 95–101. [CrossRef]
90. Rey, I.; Vallejo, C.; Santiago, G.; Iturrondobeitia, M.; Lizundia, E. Environmental Impacts of Graphite Recycling from Spent Lithium-Ion Batteries Based on Life Cycle Assessment. *ACS Sustain. Chem. Eng.* **2021**, *9*, 14488–14501. [CrossRef]
91. Mamidi, S.; Pandey, A.K.; Pathak, A.D.; Rao, T.N.; Sharma, C.S. Pencil Lead Powder as a Cost-Effective and High-Performance Graphite-Silica Composite Anode for High Performance Lithium-Ion Batteries. *J. Alloys Compd.* **2021**, *872*, 159719. [CrossRef]

92. Lithium Ion Battery Recycling. Available online: https://www.cas.org/resources/cas-insights/lithium-ion-battery-recycling (accessed on 7 August 2024).
93. Recycling Facts from around the World—Purpose Rising Blog. Available online: https://repurpose.global/blog/post/recycling-facts-from-around-the-world (accessed on 7 August 2024).
94. EU: Recycling Rate of Municipal Waste. Available online: https://www.statista.com/statistics/632839/municipal-waste-recycling-eu/ (accessed on 7 August 2024).
95. Kappalman, S. Updated Report: The West Coast Contamination Initiative: Results from California, Oregon, and Washington. Available online: https://recyclingpartnership.org/blog-the-west-coast-contamination-initiative-results-from-california-oregon-and-washington/ (accessed on 7 August 2024).
96. New Survey Discovers Why Most People Don't Recycle | WANE 15. Available online: https://www.wane.com/news/new-survey-discovers-why-most-people-dont-recycle/ (accessed on 7 August 2024).
97. 50 States of Recycling. Available online: https://www.ball.com/sustainability/real-circularity/50-states-of-recycling (accessed on 10 August 2024).
98. Lithium Mining Projects May Not Be Green Friendly—The New York Times. Available online: https://www.nytimes.com/2021/05/06/business/lithium-mining-race.html (accessed on 7 August 2024).
99. Williams, B.D.; Lipman, T.E. Strategy for Overcoming Cost Hurdles of Plug-In–Hybrid Battery in California: Integrating Post-Vehicle Secondary Use Values. *Transp. Res. Rec.* **2010**, *2191*, 59–66. [CrossRef]
100. Ng, M.-F.; Zhao, J.; Yan, Q.; Conduit, G.J.; Seh, Z.W. Predicting the State of Charge and Health of Batteries Using Data-Driven Machine Learning. *Nat. Mach. Intell.* **2020**, *2*, 161–170. [CrossRef]
101. Sun, S.; Jin, C.; He, W.; Li, G.; Zhu, H.; Huang, J. Management Status of Waste Lithium-Ion Batteries in China and a Complete Closed-Circuit Recycling Process. *Sci. Total Environ.* **2021**, *776*, 145913. [CrossRef]
102. Ozymy, D.J.; Ozymy, D.M.J. Toxic Criminals: Prosecuting Individuals for Hazardous Waste Crimes Under the United States Resource Conservation and Recovery Act. *Sustain. Dev. Law Policy* **2023**, *23*, 5.
103. House, T.W. Fact Sheet: The Bipartisan Infrastructure Deal. Available online: https://www.whitehouse.gov/briefing-room/statements-releases/2021/11/06/fact-sheet-the-bipartisan-infrastructure-deal/ (accessed on 7 August 2024).
104. Here's What the Future of Battery Recycling Is Going to Look like for EV Owners | Electrek. Available online: https://electrek.co/2022/03/21/heres-what-the-future-of-battery-recycling-is-going-to-look-like-for-ev-owners/ (accessed on 7 August 2024).
105. Heath, G.A.; Ravikumar, D.; Hansen, B.; Kupets, E. A Critical Review of the Circular Economy for Lithium-Ion Batteries and Photovoltaic Modules—Status, Challenges, and Opportunities. *J. Air Waste Manag. Assoc.* **2022**, *72*, 478–539. [CrossRef]
106. Slattery, M.; Dunn, J.; Kendall, A. Charting the Electric Vehicle Battery Reuse and Recycling Network in North America. *Waste Manag.* **2024**, *174*, 76–87. [CrossRef]
107. Meng, K.; Xu, G.; Peng, X.; Youcef-Toumi, K.; Li, J. Intelligent Disassembly of Electric-Vehicle Batteries: A Forward-Looking Overview. *Resour. Conserv. Recycl.* **2022**, *182*, 106207. [CrossRef]
108. All Our Patent Are Belong to You. Available online: https://www.tesla.com/blog/all-our-patent-are-belong-you (accessed on 7 August 2024).
109. Chirumalla, K.; Kulkov, I.; Parida, V.; Dahlquist, E.; Johansson, G.; Stefan, I. Enabling Battery Circularity: Unlocking Circular Business Model Archetypes and Collaboration Forms in the Electric Vehicle Battery Ecosystem. *Technol. Forecast. Soc. Chang.* **2024**, *199*, 123044. [CrossRef]
110. Kia Establishes Multi-Stakeholder Partnership to Drive Circular Economy in EV Battery Industry. Available online: https://www.kianewscenter.com/images-and-videos/image/kia-establishes-multi-stakeholder-partnership-to-drive-circular-economy-in-ev-battery-industry/a/5013cf97-2b59-43bd-a4a1-bd803cb3ced3 (accessed on 7 August 2024).
111. Zhang, Q.; Tang, Y.; Bunn, D.; Li, H.; Li, Y. Comparative Evaluation and Policy Analysis for Recycling Retired EV Batteries with Different Collection Modes. *Appl. Energy* **2021**, *303*, 117614. [CrossRef]
112. Balachandran, S.; Forsberg, K.; Lemaître, T.; Vieceli, N.; Lombardo, G.; Petranikova, M. Comparative Study for Selective Lithium Recovery via Chemical Transformations during Incineration and Dynamic Pyrolysis of EV Li-Ion Batteries. *Metals* **2021**, *11*, 1240. [CrossRef]
113. Boyo, S. Here's How EV Batteries Can Be Given a Second Life. Available online: https://www.cnbc.com/2023/03/29/heres-how-ev-batteries-can-be-given-a-second-life.html (accessed on 7 August 2024).
114. Bajolle, H.; Lagadic, M.; Louvet, N. The Future of Lithium-Ion Batteries: Exploring Expert Conceptions, Market Trends, and Price Scenarios. *Energy Res. Soc. Sci.* **2022**, *93*, 102850. [CrossRef]
115. Identifying Government Grants and Funding Opportunities. Available online: https://fastercapital.com/keyword/identifying-government-grants-and-funding-opportunities.html (accessed on 10 August 2024).
116. Meticulous Market Research Pvt. Ltd. EV Battery Recycling Market to Be Worth $15.8 Billion by 2030—Exclusive Report by Meticulous Research®. Available online: https://www.globenewswire.com/en/news-release/2023/03/23/2633129/0/en/EV-Battery-Recycling-Market-to-be-Worth-15-8-Billion-by-2030-Exclusive-Report-by-Meticulous-Research.html (accessed on 7 August 2024).
117. Luo, D. Research on the Influence of Consumers' Environmental Preference and Government Policy on EV Battery Recycling. *IOP Conf. Ser. Earth Environ. Sci.* **2021**, *651*, 042057. [CrossRef]

118. Cox Automotive Mobility Takes Charge of EV Battery Lifecycle Services with Global Acquisition—Cox Automotive Inc. Available online: https://www.coxautoinc.com/news/cox-automotive-mobility-takes-charge-of-ev-battery-lifecycle-services-with-global-acquisition/ (accessed on 7 August 2024).
119. The US Lags on Battery Sustainability and Supply Chain Transparency—Can a Digital 'Passport' Be the Answer? Available online: https://www.utilitydive.com/news/battery-sustainability-supply-chain-passport-tesla/642541/ (accessed on 7 August 2024).
120. Policies to Promote Electric Vehicle Deployment—Global EV Outlook 2021—Analysis. Available online: https://www.iea.org/reports/global-ev-outlook-2021/policies-to-promote-electric-vehicle-deployment (accessed on 7 August 2024).
121. Michigan Workforce Hub Will Focus on EVs, Training. Available online: https://www.govtech.com/transportation/michigan-workforce-hub-will-focus-on-evs-training (accessed on 10 August 2024).
122. House, T.W. FACT SHEET: The Biden-Harris Electric Vehicle Charging Action Plan. Available online: https://www.whitehouse.gov/briefing-room/statements-releases/2021/12/13/fact-sheet-the-biden-harris-electric-vehicle-charging-action-plan/ (accessed on 7 August 2024).

recycling

MDPI

Review

Sustainable Recycling of End-of-Life Electric Vehicle Batteries: EV Battery Recycling Frameworks in China and the USA

Amjad Ali [1,2,*], Mujtaba Al Bahrani [2], Shoaib Ahmed [1,3], Md Tasbirul Islam [1], Sikandar Abdul Qadir [1] and Muhammad Shahid [1]

1 Interdisciplinary Research Center for Sustainable Energy Systems (IRC-SES), King Fahd University of Petroleum & Minerals, Dhahran 31261, Saudi Arabia; shoaib.ahmed@unibas.it (S.A.); mdtasbirul.islam@kfupm.edu.sa (M.T.I.); sikandar.abdulqadir@kfupm.edu.sa (S.A.Q.); muhammad.shahid.1@kfupm.edu.sa (M.S.)
2 Electrical Engineering Department, King Fahd University of Petroleum & Minerals, Dhahran 31261, Saudi Arabia; s201918850@kfupm.edu.sa
3 Department of Engineering, University of Basilicata, Via dell'Ateneo Lucano, 10, 85100 Potenza, Italy
* Correspondence: amjad.ali@kfupm.edu.sa

Abstract: The increasing adoption of electric vehicles (EVs) has led to a surge in end-of-life (EOL) lithium-ion batteries (LIBs), necessitating efficient recycling strategies to mitigate environmental risks and recover critical materials. This study compares the EV battery recycling frameworks in China and the United States, focusing on policy effectiveness, technological advancements, and material recovery efficiencies. China's extended producer responsibility (EPR) policies and 14th Five-Year Plan mandate strict recycling targets, achieving a 40% battery recycling rate with 90% material recovery efficiency. Hydrometallurgical methods dominate, reducing energy consumption by 50% compared to virgin material extraction. The US, leveraging incentive-based mechanisms and private sector innovations, has a 35% recycling rate but a higher 95% resource recovery efficiency, mainly due to direct recycling and AI-based sorting technologies. Despite these advancements, challenges remain, including high recycling costs, inconsistent global regulations, and supply chain inefficiencies. To enhance sustainability, this study recommends harmonized international policies, investment in next-generation recycling technologies, and second-life battery applications. Emerging innovations, such as AI-driven sorting and direct cathode regeneration, could increase recovery efficiency by 20–30%, further reducing lifecycle costs. By integrating synergistic policies and advanced recycling infrastructures, China and the US can set a global precedent for sustainable EV battery management, driving the transition toward a circular economy. Future research should explore life cycle cost analysis and battery reuse strategies to optimize long-term sustainability.

Keywords: electric vehicle; battery; battery waste; battery waste recycling; battery waste recycling policies and regulation

1. Introduction

In today's world, batteries are indispensable, powering everything from massive industrial machinery to tiny consumer electronics. Over the past few decades, the global battery market has experienced unprecedented growth, driven by technological advancements by the increasing adoption of electric vehicles (EVs), and the proliferation of consumer electronics. As shown in Figure 1 [1], the global battery market is poised for substantial growth from 0.3 terawatt-hours (TWh) in 2020 to 1.8 TWh in 2024. With the accelerating energy transition and the push towards decarbonization by countries and consumers, global battery demand could surge exponentially, potentially reaching nearly nine terawatt-hours by 2030, five times the levels seen in 2024. Therefore, for stakeholders hoping to take advantage of the opportunities provided by this quickly changing market, knowing the fundamental drivers of this demand and the ramification of diverse businesses is vital.

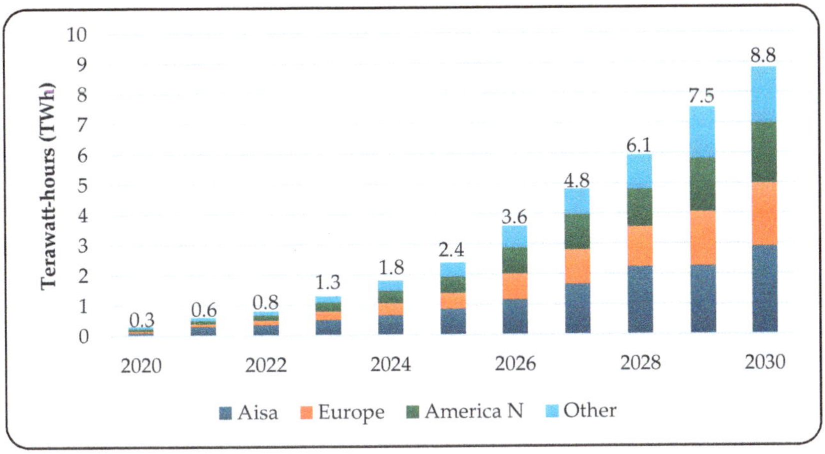

Figure 1. The expansion of batteries worldwide.

The global usage of batteries is extensive and varied, reflecting their importance across multiple aspects of modern life. Various types of batteries cater to specific applications, each offering distinct benefits. Batteries are essential for numerous technological advancements, from improving the convenience of consumer electronics to facilitating switching to electric vehicles.

According to Figure 2 [2], Lithium-ion (Li-ion) batteries dominate the market, accounting for approximately 75 to 80 percent of usage in various applications. Their high energy density, long life cycle, and efficiency account for this widespread use. The characteristics of (Li-ion) batteries make them ideal for (EVs), and a wide range of other applications.

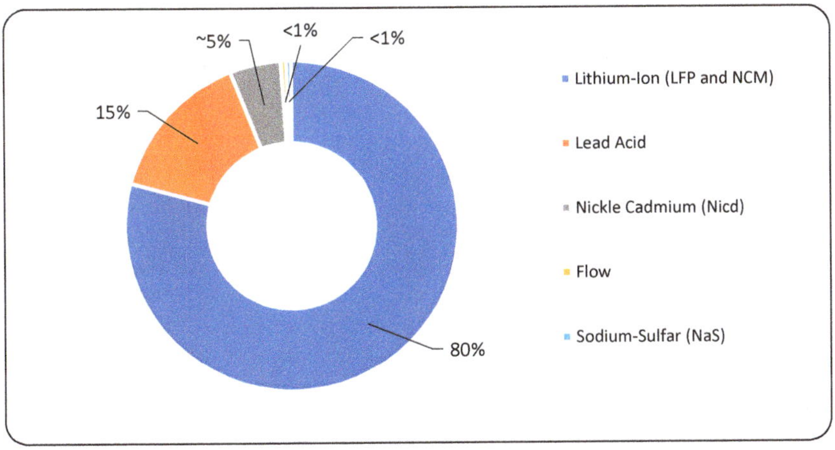

Figure 2. Percentage usage of batteries in various fields.

In the transportation sector, batteries are revolutionizing the use of (EVs) in various forms. (Li-ion) batteries are critical in replacing traditional gasoline-powered vehicles with electric models. Figure 3 [3] shows that (Li-ion) batteries are primarily used in five crucial applications, with the largest share going to passenger electric vehicles. The demand for these batteries used in (EVs) has increased dramatically, from nearly 12 GWh in 2015 to an anticipated 1400 GWh by 2030. Overall, the need for (Li-ion) batteries has surged from 20 GWh in 2010 to 290 GWh in 2019, with forecasts suggesting it will reach 2000 GWh by 2030, accounting for roughly 10% of the global energy supply.

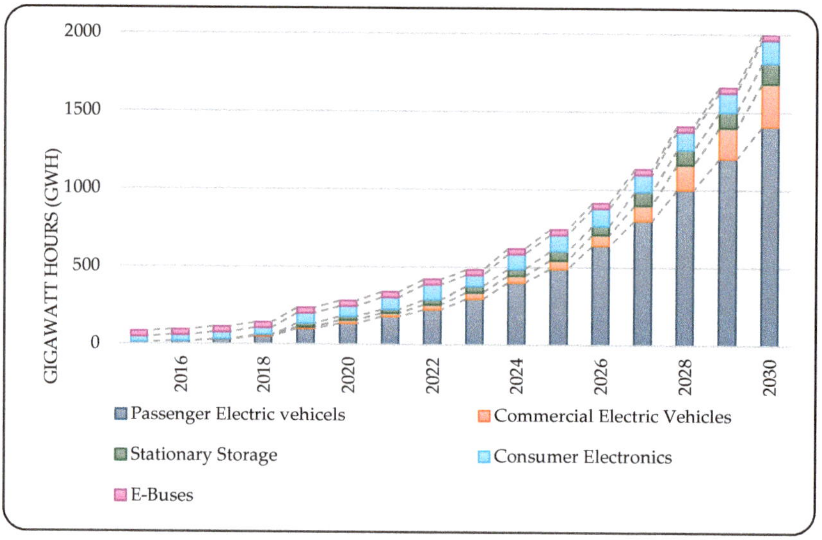

Figure 3. Global applications of lithium-ion batteries across various sectors.

The widespread adoption of Li-ion batteries is fueled by their high energy density, long life cycle, and decreasing costs, making them the preferred choice across various industries. These batteries are vital for the growth of EVs, portable electronics, and industrial applications. Li-ion batteries excel in these fields due to three key advantages: they provide more energy per unit weight, ensuring efficiency in portable and space-constrained applications; they have a long life cycle, allowing for repeated charging and discharging with minimal capacity loss, ensuring longevity and reliability; and they have a low self-discharge rate, making them ideal for devices requiring long standby times [4,5]. Understanding the diverse applications and increasing demand for Li-ion batteries is crucial for stakeholders navigating this evolving industry. Ongoing advancements in battery technology continue to improve performance, safety, and environmental impact, further accelerating their adoption.

Governments and regulatory organizations are implementing policies to control battery lifecycles more frequently, such as mandated recycling, financial incentives for recycling infrastructure, and safe disposal guidelines. As illustrated in Figure 4 [6], the percentage of EOL batteries, particularly from EVs, is projected to grow from 52% in 2020 to 94% by 2040, addressing environmental pollution concerns due to the rise in EOL batteries—technological advancements in recycling aim to improve efficiency and cost-effectiveness while fostering sustainable growth. Global efforts are essential to maximizing the benefits of Li-ion batteries while minimizing their environmental and economic impacts.

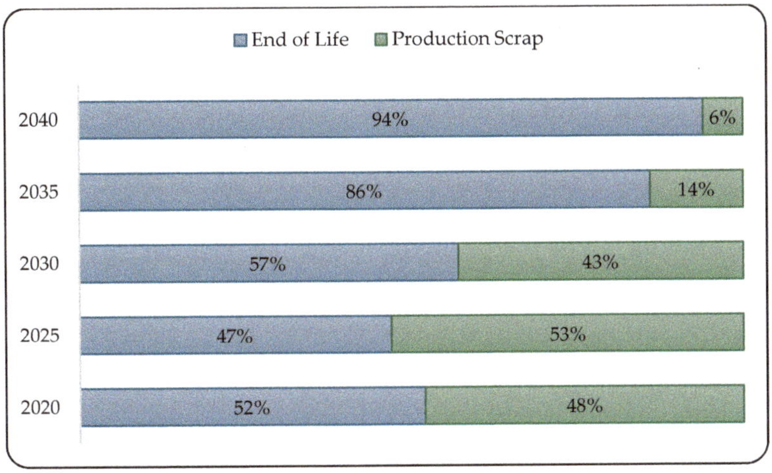

Figure 4. The global supply of EV batteries for recycling.

As shown in Figure 5, the number of electric vehicles (EVs) in use was substantial in 2023 and is expected to increase. China holds the largest share with 6.2 million EVs, while the U.K. has the lowest among the mentioned countries, with 370,000 EVs. Driven by goals to reduce fuel demand and lower pollution from car emissions, China and the US have implemented stringent rules and regulations to advance EVs development. The US encourages EV adoption through federal tax credits, state-level mandates, and emissions standards, with policies like the Clean Vehicle Tax Credit and California's ZEV program. China's 14th Five-Year Plan includes significant investments in charging infrastructure, NEV mandates, and incentives like purchase subsidies and tax exemptions. These initiatives aim to reduce carbon emissions, enhance transportation sustainability, and improve grid flexibility.

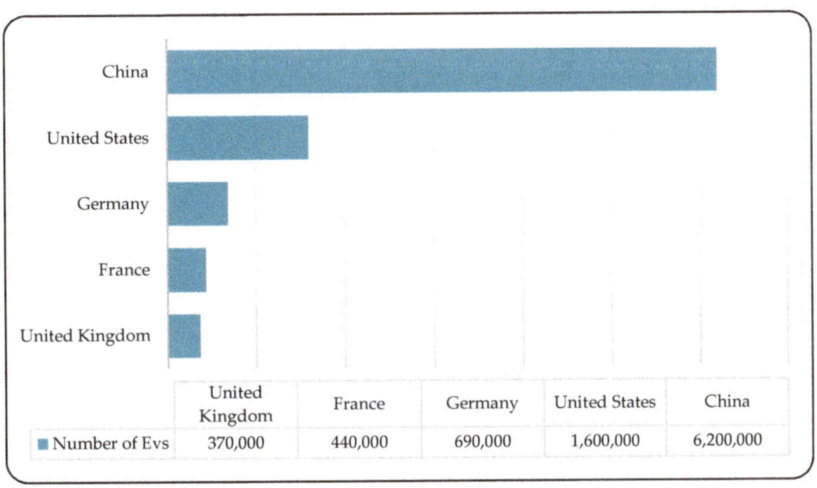

Figure 5. Leading countries in electric vehicle adoption.

This paper investigates the increasing need for energy storage in various applications, especially in EVs, highlighting the extensive application of Li-ion batteries because of their efficiency, long life cycle, and high energy density. Significant environmental issues arise from handling EOL waste and sourcing raw materials like cobalt and lithium.

This paper assesses the financial and technical obstacles to Li-ion battery recycling and explores recent technological advancements designed to enhance recycling profitability and efficiency. It also scrutinizes international laws and regulations governing the life cycle of Li-ion batteries used in EVs, such as mandatory recycling programs and financial incentives. The focus is on the measures adopted by the US and China to mitigate adverse environmental and economic impacts, instilling a sense of optimism about the future of battery recycling.

2. Literature Review

In managing EOL EV battery waste, the recent literature reveals a growing body of research addressing various aspects of waste recycling. EV batteries is a multi-step process that typically involves pre-processing (disassembly and materials separation) followed by one of three main metallurgical approaches. Key recycling methods include:

- Pyrometallurgical recycling (smelting): a high-temperature process where battery modules or cells are melted in a furnace. Pyrometallurgy can efficiently recover valuable metals like nickel, cobalt, and copper by incorporating them into an alloy or slag, but it burns off lighter components, notably lithium, electrolytes, plastics, and aluminum, which are lost or end up in waste slag.
- Hydrometallurgical recycling (chemical leaching): a wet chemical process that has become the state-of-the-art in battery recycling due to its high recovery efficiencies. After mechanical pre-processing (disassembling packs and shredding cells to produce a concentrated "black mass" rich in battery metals), hydrometallurgy uses aqueous chemistry to dissolve metals into solution via acid or base leaching. Subsequent purification and precipitation steps then selectively recover metals as salts or oxides (e.g., lithium carbonate, nickel, and cobalt sulfates). Modern hydrometallurgical processes can recover 90–99% of critical metals like cobalt, nickel, and lithium.
- Direct recycling (cathode-to-cathode): an emerging technique that seeks to recover the battery's cathode material intact for direct re-use, rather than breaking it down into elemental constituents. In direct recycling, the electrode (especially the cathode coating, such as lithium nickel manganese cobalt oxide or lithium iron phosphate) is physically separated and treated to restore its performance—for example, by removing degradation products and re-lithiating the cathode to replenish lost lithium. This preserved cathode powder can potentially be reintroduced into new battery production without the intensive energy or chemical processes of smelting or leaching. The promise of direct recycling is a lower energy, lower emission route which avoids the multiple conversion steps of other methods, and thus can be the most environmentally benign.

However, direct recycling also faces several challenges, including the complexity of separating and identifying different cathode chemistries from mixed battery waste streams. It requires battery designs that allow for easier disassembly, and standardized cell formats to be more effective. In addition, the quality and performance of the recycled cathode material may vary depending on the degree of degradation, which can affect its reuse potential in high-performance applications. These technical and logistical limitations may restrict the scalability of this method.

Over the past five years, from 2019 to 2024, approximately 1440 papers have been published on EV battery waste recycling, highlighting this issue's global relevance and importance. These papers encompass diverse topics, including technological advancements in recycling processes, strategies for improving recycling efficiency, and comprehensive reviews of international trends and future perspectives.

As shown in Figure 6, the distribution of research papers over time demonstrates the growing emphasis on EV battery waste management. However, it also highlights a substantial gap in understanding regulatory regimes. This gap is further underscored in Table 1, which contains fourteen significant articles that address laws and regulations related to EV batteries' (EOL) management. A more concentrated study demonstrates the urgent need for thorough policy analysis and development. Such efforts could have a revolutionary impact, promoting sustainable EV battery waste management methods and aiding countries like the Middle East region, which is increasingly adopting electric vehicles.

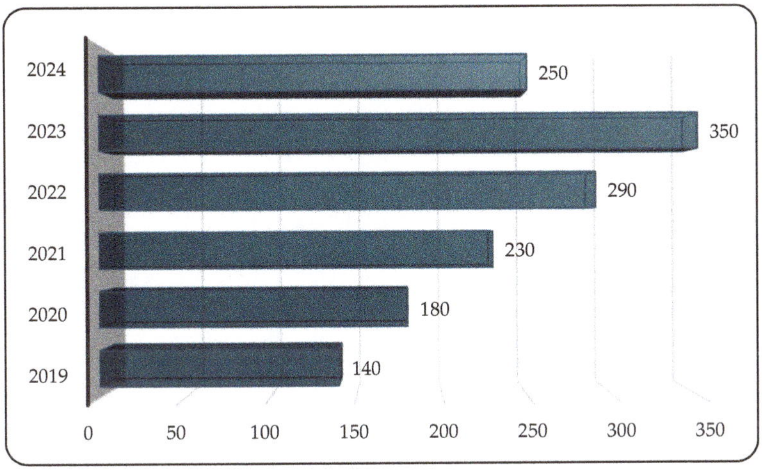

Figure 6. Research articles published on EV battery waste recycling in different areas (2019–2024).

Table 1. Research articles published on EV battery EOL waste recycling policies and regulations (2019–2024).

S No.	Paper Title	Journal	Year of Publication
1	Life cycle assessment of secondary use and physical recycling of lithium-ion batteries retired from electric vehicles in China [1]	Waste Management	2024
2	A Comprehensive Review of Lithium-Ion Battery (LiB) Recycling Technologies and Industrial Market Trend Insights [2]	Recycling	2024
3	Study on the impact of government policies on power battery recycling under different recycling models [3]	Journal of Cleaner Production	2023
4	Recycling technologies, policies, prospects, and challenges for spent batteries [4]	i-Science	2023
5	A review on comprehensive recycling of spent power lithium-ion battery in China [5]	e-Transportation	2022
6	Challenges and Recent Developments in Supply and Value Chains of Electric Vehicle batteries: a Sustainability Perspective [6]	Conservation and Recycling	2022
7	Treatment of electric vehicle battery waste in China: A review of existing policies [7]	Journal of Environmental Engineering and Landscape Management	2021
8	Implication viability assessment of electric vehicles for different regions: An approach of life cycle assessment considering exergy analysis and battery degradation [8]	Energy Conversion and Management	2021

Table 1. *Cont.*

S No.	Paper Title	Journal	Year of Publication
9	Recycling and environmental issues of lithium-ion batteries: Advances, challenges and opportunities [9]	Energy Storage Materials	2021
10	Recycling of mixed cathode lithium-ion batteries for electric vehicles: Current status and future outlook [10]	Carbon Energy	2020
11	Recycling lithium-ion Batteries from Electric Vehicles [11]	Nature	2019
12	Life Cycle Analysis of Lithium-Ion Batteries for Automotive Applications [12]	Batteries	2019

3. Research Methodology

3.1. Research Design

This study employs a comparative analysis of policy frameworks governing end-of-life (EOL) electric vehicle (EV) battery recycling in China and the United States. The primary objective is to extract valuable insights and best practices that could serve as guiding principles for other nations aiming to refine their own EOL battery management strategies in line with broader environmental, economic, and societal goals.

3.2. Data Collection

The investigation draws upon both primary and secondary sources. The primary sources encompass legal texts, national recycling mandates, and technical standards relevant to battery disposal and reuse in the selected countries. Secondary information was gathered from peer-reviewed literature, technical reviews, industrial reports, and prior policy analyses that document the practical outcomes of these regulations within the EV context.

3.3. Analytical Framework

A qualitative content analysis technique was applied to systematically review and contrast the essential features of the recycling policies in both nations. The examination concentrated on core elements such as legal structures, roles and duties of key stakeholders, compliance systems, and technical procedures for battery recovery and disposal.

3.4. Criteria for Analysis

The assessment of the regulatory schemes was based on four principal dimensions critical to effective EOL battery recycling:

- Regulatory scope: the extent to which each framework addresses the full battery lifecycle—from collection to recycling and disposal.
- Stakeholder participation: the degree of involvement and responsibility shared among producers, consumers, recyclers, and government agencies.
- Compliance mechanisms: the availability and rigor of systems that monitor and enforce adherence to the policies.
- Sustainability outcomes: the environmental and economic performance of the recycling practices encouraged by the respective frameworks.

3.5. Link to Objective and Research Questions

This methodology is directly aligned with the study's aim of understanding how leading EV markets handle battery recycling at the policy level. By evaluating real-world approaches and results, the study seeks to propose informed strategies that can be adapted globally. Specifically, the research aims to answer the following questions:

- What are the most effective policy-driven practices in EOL battery recycling among leading nations?
- What types of legislative and regulatory instruments are in place in China and the USA?
- What operational procedures are being used in both countries for managing used EV batteries?

This structured approach allows for a well-rounded assessment of international experiences in EV battery recycling and supports the formulation of context-sensitive recommendations for improved regulatory implementation. A summary of the research process is depicted in Figure 7.

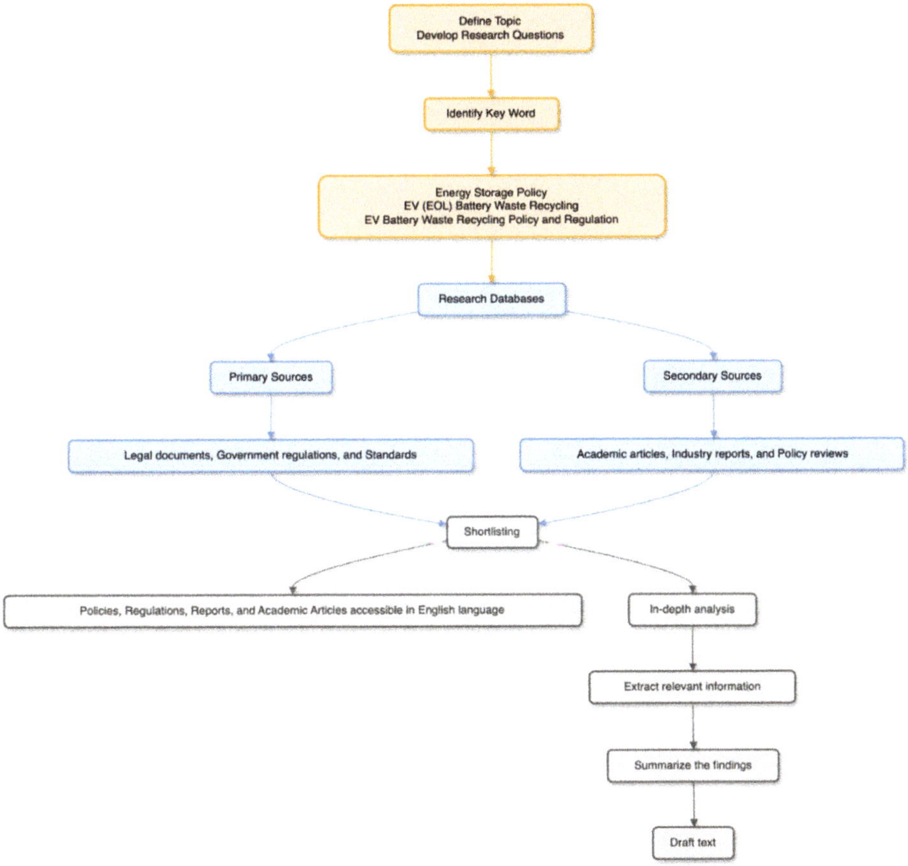

Figure 7. Research methodology flowchart.

4. EV Batteries Waste at End-of-Life Waste Recycling and Its Environmental Impacts

EV batteries have emerged as a cornerstone of the global shift toward cleaner and more sustainable transportation, offering a promising solution to reducing greenhouse gas emissions and dependence on fossil fuels [13]. As the adoption of electric vehicles continues to surge, the production and eventual disposal of EV batteries are becoming increasingly significant issues. While instrumental in the transition to electric mobility, these batteries present unique challenges at the end of their lifecycle, particularly in waste management and environmental stewardship [14,15].

The growing reliance on EV batteries has intensified concerns about effectively managing the waste they generate once they reach the end of their useful life. Unlike conventional vehicle batteries, EV batteries are complex, containing various valuable but challenging-to-recycle materials, such as lithium, cobalt, nickel, and graphite. These materials are critical to the batteries' functioning and are also finite resources, making their recovery and recycling a priority for economic and environmental reasons. Therefore, the potential environmental impacts associated with EV battery waste include:

Composition of EV Battery Waste

EV battery waste comprises a complex mix of materials, including metals, plastics, and electrolytes. As shown in Figure 8, research indicates that a typical Li-ion EV battery primarily consists of 40% cathode materials, which include 10–20% nickel, 5–15% cobalt, 10–20% manganese, and 5–10% lithium. The anode materials, mainly graphite, account for 15% of the battery. The electrolyte, which includes lithium salts and solvents, makes up 10–15%, while separators (usually polyethylene or polypropylene) contribute another 5–10%. Additionally, 5–10% of the battery consists of aluminum and 10–20% of copper, both of which are used as current collectors, with 10–15% comprising steel and other structural materials. The remaining 5% consists of plastics and other components used for encasing and insulation. It is important to note that these compositions can vary depending on the battery type and manufacture.

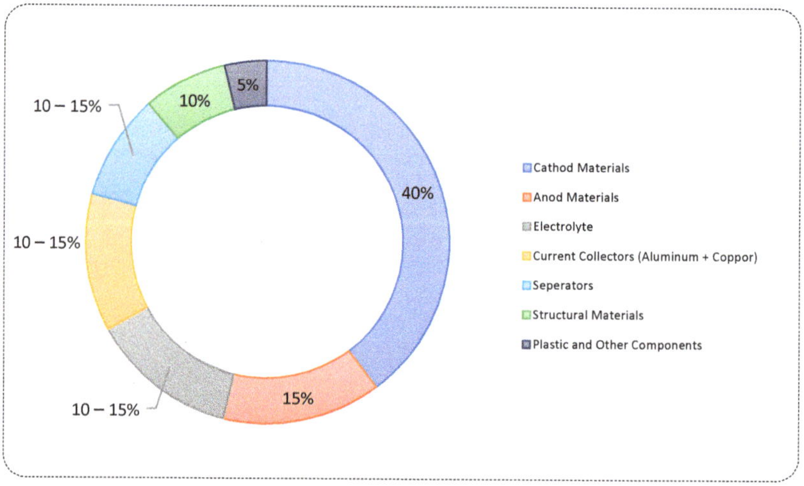

Figure 8. Weight percentage (wt%) composition of EV battery waste by material type.

1. Challenges in Recycling EV Battery Waste

The improper handling of EV battery waste contributes significantly to environmental hazards. Issues such as limited recycling infrastructure, technological barriers, and inconsistent regulations hinder efficient recycling. These challenges result in batteries ending up in landfills or being processed inefficiently, increasing pollution, and reducing resource recovery rates.

Furthermore, recycling (EV) battery waste presents several significant challenges that complicate the recovery of valuable materials and the safe disposal of hazardous components:

(a) Technological and Economic Barriers

The recycling of (EV) batteries is technologically complex due to their intricate design and composition, including various metals like lithium, cobalt, nickel, manganese, plastics, and electrolytes. Separating and recovering these materials requires advanced recycling technologies that are still in development and have yet to become widely available or economically feasible. The high costs associated with establishing and operating specialized recycling plants capable of processing (EV) batteries can significantly deter investment in the necessary infrastructure.

(b) Regulatory and Logistical Issues

There needs to be more consistent regulatory frameworks across different regions regarding the disposal and recycling of (EV) batteries. This inconsistency leads to variations in recycling practices and standards, complicating global efforts to manage (EV) battery waste effectively. Additionally, the logistics of collecting, transporting, and storing (EOL) batteries pose significant challenges, particularly in areas where dedicated recycling facilities are limited or nonexistent. These logistical hurdles add to the overall complexity and cost of (EV) battery recycling, further hindering progress in this critical area.

2. Potential Environmental Impacts

The environmental impacts of (EV) battery waste are primarily categorized into resource depletion, pollution, and health and safety risks:

(a) Resource Depletion

Producing (EV) batteries requires substantial raw materials such as lithium, cobalt, nickel, and graphite. These materials are critical for battery manufacturing and other high-tech industries, making their recovery from (EOL) batteries crucial. The extraction and processing of these materials can lead to resource depletion, particularly given their limited availability. Without effective recycling, the depletion of these resources could be exacerbated, potentially leading to scarcity, higher costs, and negative impacts on other industries that rely on these materials. According to various studies, the limited supply of these critical materials underscores the importance of developing efficient recycling processes to ensure their sustainable use.

(b) Pollution

Improper disposal of electric vehicle (EV) batteries can lead to significant environmental pollution. These batteries contain hazardous materials such as heavy metals (e.g., cobalt, nickel, and manganese) and toxic electrolytes, which can leach into the soil and water if not managed properly. This contamination poses risks to ecosystems and human health, as these poisonous substances can accumulate in the environment and cause long-term damage. Additionally, the breakdown of battery components can release harmful substances into the air, soil, and water, leading to widespread pollution.

Furthermore, incinerating EV batteries, especially those containing hazardous materials, can result in air pollution. The incineration process can release toxic gases and particles, including dioxins and heavy metals, which harm human health and degrade air quality. These pollutants can have severe environmental and public health consequences, highlighting the importance of developing safe and effective recycling and disposal methods for EV batteries.

Another environmental concern is the presence of per- and polyfluoroalkyl substances (PFAS) in lithium-ion batteries. PFAS, often referred to as "forever chemicals" due to their persistence, are used in battery components to enhance safety and performance. However, improper disposal or recycling of these batteries can lead to PFAS contamination in air, soil, and water. Studies have detected PFAS pollution near battery manufacturing plants and in landfill leachates, indicating that these substances can enter the environment during both production and disposal processes. Given their resistance to degradation and potential health risks, the release of PFAS from EV batteries underscores the need for stringent waste management practices and the development of PFAS-free battery technologies.

5. Policies and Regulations in China and the USA

Effective reuse and recycling of end-of-life (EOL) electric vehicle (EV) batteries are critical components of sustainable battery lifecycle management. Both China and the United States have implemented comprehensive regulatory frameworks aimed at addressing the growing volume of battery waste. These policies emphasize material recovery, environmental protection, and circular economy practices. China's 14th Five-Year Plan and related recycling mandates, as well as the USA's incentive-based approaches and extended producer responsibility (EPR) programs, illustrate contrasting, yet complementary, strategies to manage EOL batteries. This section focuses specifically on the policy mechanisms, regulatory initiatives, and enforcement practices in these two leading markets that directly relate to the reuse and recycling of EV batteries [16,17].

5.1. China

China has rapidly emerged as the global leader in electric vehicle (EV) adoption, driven by supportive policies and infrastructure development. As illustrated in Figure 9, nearly 9 million new EVs were sold in China in 2023, representing 34% of the global EV market and marking a 37% increase from the previous year [18]. This surge in EV deployment, while commendable for decarbonization, is directly contributing to a growing volume of end-of-life (EOL) batteries, necessitating urgent and efficient recycling solutions.

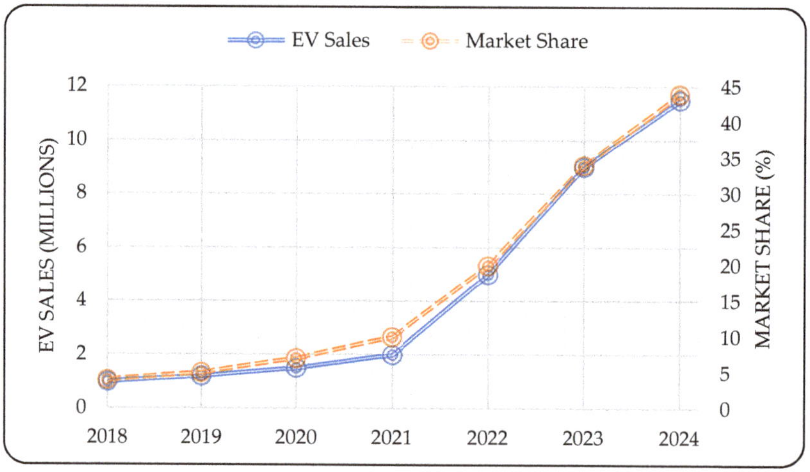

Figure 9. The growth of EVs in China.

To address this challenge, China has implemented a series of regulatory measures targeting the reuse, recycling, and traceability of EV batteries. According to projections, battery waste is expected to reach nearly 360,000 metric tons by 2030, over three times the estimated level in 2025 Figure 10. This alarming trend has prompted the government to enforce new national standards effective from December 2023, requiring EV manufacturers and battery suppliers to adopt robust recycling and traceability mechanisms [19,20].

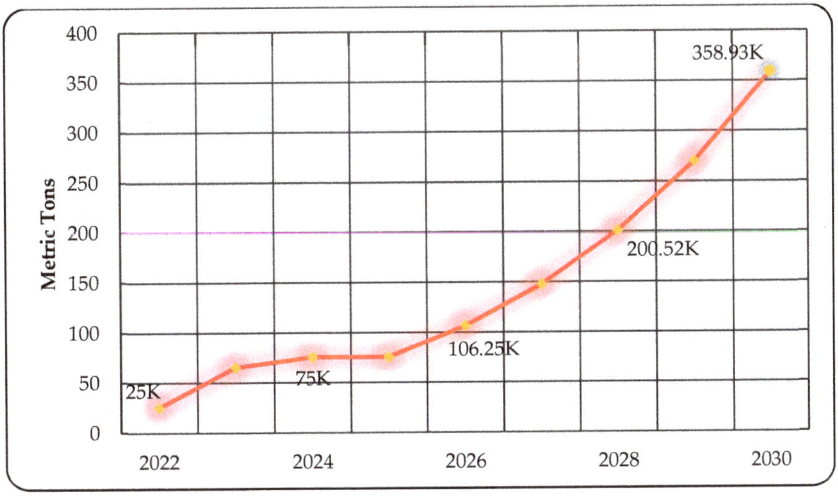

Figure 10. Battery waste in China.

China's evolving framework now mandates closed-loop systems where used batteries are disassembled and processed to recover critical materials such as lithium, cobalt, and nickel. These recovered resources are then reintegrated into the production cycle, supporting a circular economy model. Policies also require digital tracking of battery lifecycles, enabling better monitoring of EOL batteries and improving compliance.

By embedding reuse and recycling strategies into its broader new energy and environmental policies, China demonstrates a proactive approach toward sustainable battery lifecycle management. These efforts not only support environmental protection and resource efficiency, but also position China as a regulatory leader in EV battery circularity.

5.1.1. Policies

China has revised its policies on energy storage batteries to support the NEV market and technological advancements. Early policies, like the 2016 subsidy notice [Caijian (2021) No. 466], boosted NEV adoption and battery tech. These are being replaced by newer frameworks, such as the 14th Five-Year Plan (2021–2025), which targets a 30% reduction in battery costs by 2025 and 100 GW of new storage capacity by 2030. The 2021 Guidance on New-Type Energy Storage plans large-scale energy storage development, and the 2023 Policies for Power Battery Industry Management [Caishui [2023] No. 13] focus on industry optimization and safety, as shown in Table 2. The 2024 Draft Regulations on the Lithium Battery Industry aim for sustainable development with stringent standards and significant R&D investment.

Table 2. Recycling methods with their advantages and disadvantages.

Old Policy/Plan	New Policy/Plan	Year of Effective
13th Five-Year Plan	14th Five-Year Plan	2021
Caijian (2021) No. 466	Caishui (2023) No. 13	2023

In Beijing and Shanghai, national policies are implemented with enhancements through local partnerships and projects. Beijing collaborates with recycling firms, while Shanghai has projects like the BYD recycling plant and partnerships with Volvo and CATL for closed-loop recycling. In Shenzhen and Guangdong, national guidelines are followed, with local initiatives like partnerships with BYD and major recycling facilities by Guangdong Brunp and CATL. These policies underscore China's commitment to advancing energy storage capabilities aligned with technological progress and market needs. The following sections provide detailed insights into China's evolving energy storage battery policies.

- China's Five-Year Development Plans

China's five-year development plans focus on successfully developing and deploying renewable energy technologies, including waste recycling and management, as shown in Table 3. The main goals of these plans are centered around promoting a cleaner, more efficient, and sustainable energy system. The 13th Five-Year Plan for Energy Development (2016–2020) aimed to reduce coal consumption, increase the share of renewable energy, improve energy efficiency, and promote technological innovation in the energy sector [21]. Building on this, the 14th Five-Year Plan for New Energy Storage Development Implementation Plan (2021–2025) outlines China's strategy to enhance its energy storage capabilities, which is crucial for integrating renewable energy sources like solar and wind into the grid [22]. This plan emphasizes the development of advanced energy storage technologies, establishing a supportive regulatory framework, and creating robust infrastructure to ensure a stable and resilient energy supply system. These plans reflect China's commitment to transitioning towards a low-carbon economy and addressing the challenges of climate change.

Table 3. Thirteenth and fourteenth five-year development plans primary goals related to energy storage and waste recycling.

Key Points	Description
Energy Storage Technologies	Emphasizes the development of advanced energy storage technologies, including batteries, to support the integration of renewable energy and enhance grid stability. This article highlights the need to improve the performance and reduce the costs of battery storage systems.
Support for Electric Vehicles (EVs)	Addresses the need for extensive electric vehicle charging infrastructure, which indirectly supports the development and utilization of battery technologies. This includes constructing a national network of fast-charging stations and decentralized charging piles.
Enhance Resource Recycling	Establish a comprehensive power battery recycling and utilization system to manage retired power batteries effectively by 2025.
Promote Circular Economy	Develop a resource recycling industrial system with an output value reaching RMB 5 trillion (approximately USD 773 billion) by 2025.
Increase Utilization of Recycled Materials	Produce 20 million tonnes of recycled non-ferrous metals and utilize 320 million tonnes of scrap steel annually by 2025.

- 13th Five-Year Plan for Energy Development (2016–2020) [Old]:

The "13th Five-Year Plan for Energy Development (2016–2020)" aimed to enhance energy efficiency, increase the use of non-fossil energy, and reduce carbon emissions, laying the groundwork for future advancements in the energy sector. The key objectives included improving energy efficiency, promoting energy-saving technologies, expanding renewable energy sources like wind, solar, and hydroelectric power, setting specific capacity targets, reducing carbon emissions through clean energy technologies, and fostering technological innovation in advanced coal, nuclear power, and energy storage [21], and these are summarized in Table 4. The plan succeeded in significantly boosting renewable energy capacity and efficiency.

Table 4. Some of the 13th Five Year Plan's primary goals.

Article No.	Description
8	Enhance Energy Storage Capacity: Sets an ambitious target of achieving 30 GW of new energy storage capacity by 2025, with a long-term goal of 100 GW by 2030. This is crucial for balancing supply and demand fluctuations, especially with increased use of intermittent renewable energy sources like wind and solar power.
11	Support for Renewable Energy Integration: Focuses on enhancing the role of energy storage in balancing the grid and supporting the integration of intermittent renewable energy sources. This involves developing independent battery storage facilities to store excess energy generated during peak production periods and release it during high-demand periods.
14	Technological Advancements and Innovation: Emphasizes the importance of supporting research and development (R&D) in new energy storage technologies. The goal is to improve performance, reduce costs, and enhance the safety of energy storage systems. This includes promoting pilot projects and demonstrations of advanced energy storage systems.
17	Policy and Regulatory Support: Aims to develop market-oriented policies that encourage investment in energy storage technologies. Establishes pricing mechanisms and financial incentives to make energy storage projects economically viable and attractive to private investors.
18	Environmental and Economic Impact: Contributes to China's carbon neutrality goals by 2060 by enhancing the role of clean energy storage solutions. This article addresses reducing reliance on fossil fuels, supporting renewable energy integration, lowering carbon emissions, and improving air quality. Additionally, it highlights the economic benefits, such as stimulating growth, creating jobs, and enhancing energy security.

14th Five-Year Plan for New Energy Storage Development Implementation Plan (2021–2025) [New]

The "14th Five-Year Plan for New Energy Storage Development Implementation Plan (2021–2025)" is a strategic initiative by the Chinese government to significantly advance the country's energy storage capabilities. This plan is essential for integrating renewable energy sources and ensuring the stability of the national grid [23,24]. It builds upon the foundations of previous Five-Year Plans, setting more ambitious targets and focusing on technological innovations to achieve these goals.

Implementation Details

- Official Name: 14th Five-Year Plan for New Energy Storage Development Implementation Plan (2021–2025).
- Policy reference: Part of China's broader 14th Five-Year Plan framework.
- Effective Date: 1 January 2021 to 31 December 2025.

Description of the policy

(a) Energy Storage Technologies:

The plan focuses on various energy storage technologies, including lithium-ion, sodium-ion, lead–carbon, and redox flow batteries. Emphasis is placed on developing long-duration energy storage systems capable of providing backup power and stabilizing the grid during peak demand. This technological diversity ensures that multiple storage solutions can be created and optimized for different applications and conditions.

(b) Infrastructure Development:

Expanding the infrastructure for energy storage systems is a critical plan component. This includes constructing large-scale storage facilities and integrating energy storage with existing power plants and renewable energy installations. By enhancing the infrastructure, the plan aims to optimize the operation and efficiency of the entire energy system, ensuring reliable and cost-effective energy delivery.

(c) Research and Development (R&D):

Increased funding for R&D is allocated to develop advanced energy storage materials and technologies. The plan encourages collaboration with international partners to accelerate technological advancements and share best practices. This collaborative approach ensures that China remains at the forefront of energy storage innovation and can adopt the latest and most effective solutions.

(d) Regulatory Framework:

Establishing clear regulatory guidelines for deploying and operating energy storage systems is essential for the plan's success. These guidelines ensure compliance with safety and performance standards, protecting consumers and the environment. The regulatory framework provides the necessary oversight and governance to support the sustainable development of the energy storage sector.

(e) Impact and Future Directions:

The plan is expected to drive substantial market growth in the energy storage sector, fostering innovation and attracting significant investment. By supporting R&D and pilot projects, the plan encourages technological breakthroughs that can reduce costs and improve the efficiency and safety of energy storage systems. Enhanced energy storage capabilities support integrating renewable energy and contributing to carbon emission reduction goals. Additionally, the development of the energy storage industry is anticipated to create new jobs, stimulate economic growth, and enhance energy security, positioning China as a global leader in clean energy technologies.

The comparative differences between the 13th and 14th Five Year Plans are given in Table 5.

Table 5. Comparing China's 13th and 14th Five-Year Plans reveals a significant escalation in targets related to resource productivity, recycling, and the circular economy.

Metric	13th Five-Year Plan	14th Five-Year Plan
Resource Productivity Increase (%)	Increase by 15% from 2015 levels [25].	Increase by 20% compared to 2020 levels [26].
Recycled Non-Ferrous Metals Production (Million Tons)	Recycled non-ferrous metals reached approximately 14.5 million tons [25].	Achieve 20 million tons [26].
Resource Recycling Industry Output (RMB Trillion)	Recycling industry achieved an output value of RMB 2.8 trillion (approximately USD 406 billion) in 2020 [25].	Reach a value of RMB 5 trillion (approximately USD 773 billion) [26].
Battery Swapping Stations	Around 555 battery swapping stations were built in 2020 [27].	Plan to build over 1000 battery swapping stations by 2025 [28].

Policies Implemented from These Plans

The "Notice on the Promotion and Application of Financial Subsidy Policy for New Energy Vehicles (2022)" [Caijian (2021) No. 466] aimed to support the NEV industry's growth by gradually reducing subsidies while maintaining high technical standards to encourage innovation, and was effective from 1 January 2022. Key provisions included a 30% subsidy reduction for 2022, stringent performance and technical standards, and support for battery technology advancements and infrastructure development. This policy contributed to market growth, technological advancements, and environmental benefits. Replacing this, the "Notice on Extending and Adjusting the Purchase Tax Exemption Policy for New Energy Vehicles (2023)" [Caishui [2023] No. 13], effective from 1 January 2024 to 31 December 2027, shifts the focus to tax exemptions, providing up to RMB 30,000 per vehicle for 2024–2025 and RMB 15,000 for 2026–2027. This new policy aims to promote NEV adoption, support technological advancements, and expand charging infrastructure, contributing to China's carbon neutrality goals by reducing greenhouse gas emissions and improving air quality. The key feature points of Caijian (2021) No. 466 and Caishui [2023] No. 13 are given in Table 6.

Table 6. The criteria of each policy in China.

Caijian (2021) No. 466 [Old]	Caishui (2023) No. 13 [New]
▪ Reducing subsidies by 30% in 2022. ▪ The transport sector received a 20% reduction. ▪ High technical standards. ▪ Support development for charging stations. ▪ Battery recycling regulations ▪ Environmental impact.	▪ Provide tax exemption to lower the upfront cost of NEV. ▪ Invest in research and development on batteries and NEV technologies. ▪ Support for battery swap models. ▪ Supporting the growth of the NEV market ▪ Reducing greenhouse gas emission. ▪ NEV must perform at a high standard.

Notice on the Promotion and Application of Financial Subsidy Policy for New Energy Vehicles (NEVs) (2022) [OLD]

The "Notice on the Promotion and Application of Financial Subsidy Policy for New Energy Vehicles (2022)" [Caijian (2021) No. 466] aims to support the NEV industry's growth by gradually reducing subsidies while maintaining high technical standards to encourage innovation [29]. Issued by the Ministry of Finance, MIIT, MOST, and NDRC, it became effective on 1 January 2022. Key provisions include a 30% subsidy reduction for 2022, with more minor cuts for public service vehicles and stringent performance and technical standards for eligibility. The policy promotes advancements in battery technology, supports infrastructure development for charging stations, and integrates renewable energy. It also emphasizes battery recycling and includes mechanisms for monitoring and evaluation. Public awareness campaigns and industry collaboration are encouraged to foster innovation. This policy has significantly contributed to market growth, technological advancements, and environmental benefits, helping to position China as a global leader in the NEV sector.

Notice on Extending and Adjusting the Purchase Tax Exemption Policy Foe New Energy Vehicles (2023) [NEW]

The "Notice on Extending and Adjusting the Purchase Tax Exemption Policy for New Energy Vehicles (2023)" [Caishui [2023] No. 13] is an updated policy aimed at promoting the growth and technological advancement of the new energy vehicle (NEV) market through tax exemptions, replacing direct subsidies with tax incentives [30]. This approach is intended to sustain market momentum while encouraging innovation in NEV technology and infrastructure.

Implementation Details

- Official Name: Notice on Extending and Adjusting the Purchase Tax Exemption Policy for New Energy Vehicles (2023).
- Policy reference: [Caishui [2023] No. 13].
- Effective Date: 1 January 2024 to 31 December 2027.

Description of the policy

(a) Tax Exemptions:

There are two tax exemption plans, one from 2024 to 2025. NEV owners will receive a purchase tax exemption of up to RMB 30,000 per car, providing significant financial relief and encouraging NEV purchases [31]. This is projected to drive significant sales growth and aid the industry's recovery following COVID-19. From 2026 to 2027, the second plan will cut the tax exemption to RMB 15,000 per car to gradually phase out tax breaks while providing continuous market support, preserving momentum while the industry transitions to a future with fewer government incentives.

(b) Eligibility Criteria:

NEVs must meet specific performance standards, such as a minimum driving range of 200 km for pure electric vehicles and a battery energy density of at least 125 Wh/kg, to ensure that only high-performance vehicles receive tax breaks, encouraging manufacturers to produce technologically advanced and efficient vehicles [32]. Furthermore, vehicles must conform to tight battery performance and safety criteria, ensuring high levels of safety and efficiency in the NEV industry and boost consumer confidence and market stability. To ensure compliance, vehicles must undergo a rigorous certification procedure that validates their eligibility for tax breaks and protects policy integrity by ensuring that only compliant vehicles receive benefits.

(c) Support for Battery Swap Models:

Vehicles that feature battery swap technology receive significant tax breaks. The goal is to promote the battery swap model to minimize range anxiety and increase convenience. This project encourages the use of novel battery technology and the development of related infrastructure, helping to advance the NEV industry.

(d) Impact and Future Directions:

The strategy is intended to sustain the NEV market's rapid growth by encouraging both domestic and foreign sales and maintaining market momentum through financial incentives that make NEVs more accessible and appealing to customers. By establishing rigorous technical criteria, the policy promotes breakthroughs in battery technology and overall vehicle performance, pushing manufacturers to develop and improve the quality and efficiency of their NEVs. In addition, the regulation improves the environment by lowering greenhouse gas emissions and increasing air quality, which is consistent with China's overall environmental, and sustainability aims. It encourages using NEVs as part of the transition to greener transportation options, thereby mitigating the effects of climate change.

Lithium-Ion Battery Policies

China has implemented a national policy to promote the development, use, safety, and recycling of Li-ion batteries in EVs. This policy aims to ensure the sustainable growth of the EV industry while minimizing environmental impacts and enhancing technological advancements. Below is a critical national policy:

Policies for Power Battery Industry Management (2023):

The Policies for Power Battery Industry Management (2023) include comprehensive measures explicitly targeting the end-of-life management and recycling of lithium-ion (Li-ion) batteries used in electric vehicles (EVs) [33]. This policy ensures environmental safety and resource recovery through a structured regulatory framework, financial incentives, and extended producer responsibility (EPR). Below is an expanded and detailed description of each aspect of the recycling policy concerning EVs.

Implementation Details

- Regulatory Framework: MIIT's Technical Standards for Li-ion Batteries.
- National Standard: GB/T 31574-2015
- Effective Date: 1 January 2023

Description of the policy

End-of-Life Battery Management and Recycling Policy

(a) Regulatory Framework for Recycling:

The goal is to provide a comprehensive regulatory framework that oversees the recycling and disposal of (EOL) EV batteries while ensuring environmental safety and resource recovery. To do this, the policy imposes strict rules on all EV battery recycling phases, including collecting, transportation, storage, and dismantling. Standards like GB/T 31574-2015 [34] define safe, efficient, and environmentally friendly recycling procedures, including how to handle and process batteries to maximize material recovery. Furthermore, recycling facilities must obtain permission and certifications and follow government rules to ensure that their operations are responsible and sustainable. The End-of-Life Battery Management and Recycling Policy's key metrics align with the targets set in the 14th Five-Year Plan, as tabulated in Table 7.

Table 7. End-of-Life Battery Management and Recycling Policy's key metrics align with the targets set in the 14th Five-Year Plan.

Metric	Value (End of 2024)
Market Size	Approximately RMB 40 billion (approximately USD 5.8 billion) [35]
Decommissioned Batteries	Estimated at 52.29 GWh in 2022; projected to reach 134.49 GWh by 2025 [36]
Recycling Capacity	Approx. 337.5 GWh (75% of global battery recycling capacity) [37]
Recycling Enterprises	162,000 enterprises [38]
Circular Economy Leadership	Release of 2024 Edition of Battery Recycling Regulations [39]

(b) Incentives for Recycling and Reuse:

The objective is to encourage recycling and reuse of battery materials through financial incentives and support. To implement this, the legislation provides subsidies and tax breaks to EV battery recycling companies, which help to offset recycling costs and increase participation. These financial incentives include financing for constructing or upgrading recycling facilities to improve efficiency and environmental compliance, making recycling infrastructure investments more economically viable for enterprises [35]. Furthermore, the policy supports R&D projects to improve recycling technology and increase material recovery efficiency.

(c) Extended Producer Responsibility (EPR):

The goal is to hold EV producers responsible for their products' whole lifecycle, including end-of-life (EOL) management. To accomplish this, manufacturers must provide take-back procedures for spent EV batteries, assuring proper collection, recycling, and responsible disposal. Companies such as BYD and NIO have set up collecting points and recycling programs to keep batteries out of landfill sites. Producers must also disclose their collection and recycling rates regularly, and authorities will monitor compliance with extended producer responsibility (EPR) requirements. This involves presenting annual reports specifying the number of batteries collected and recycled and the percentage of material recovered, ensuring transparency and accountability in recycling.

In addition to support for China's national policies, different provinces have formulated and implemented their policies, as shown in Table 8.

Table 8. Some of the provincial policies in China.

Province	Policy
Beijing [40]	• Adherence to National Policies: Beijing follows MIIT regulations for Li-ion battery recycling, ensuring environmental and safety standards., e.g., "Draft Regulations on the Lithium Battery Industry (2024)" and "Policies for Power Battery Industry Management (2023)". • Local Implementations: Beijing enhances national policies through public–private partnerships, collaborations with recycling firms, and pilot programs., e.g., initiatives involving government, private companies, and research institutions to manage used batteries, streamline processes, and test new recycling technologies.

Table 8. *Cont.*

Province	Policy
Shanghai [41]	• Adherence to National Policies: Shanghai follows MIIT regulations for Li-ion battery recycling, ensuring compliance with national environmental and safety standards., e.g., policies like the "Draft Regulations on the Lithium Battery Industry (2024)" and "Policies for Power Battery Industry Management (2023)" guide Shanghai's recycling processes. • Local Projects: Shanghai's battery recycling efforts include the BYD recycling plant for efficient Li-ion battery recycling and material recovery, partnerships with Volvo Cars and CATL for closed-loop recycling, and sustainability initiatives using renewable energy and energy-efficient technologies.
Shenzhen [42]	• Local Implementation of National Policies: Shenzhen follows MIIT guidelines for Li-ion battery recycling, ensuring strict environmental and safety standards. For example, enhanced local execution through partnerships and advanced recycling facilities to address the city's rapid urbanization and high volume of EVs. • Strategic Collaborations: Shenzhen partners with companies like BYD for efficient battery recycling, setting up collection points and advanced facilities., e.g., BYD and local authorities have citywide collection points and a recycling plant that uses automated systems to recover materials and dispose of hazardous components.
Guangdong Province [43]	• Guangdong Brunp Recycling Technology: A CATL subsidiary with major facilities in Foshan and Changsha, Brunp uses advanced technology to recover valuable materials from used batteries., e.g., the Foshan facility processes thousands of tons annually, highlighting Guangdong's commitment to sustainable recycling. • CATL Investments: CATL has invested significantly in Guangdong to develop advanced battery recycling plants., e.g., a new lithium battery production and recycling base in Zhaoqing aims to boost capacity, support sustainable battery life cycles, and maximize material recovery with minimal environmental impact.

5.2. The United States of America

The increasing adoption of electric vehicles (EVs) in the United States has raised critical concerns regarding the management of end-of-life (EOL) lithium-ion (Li-ion) batteries. As shown in Figure 11, the number of EVs grew from 360,000 in 2020 to nearly 900,000 by 2023, with market share rising from under 2% in 2018 to over 6% in 2023 [44,45]. This rapid expansion, while beneficial for reducing emissions and boosting technological innovation, also contributes to a growing volume of battery waste.

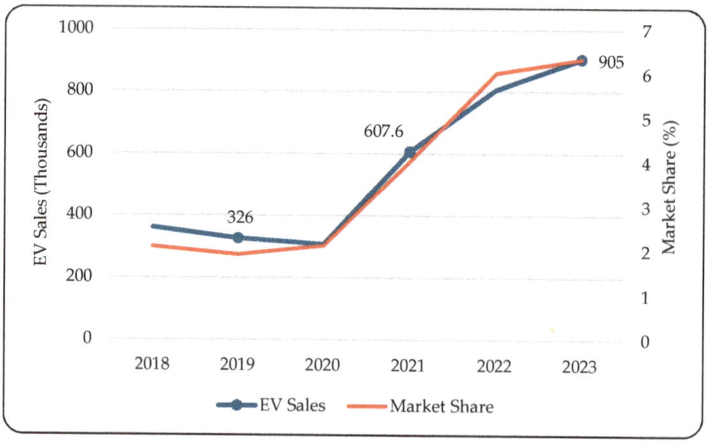

Figure 11. The growth of EVs in the US.

energy and contributing to the 100% clean electricity target by 2035. The policy promotes technologies that allow for efficient energy storage from renewable sources like solar and wind to ensure a steady and predictable energy supply [59]. The policy also intends to lower the carbon footprint of the transportation and energy sectors by developing battery technology for electric vehicles (EVs) and energy storage systems. Significant investments have been made in initiatives such as large-scale battery storage systems and expanding EV charging infrastructure, both critical components of the clean energy transition.

General FEMP Energy Efficiency Initiatives [Public Law 110-140, Title IV, Subtitle C, Section 432, Public Law 95-619, Amended Multiple Times, and Public Law 109-58, Title I, Subtitle B, Section 103] [Old]

The Federal Energy Management Program (FEMP) initially focused on broad energy efficiency initiatives across federal facilities. These initiatives were guided by various Department of Energy (DOE) directives and guidance documents [Public Law 110-140, Title IV, Subtitle C, Section 432, Public Law 95-619, amended multiple times, and Public Law 109-58, Title I, Subtitle B, Section 103]. The program aimed to improve energy and water efficiency, reduce greenhouse gas emissions, and ensure energy resilience in federal operations [55,60]. Key strategies included optimizing building performance, enhancing lighting and HVAC systems, and implementing renewable energy projects.

Executive Order 14008 and the Infrastructure Investment and Jobs Act [Executive Order 14008, and Public Law 117-58, Title IX, Subtitle B] [New]

The Federal Energy Management Program (FEMP) has expanded its focus to include battery storage projects as part of broader initiatives under [Executive Order 14008], "Tackling the Climate Crisis at Home and Abroad", and the Infrastructure Investment and Jobs Act [Public Law 117-58, Title IX, Subtitle B]. These initiatives aim to enhance energy resilience, integrate renewable energy sources, and reduce greenhouse gas emissions across federal facilities [56,61]. By providing technical assistance, funding, and guidance for deploying advanced battery storage systems, FEMP supports the US government's sustainability and climate goals, ensuring that federal operations lead by example in adopting clean energy technologies.

Implementation Details

- Official name: Executive Order 14008: Tackling the Climate Crisis at Home and Abroad.
- Reference number: Executive Order 14008
- Date effective: 27 January 2021
- Official name: Infrastructure Investment and Jobs Act.
- Reference number: Public Law 117-58, Title IX, Subtitle B
- Date effective: 15 November 2021

Description of the Policy

(a) Executive Order 14008:

The goal of President Biden's executive order, issued on 27 January 2021, is to prioritize addressing the climate crisis through a whole-of-government approach, directing federal agencies to focus on clean energy technologies, including energy storage, to reduce carbon emissions and improve resilience. The directive focuses on establishing innovative energy storage systems across government sites to help integrate renewable energy sources and increase grid stability. To carry out this instruction, federal agencies, including the Department of Energy (DOE), are responsible for identifying opportunities for implementing battery storage and incorporating these systems into their energy management programs. For instance, the United States Department of Defense has implemented battery storage initiatives at military stations such as the Marine Corps Air Station Miramar in California, where a microgrid with battery storage improves energy security and promotes the use of renewable energy.

(b) Infrastructure Investment and Jobs Act (IIJA):

The Infrastructure Investment and Jobs Act (IIJA), passed on 15 November 2021, seeks to update the nation's infrastructure, particularly energy infrastructure [62]. The act encourages the development and implementation of renewable energy technology, particularly improvements to energy storage capacity. To do this, the IIJA provides significant money for energy storage projects, such as grants and technical help installing battery storage systems at federal sites. For example, the IIJA provided funding for building a large-scale battery storage system at Fort Carson Army Base in Colorado, improving energy resilience and promoting renewable energy integration. The act requires examining and implementing energy storage options to increase the strength and efficiency of the federal energy systems. The Department of Energy (DOE), through the Federal Energy Management Program (FEMP), provides technical help, funding possibilities, and project advice. The DOE, for example, has provided grants to NASA Johnson Space Center to deploy a lithium-ion battery system, improving energy reliability and sustaining mission-critical operations.

(c) Technical Assistance and Evaluation:

The Federal Energy Management Program (FEMP) aims to provide technical help to federal agencies in evaluating and executing energy storage projects. This service includes determining the viability of battery storage systems, designing and installing them, and optimizing their performance. FEMP's services include conducting site surveys, generating project specifications, performing financial analysis, and verifying compliance with federal energy regulations. For example, FEMP conducted a site evaluation for the United States Postal Service's facility in Los Angeles, California, to determine the viability of adding a battery storage system to manage peak energy demand and save electricity costs. Several government locations, including military bases and national labs, are being assessed for prospective energy storage installations as part of efforts to cut energy costs, improve energy security, and promote federal sustainability objectives. One example is the Oak Ridge National Laboratory in Tennessee, which developed a battery storage system to enable grid integration and energy storage research, lowering energy costs and increasing research capacities.

(d) Coordination and Collaboration:

FEMP actively works with other federal agencies, including the Department of Defense (DoD) and the General Services Administration (GSA), to encourage the use of energy storage technologies. For example, FEMP and the DoD collaborated to establish a battery storage system at the United States Army's Fort Hood in Texas, increasing the base's energy resilience and supporting its sustainability objectives. Furthermore, FEMP promotes public–private partnerships to utilize private sector expertise and money to deploy sophisticated energy storage technologies. The GSA and a commercial energy business collaborated to install a battery storage system at the Ronald Reagan Building and International Trade Center in Washington, D.C., resulting in lower energy costs and better grid stability.

Lithium–Ion Battery Policies

The US has established comprehensive policies to develop a robust domestic supply chain for lithium batteries and ensure efficient recycling of critical minerals. The National Blueprint for Lithium Batteries 2021–2030, published by the Federal Consortium for Advanced Batteries (FCAB), aims to cover the entire lifecycle of batteries, from raw material extraction to end-of-life recycling and reuse. It emphasizes the development of technologies and processes for efficient recycling, infrastructure development, public–private partnerships, and international collaboration to promote a circular economy for battery materials.

The Battery and Critical Mineral Recycling Act of 2021, introduced as S.1918, supports research, development, and demonstration projects to improve battery recycling technologies and critical mineral recovery. The Act establishes grant programs to incentivize recycling and reuse and funds research initiatives to enhance recycling efficiency. It mandates the Department of Energy (DOE) to oversee the implementation of these initiatives.

The major points of the Federal Consortium for Advanced Batteries (FCAB) and Battery and Critical Mineral Recycling Act are encapsulated in Table 11.

Table 11. The main points between [FCAB] and [S.1918] policies in lithium-ion batteries.

Federal Consortium for Advanced Batteries (FCAB)	Battery and Critical Mineral Recycling [S.1918]
▪ Develop a robust domestic supply chain for lithium batteries. ▪ Recycle the (EOL) batteries and reuse the valuable materials in the production of new ones. ▪ Invest in R&D to improve the efficiency and cost-effectiveness of lithium-ion battery recycling processes. ▪ Implement policies and regulations that incentivize battery recycling and ensure environmentally sound disposal practices. ▪ Encourage battery manufacturers to design batteries with recycling in mind. ▪ Develop markets for recycled battery materials by ensuring that they meet the quality standards required for new battery production.	▪ Support the research, development, and demonstration of advanced battery recycling and critical mineral recovery. ▪ Provide financial assistance to companies and research institutions working on innovative recycling technologies. ▪ Enhancing the efficiency and cost-effectiveness of battery recycling. ▪ Implement programs aim to lower the costs associated with recycling and increase the recovery rates of valuable materials such as lithium. ▪ Improving the processes for separating and purifying critical minerals from used batteries. ▪ The Act aims to ensure that valuable materials are not wasted, and environmental harm is minimized.

(a) National Blueprint for Lithium Batteries 2021–2030 [Federal Consortium for Advanced Batteries (FCAB)]

The Federal Consortium for Advanced Batteries (FCAB) released the National Blueprint for Lithium Batteries 2021–2030, which attempts to address every stage of the battery lifetime, from extraction of raw materials to recycling and re-use at the end of life [63]. To support a circular economy for battery materials, it strongly emphasizes the creation of technologies and procedures for effective recycling, infrastructure development, public–private partnerships, and international cooperation. The primary objective of the National Blueprint for Lithium Batteries 2021–2030 is to develop a robust domestic supply chain for lithium batteries. This includes the entire lifecycle of batteries from raw material extraction through production, utilization, and (EOL) recycling and reuse.

Implementation Details

- Official name: National Blueprint for Lithium Batteries 2021–2030.
- Reference number: It is a strategic framework published by the Federal Consortium for Advanced Batteries (FCAB), which includes various federal agencies.
- Date effective: June 2021

Description of the Policy

(a) Recycling and Reuse:

The goal is to promote the development of technology and procedures for efficiently recycling lithium-ion batteries, ensuring that valuable materials may be recovered and reused to make new batteries. To accomplish this, investments are made in research and development (R&D) to improve the efficiency and cost-effectiveness of lithium-ion battery recycling processes, focusing on advanced separation and extraction technologies for recovering critical materials like lithium, cobalt, and nickel. Furthermore, the policy prioritizes creating and growing the infrastructure required for large-scale battery recycling, such as recycling facilities and logistics networks for collecting and transporting end-of-life batteries.

To assist these efforts, legislation, and regulations are in place to encourage battery recycling and ensure environmentally sound disposal procedures. This includes extended producer responsibility (EPR) schemes, which hold manufacturers accountable for managing their products at the end of life. Furthermore, public–private partnerships are encouraged to speed up the development and deployment of recycling technologies by leveraging government and industry's necessary money and knowledge. Consumer awareness and education campaigns are also encouraged to educate consumers and companies on the need for battery recycling and the availability of recycling schemes.

(b) Circular Economy:

The goal is to create a circular economy for battery materials, reducing environmental impact and dependency on new raw materials by reintegrating spent battery materials into supply chains. To facilitate this, battery makers are encouraged to design goods with recycling in mind, making them easier to disassemble and utilizing components that can be quickly recovered and reused. Comprehensive lifecycle evaluations are undertaken to examine the environmental and economic implications of battery production, usage, and disposal, assisting in identifying potential for enhancing sustainability throughout the battery's lifecycle.

Setting targets for material recovery and reuse from end-of-life batteries is critical for stimulating innovation and investment in recycling technology. Furthermore, promoting second-life applications for retired EV batteries, such as stationary energy storage devices, increases battery life and delays disposal. To further promote the circular economy, markets for recovered battery materials are being formed, guaranteeing they fulfill the quality criteria necessary for manufacturing new batteries. This results in stable demand and makes recycling economically viable.

International collaboration is crucial in harmonizing standards and practices for battery recycling and material recovery. This global approach is essential in addressing the environmental impact of battery production and disposal worldwide. Working with international partners enhances the efficiency and effectiveness of recycling efforts, helping to reduce the environmental impact of battery production and disposal worldwide. The key economic data and funding allocations under the Federal Consortium for Advanced Batteries (FCAB) are summarized in Table 12.

Table 12. The key economic data and funding allocations under the Federal Consortium for Advanced Batteries (FCAB).

Matrix	Economic Data
Battery Recycling and Second-Life Applications Program (2021–2026)	• Authorized under the Bipartisan Infrastructure Law, this program allocates USD 200 million over five years to support research, development, and demonstration projects focused on EV battery recycling and second-use applications [64].
Advanced Battery Research and Development Consortium (2023)	• The US Department of Energy announced more than USD 192 million in new funding for recycling batteries from consumer products [65].
DOE Funding for Battery Manufacturing and Recycling (2022)	• USD 2.8 billion for domestic manufacturing of batteries for electric vehicles (EVs) and the electrical grid [66].

Table 12. *Cont.*

Matrix	Economic Data
Projected US Battery Market Growth	• Expected to reach USD 100 billion in the coming decades [67].
Recent DOE Investment in Battery Supply Chain (2024)	• Over USD 3 billion allocated to 25 projects across 14 states, creating approximately 12,000 jobs [68].
Loan for EV Battery Factories (2024)	• USD 7.54 billion loan to Stellantis and Samsung SDI for two EV battery factories in Indiana [69].

Battery and Critical Mineral Recycling Act of 2021 [S.1918]

The Battery and Critical Mineral Recycling Act of 2021 supports the research, development, and demonstration of advanced battery recycling and critical mineral recovery [70]. This aims to ensure that the United States develops a sustainable and efficient approach to managing the EOL of Li-ion batteries, which are crucial for electric vehicles.

Implementation Details

- Official name: Battery and Critical Mineral Recycling Act of 2021.
- Reference number: S.1918
- Date effective: introduced on 27 May 2021

Description of the Policy

(a) Grants and Funding:

The Act provides substantial funding for projects that improve battery recycling technologies and processes. It establishes three grant programs to incentivize the reuse and recycling of batteries and their critical minerals. These grants aim to enhance the overall efficiency of recycling operations and support the creation of a robust recycling infrastructure. For instance, one of the key initiatives is the establishment of grant programs that provide financial assistance to companies and research institutions working on innovative recycling technologies. These programs aim to lower the costs associated with recycling and increase the recovery rates of valuable materials such as lithium, cobalt, and nickel.

(b) Research and Development:

The Act supports various R&D initiatives aimed at enhancing the efficiency and cost-effectiveness of battery recycling. It encourages the development of new technologies and methods that can make the recycling process more sustainable and economically viable. This includes improving the processes for separating and purifying critical minerals from used batteries. For instance, continuing the Lithium-Ion Battery Recycling Prize competition is a significant component of this Act. This competition incentivizes innovation by awarding prizes to individuals or teams that develop breakthroughs solving problems in collecting, storing, and recycling Li-ion batteries. The goal is to stimulate the creation of scalable and commercially viable recycling methods.

Implementation and Impact:

The Act is designed to steer the Department of Energy (DOE) towards funding programs that foster the development of battery recycling infrastructure and technology. This forward-thinking strategy promotes a circular economy by recovering and reusing materials from end-of-life batteries, reducing the environmental impact and the need for additional raw materials. For instance, the DOE's Lithium-Ion Battery Recycling Prize is a testament to this, as it encourages innovative disassembly and material recovery options, thereby accelerating recycling efforts.

This legislation is critical for resolving the developing issues with lithium-ion batteries in electric vehicles. The Act aims to reduce waste and pollution by improving recycling technologies and infrastructure. Notably, DOE support enabled the American Battery Technology Company (ABTC) to develop a Nevada recycling facility to recover and recombine essential elements, minimizing the demand for new raw materials.

Key matrix and data points for the Battery and Critical Mineral Recycling Act (S.1918) are tabulated in Table 13.

Table 13. The key economic data and initiatives under the Battery and Critical Mineral Recycling Act (S.1918) [70].

Metrix	Economic Data
Lithium-Ion Battery Recycling Prize	• The Department of Energy (DOE) is authorized to continue this competition, with an additional USD 10 million appropriated for fiscal year 2021 to support Phase III, focusing on innovative solutions for battery recycling.
State and Local Battery Collection Programs	• The Act establishes a grant program for states and local governments to develop or enhance battery collection, recycling, and reprocessing programs. The non-federal cost share for these projects is 50% of the total cost.
Retailer Battery Collection Systems	• Grants are available for retailer's selling batteries or battery-containing products to implement systems for accepting and collecting used batteries for reuse, recycling, or proper disposal. These systems must offer take-back services at no cost to consumers.
Voluntary Labeling Program	• The Act directs the DOE and the Environmental Protection Agency (EPA) to establish a voluntary labeling program to improve battery collection and reduce waste, facilitating consumer awareness and participation in recycling efforts.

Recycling Steps

(a) Collection:

Manufacturers, dealerships, and recycling companies set up EV battery collecting locations, which are administered in collaboration with automakers and third-party recyclers such as Redwood Materials and Li-Cycle [71,72]. Ads, social media, and events boost consumer awareness, while collection stations stress safety by training staff and utilizing appropriate containers to prevent spills or fires.

(b) Transportation:

Transportation companies adhere to federal safety laws such as DOT guidelines [73], using fireproof containers and real-time tracking to ensure safe battery shipments to recycling sites. They also document each shipment, including the battery's origin, condition, and destination, to guarantee correct handling and environmental compliance during transit.

Storage

Recyclers and waste management facilities create secure, climate-controlled storage chambers for incoming batteries, ensuring stability and safety until processing. Advanced inventory systems monitor each battery's progress from collecting to dismantling, optimizing timing and procedures. Furthermore, regulators such as the Environmental Protection Agency (EPA) regularly audit these locations to guarantee compliance with environmental standards [74].

Dismantling

Batteries are dismantled manually or automatically to separate components such as modules and cells, and businesses like Tesla and Redwood Materials are developing automated procedures to improve material recovery efficiency. The main goal is to extract valuable metals such as lithium, cobalt, nickel, and graphite. Recycling companies established in the United States, such as Li-Cycle and Ascend Elements [75], use patented technology to increase recovery rates for these critical minerals. Hazardous waste management standards closely adhere to adequately handling and neutralizing harmful components such as electrolytes to prevent environmental contamination [76].

Recycling and Processing

Depending on the battery chemistry, recovered components are chemically processed using procedures such as hydrometallurgy or pyrometallurgy to make them suitable for reuse. Recycling facilities follow EPA rules and state requirements to reduce emissions and waste during recycling, assuring environmental compliance. Furthermore, the recycled materials are subjected to rigorous quality control testing before being reintroduced into the supply chain to manufacture new batteries.

Reporting and Compliance

Companies regularly submit detailed reports to regulatory bodies like the EPA, outlining the amount of material recovered and recycled. Internal and external audits are conducted to ensure that recycling programs comply with current safety and environmental regulations. Additionally, companies maintain transparency by sharing progress reports and collaborating with stakeholders to improve recycling practices continuously.

6. Empirical Data: Effectiveness of Recycling Technologies and Policies

6.1. Case Studies on Recycling Success

6.1.1. China

CATL's Recycling Initiatives: Contemporary Amperex Technology Co., Limited, which is by far the leading battery manufacturer, has constructed state-of-the-art facilities in Guangdong province in which over 90% of lithium and cobalt can be recovered from used batteries. This facility is hydrometallurgical in modes of modern application and focused on minimal environmental impact in the process of achieving high recovery rates. Recycled material is chained into production cycles again for making new batteries, thereby limiting the reliance on raw virgin materials. Further, the company invests in research and development to optimize recycling efficiency. Policies such as the 14th Five-Year Plan have played an important role in formulating recycling legislatures, propelling infrastructure development, and setting ambitious sustainability targets [77–84].

Empirical Evidence: According to the Ministry of Industry and Information Technology (MIIT), China's recycling rate for EV batteries is now 40% in 2023, compared to just 25% in 2020 [85–88]. This phenomenal growth is because of adopting extended producer responsibility (EPR), which holds manufacturers accountable to ensure the collection and recycling of used batteries. Furthermore, the new policy encourages innovation and creates a need for companies to give more than the cost of collection manufacturing recycling target incentives through grants and tax breaks. A nationwide battery tracking infrastructure has significantly improved the capabilities of collection and recycling operations and ensured that a higher percentage of EOL batteries are disposed of well.

6.1.2. United States

(a) Redwood Materials, based in Nevada, United States, is a pioneering company in the domestic recovery of critical metals such as nickel, cobalt, and lithium from end-of-life EV batteries. With advanced hydrometallurgical processes, Redwood can achieve recovery efficiencies of over 95% for its key metals. Its processes enable environmentally friendly and economically viable extraction with minimal reliance on energy-intensive mining operations. The company also introduces innovations in battery disassembly through automated sorting and separation technologies, enhancing material recovery and streamlining the recycling process. By returning recovered materials to the EV battery supply chain, Redwood contributes to building a circular economy within the US and improves the overall sustainability of domestic EV production [89–93]. While several international companies have established large-scale battery recycling operations, Redwood Materials represents a leading example of such efforts within the United States context.

(b) Policy Impact: S.1918, Battery and Critical Mineral Recycling Act of the US, allocates a large part of funds to the construction of facilities such as Redwood Materials. The grant from Greentech between 2021 and 2024 is aimed at funding infrastructure development and new technology that will reduce virgin raw materials dependency by 30 percent. Redwood has, for example, used federal dollars to scale up its operations, increasing its recycling capacity by 50 percent while introducing its gram-scale advanced processing units capable of handling higher volumes of spent batteries. Such policy also enables the cooperation between their own battery manufacturers and recycling facilities towards enabling a circular economy of EV batteries in the United States [94,95].

7. Technological Innovations in Recycling

7.1. Advanced Recycling Techniques

7.1.1. China

(a) Hydrometallurgy

Recycling lithium, cobalt, and other critical metals from old EV batteries using aqueous solutions is called hydrometallurgy. Plants in China largely use this process because it yields high recovery rates with a minimal environmental footprint. In this procedure, special leaching agents select these metals, and combined with advances in filtration systems, recycling will ensure that over 90% of metals are recovered in pure form, and ready for use in new batteries. For example, hydrometallurgical research is gaining traction so that recycling facilities in Guangdong can raise their cobalt and lithium recovery rates over 90% and even further, with the figure now being much higher than that [96–99].

(b) Automation in Recycling

This integration of robotic and AI-based systems in recycling has brought about amazing strides in efficiency and precision. Robotic arms with highly sensitive equipment for detecting items will be used in Shanghai installations to disassemble components in EV batteries such as cathode, anode, and casing. Furthermore, AI algorithms improve the sorting process of materials, thereby reducing human error while operating by much more than 40% throughput. The whole automation has been able to achieve a labor cost reduction and lesser material contamination, resulting in higher recovery yields.

7.1.2. United States

(a) Direct Recycling:

Direct recycling aims at ensuring that the cathode materials remain unchanged and that there will be minimum chemical processing in their use in new batteries. Such methods are energy-efficient, saving up to thirty percent of the energy consumed by normal methods like pyrometallurgy. Redwood Materials was the first company to directly recycle EV batteries, recovering important properties like nickel and cobalt and preserving their electrochemical properties. Such innovation greatly reduces the carbon footprint associated with battery recycling [100–103].

(b) AI-Driven Sorting:

Advanced AI algorithms are increasingly used for the sorting of EV battery components. These systems can detect precious material from waste by analyzing material properties in real-time. US startups have cited recovery improvements of up to 25%, thanks to the precision offered by AI-driven sorting technologies [104,105].

Commercialization Gap between China and the US:

While both China and the United States are advancing in EV battery recycling technologies, there remains a clear gap in commercialization maturity. Chinese companies such as BRUNP, GEM, and Huayou Cobalt have fully commercialized recycling systems capable of producing battery grade materials like nickel sulfate and cobalt sulfate, which are sold directly to cathode manufacturers. In contrast, most US-based recyclers, including Redwood Materials and Li-Cycle, are currently focused on producing black mass, an intermediate product containing a mix of critical metals. The refining infrastructure to convert black mass into battery-grade materials is still under development in the United States. However, ongoing federal support and private sector investments are rapidly closing this gap, with several US recyclers planning to operationalize full refining capabilities in the near future **see** Table 14 **for comparison**.

Table 14. Comparative overview of battery recycling companies in China and the United States.

Company	Country	Processing Capacity and Focus	Commercialization Level
BRUNP Recycling	China	• Processes approximately 120,000 tons of battery waste annually, achieving a metal recovery rate of 99.3% for nickel, cobalt, and manganese. • Engages in comprehensive recycling, producing battery-grade materials.	Fully commercialized operations, supplying refined materials directly to battery manufacturers.
GEM Co., Ltd.		• Has production capacity of 100,000 tons/year of battery-grade nickel–cobalt salt crystals. • Includes 33,000 tons of nickel sulfate.	Established commercial operations, produces and supplies nickel sulfate, cobalt sulfate, and ternary precursors to battery producers.
Redwood Materials	USA	• Processes about 20 GWh of lithium-ion batteries annually, equivalent to over 250,000 electric vehicles. • Currently produces black mass. • Refining facilities for battery-grade materials are under development.	Partially commercialized. Focused on black mass production. Currently developing refining processes; aims to supply battery materials domestically in the near future.

8. Evaluation of Recycling Efficiency (Quantitative Metrics)

In contrast, China, with its upgraded EPR policies, has increased the collection rates of batteries to attain resource recovery efficiencies of 90%, and an impressive drop in energy consumption level, as the collection now uses up to 50% less energy than that of traditional mining. In addition, advanced hydrometallurgy processes have significantly contributed to these achievements and provide an economically viable and environmentally friendly alternative to raw material extraction. Facilities such as Redwood Materials in the USA provide a remarkable 95% material recovery, as well as considerable energy savings and CO_2 improvement, as shown in Table 15. Direct recycling and AI-driven sorting are some of the innovations that further boost the performance and environmentally friendly nature of recycling operations.

Table 15. Highlights the key differences in policy impacts between China and the US.

Country	Recycling Rate (%)	Resource Recovery Efficiency (%)	Emission Reduction (Tons CO_2)
China	40%	90%	2.5 million
United States	35%	95%	1.8 million

9. Observation

China has developed a comprehensive and strategic policy framework for managing the lifespan of (EV) batteries, from manufacturing to (EOL) recycling. These regulations are closely aligned with the nation's overarching aims of encouraging the use of new energy vehicles (NEVs) and decreasing environmental impact. The country's Five-Year Plans, particularly the 13th and 14th, prioritize the development of energy storage and battery recycling technologies. These plans set lofty aims for lowering battery costs and improving recycling efficiency, establishing China as a global leader in the EV market. In addition to these broad goals, the Chinese government has adopted several incentives, such as subsidies and tax breaks, to encourage recycling (EOL) batteries. These incentives are supplemented by solid regulatory requirements that compel battery recycling programs and manufacturers to adopt take-back systems, ensuring that battery waste is effectively managed. Furthermore, municipal initiatives in key cities such as Beijing, Shanghai, and Shenzhen reinforce national policy by strengthening China's battery recycling capacity through partnerships with private enterprises and investments in recycling infrastructure.

The United States' approach to end-of-life EV battery recycling is characterized by a blend of federal and state-level programs, emphasizing public–private partnerships. At the federal level, programs like the Advanced Technology Vehicles Manufacturing (ATVM) Loan Program and the Bipartisan Infrastructure Law (BIL) aim to promote the development of innovative battery technologies and establish a domestic supply chain for crucial battery materials. These programs are supplemented by state-specific regulations in California, New York, and Washington that govern the proper handling, recycling, and disposal of lithium-ion batteries. These state regulations safeguard the environment and encourage a circular economy within their jurisdictions. US rules also promote the recycling and recovery of essential minerals required for lithium-ion battery production. Strategic frameworks such as the National Blueprint for Lithium Batteries 2021–2030 and the Battery and Critical Mineral Recycling Act of 2021 aim to improve recycling technology and reduce reliance on international supply chains. Furthermore, the United States government aggressively promotes public–private partnerships to enhance battery recycling technology, recognizing the vital role of private industry in expanding recycling operations and ensuring that recovered materials fulfill the quality standards needed to manufacture new batteries.

When comparing the two countries, both China and the United States understand the crucial need to have solid policies for recycling (EOL) electric vehicle batteries. However, their methodologies are fundamentally different. China's policy is more centralized, with specific directives embedded in national development goals. At the same time, the United States strategy is more decentralized, with state governments and private-sector partnerships playing an important role. Furthermore, China's rules are more stringent, with mandated recycling requirements and direct government intervention. In contrast, US policies prioritize incentives, technological innovation, and the creation of a sustainable supply chain for crucial minerals. These distinctions reflect each country's larger policy environment and industrial strategy, influencing their methods to control EV battery lifecycle are encapsulated in Table 16.

Table 16. Comparison of recycling policy aspects between China and the United States.

Aspect	China	USA
Policy Structure	Centralized, with clear directives integrated into national Five-Year Plans.	Decentralized, with significant autonomy given to state governments and private sector partnerships.
Regulatory Approach	Prescriptive, with mandatory recycling requirements and direct government intervention.	Emphasizes incentives, technological innovation, and voluntary measures to encourage compliance.
Government Involvement	Strong government role in setting and enforcing recycling standards.	Federal government sets broad guidelines, but state governments significantly influence specifics.
Economics Incentives	Limited economic incentives for recycling, focusing more on regulatory compliance.	Focuses on providing incentives, such as subsidies and tax credits, to encourage recycling and innovation.
Technological Development	Faces challenges due to a lack of advanced recycling technologies.	Emphasis on research and development to advance recycling technologies and processes.
Implementation	National policies reinforced by local initiatives in major cities.	Implementation varies widely by state, leading to inconsistencies in recycling practices.
Supply Chain Focus	Less emphasis on developing a sustainable supply chain for critical minerals.	Strong focus on building a domestic supply chain for critical minerals essential for battery production.
Stakeholder Engagement	Limited focus on public awareness and industry collaboration.	Encourages public–private partnerships and industry engagement to foster innovation and compliance.
Overall Strategy	Centralized control with mandatory compliance measures.	Decentralized, incentive-based approach with a focus on innovation and state-level customization

10. Conclusions

The exponential growth of electric vehicle (EV) adoption presents a pressing challenge in managing end-of-life (EOL) EV battery waste. Without effective policies and recycling strategies, the environmental, economic, and social impacts of EOL batteries could be substantial. This study provides a comprehensive evaluation of the regulatory frameworks for EV battery recycling in China and the United States, analyzing their effectiveness in mitigating waste and recovering critical materials.

Our analysis highlights that China has achieved a battery recycling rate of 40% with a material recovery efficiency of 90%, primarily through state-mandated policies and centralized regulations. In contrast, the US reports a slightly lower recycling rate of 35%, but exhibits a higher recovery efficiency of 95%, attributed to advanced technological innovations and private-sector involvement. Both countries emphasize the importance of extended producer responsibility (EPR) schemes, financial incentives, and stringent recycling mandates to ensure sustainable battery lifecycle management.

Despite these advancements, several critical gaps persist. Many nations lack standardized policies for battery collection, transportation, and disposal, leading to inefficiencies and environmental risks. Moreover, current recycling processes remain energy-intensive, with hydrometallurgical and pyrometallurgical methods still requiring optimization to reduce emissions and improve cost-effectiveness. The study also underscores the need for cross-border collaboration in material supply chains, particularly in securing lithium, cobalt, and nickel, which are essential for battery manufacturing.

To address these challenges, the study proposes the following recommendations:

1. Policy harmonization and global collaboration: establish a standardized international regulatory framework to enhance material traceability and ensure the responsible management of EOL EV batteries.
2. Technological innovation: increase research investment in direct recycling and AI-driven sorting systems, which have demonstrated up to 25% improvement in recovery efficiency.
3. Economic incentives: expand financial mechanisms such as tax credits, subsidies, and recycling grants to promote industry-wide adoption of sustainable practices.
4. Stakeholder engagement: foster public–private partnerships to streamline collection systems and integrate circular economy models into EV battery supply chains.

In conclusion, this study underscores the urgency of adopting a multi-faceted approach to EV battery waste management, incorporating regulatory frameworks, technological advancements, and economic incentives. By leveraging best practices from leading nations, policymakers can develop sustainable and scalable recycling models that not only mitigate environmental harm, but also support the growing EV industry. Future research should focus on life cycle cost analysis, second-life battery applications, and advanced material recovery techniques to further optimize battery circularity.

Author Contributions: Conceptualization, A.A.; methodology, A.A.; formal analysis, A.A. and M.A.B.; investigation, M.S., S.A. and S.A.Q.; resources, A.A.; data curation, M.T.I. and M.A.B.; writing—original draft preparation, A.A.; writing—review and editing, S.A. and M.T.I.; project administration. All authors have read and agreed to the published version of the manuscript.

Funding: The authors would like to express their profound gratitude for funding support from the Interdisciplinary Research Center for Sustainable Energy Systems (IRC-SES), King Fahd University of Petroleum and Minerals, under project No. INSE2402.

Data Availability Statement: Not applicable.

Acknowledgments: The authors would like to express their profound gratitude to Interdisciplinary Research Center for Sustainable Energy Systems (IRC-SES), King Fahd University of Petroleum and Minerals.

Conflicts of Interest: The authors declare no conflicts of interest.

References

1. Global Battery Demand to Surge by 2030, Supply Headaches on the Horizon. Saudi Gazette. 16 August 2021. Available online: https://saudigazette.com.sa/article/617999/BUSINESS/Powering-up-Global-battery-demand-to-surge-by-2030-supply-headaches-on-the-horizon (accessed on 14 December 2024).
2. Kirw, B.; Reddy, T.B.; Linden, D. *Linden's Handbook of Batteries*; McGraw-Hill: New York, NY, USA, 2019.
3. United Nations Department of Economic and Social Affairs. Frontier Technology Issues: Lithium-Ion Batteries—A Pillar for a Fossil-Fuel-Free Economy. Available online: https://www.un.org/development/desa/dpad/publication/frontier-technology-issues-lithium-ion-batteries-a-pillar-for-a-fossil-fuel-free-economy/ (accessed on 14 December 2024).
4. Battery University. BU-205: Types of Lithium-Ion. Available online: https://batteryuniversity.com/article/bu-205-types-of-lithium-ion (accessed on 14 August 2024).
5. Palacín, M.R. Advances in Battery Technologies for Electric Vehicles ResearchGate. October 2018. Available online: https://tinyurl.com/5csynmau (accessed on 14 December 2024).
6. McKinsey Company. Battery Recycling Takes the Driver's Seat. Available online: https://www.mckinsey.com/industries/automotive-and-assembly/our-insights/battery-recycling-takes-the-drivers-seat (accessed on 14 December 2024).
7. Yang, H.; Hu, X.; Zhang, G.; Dou, B.; Cui, G.; Yang, Q.; Yan, X. Life cycle assessment of secondary use and physical recycling of lithium-ion batteries retired from electric vehicles in China. *Waste Manag.* **2024**, *178*, 168–175. [CrossRef] [PubMed]
8. He, B.; Zheng, H.; Tang, K.; Xi, P.; Li, M.; Wei, L.; Guan, Q. A Comprehensive Review of Lithium-Ion Battery (LiB) Recycling Technologies and Industrial Market Trend Insights. *Recycling* **2024**, *9*, 9. [CrossRef]
9. Zhang, M.; Wu, W.; Song, Y. Study on the impact of government policies on power battery recycling under different recycling models. *J. Clean. Prod.* **2023**, *413*, 137492. [CrossRef]
10. Kang, Z.; Huang, Z.; Peng, Q.; Shi, Z.; Xiao, H.; Yin, R.; Fu, G.; Zhao, J. Recycling technologies, policies, prospects, and challenges for spent batteries. *iScience* **2023**, 108072. [CrossRef]
11. Yu, W.; Guo, Y.; Shang, Z.; Zhang, Y.; Xu, S. A review on comprehensive recycling of spent power lithium-ion battery in China. *eTransportation* **2022**, *11*, 100155. [CrossRef]
12. Rajaeifar, M.A.; Ghadimi, P.; Raugei, M.; Wu, Y.; Heidrich, O. Challenges and Recent Developments in Supply and Value Chains of Electric Vehicle batteries: A Sustainability Perspective. *Resour. Conserv. Recycl.* **2022**, *180*, 106144. [CrossRef]
13. Li, W.; Yang, M.; Long, R.; Mamaril, K.; Chi, Y. Treatmentofelectric vehicle battery waste in China: A review of existing policies. *J. Environ. Eng. Landsc. Manag.* **2021**, *29*, 111–122. [CrossRef]
14. Costa, C.M.; Barbosa, J.C.; Gonc, R.; Castro, H.; Campo, F.J.D.; Lanceros-Méndez, S. Recycling and environmental issues of lithium-ion batteries: Advances, challenges, and opportunities. *Energy Storage Mater.* **2021**, *37*, 433–465. [CrossRef]

15. Nimesh, V.; Kumari, R.; Soni, N.; Goswami, A.K.; Reddy, V.M. Implication viability assessment of electric vehicles for different regions: An approach of life cycle assessment considering exergy analysis and battery degradation. *Energy Convers. Manag.* **2021**, *237*, 114104. [CrossRef]
16. Or, T.; Gourley, S.W.D.; Kaliyappan, K.; Yu, A.; Chen, Z. Recycling of mixed cathode lithium-ion batteries for electric vehicles: Current status and future outlook. *Carbon Energy* **2020**, *2*, 6–43. [CrossRef]
17. Harper, G.; Sommerville, R.; Kendrick, E.; Driscoll, L.; Slater, P.; Stolkin, R.; Walton, A.; Christensen, P. Recycling lithium-ion Batteries from Electric Vehicles. *Nature* **2019**, *575*, 75–86. [CrossRef] [PubMed]
18. Oxford Institute for Energy Studies. China EVs Presentation, February 2024. Available online: https://www.oxfordenergy.org/wpcms/wp-content/uploads/2024/02/Presentation-China-EVs-Feb-2024.pdf (accessed on 14 August 2024).
19. Business Insider. China's EV Growth Set to Explode in 2024. 11 August 2023. Available online: https://markets.businessinsider.com/news/stocks/chinas-ev-growth-set-to-explode-in-2024-1033088764 (accessed on 16 December 2024).
20. Adamas Intelligence. Charts: China's Global Electric Car Dominance. Available online: https://www.adamasintel.com/charts-china-global-electric-car-dominance/ (accessed on 16 December 2024).
21. National Development and Reform Commission of China. Notice on Matters Related to the New Energy Vehicle Purchase Subsidy Policy. May 2021. Available online: https://en.ndrc.gov.cn/policies/202105/P020210527785800103339.pdf (accessed on 20 December 2024).
22. National Development and Reform Commission of China. Policies and Guidelines. Available online: https://en.ndrc.gov.cn/policies/index.html (accessed on 20 December 2024).
23. Fujian Provincial Government. Fujian Focus: Province Develops New High-Quality Lithium Battery Recycling Plant. 9 August 2021. Available online: https://shorturl.at/B8txl (accessed on 28 December 2024).
24. Climate Laws. Outline of the People's Republic of China 14th Five-Year Plan for National Economic and Social Development and Long-Range Objectives for 2035. 2021. Available online: https://climate-laws.org/search?q=China+14th+five+year+plan (accessed on 28 December 2024).
25. Battery Swapping in China. Available online: https://changing-transport.org/wp-content/uploads/2022_Overview_on_Battery_Swapping_and_Battery-as-a-Service_BaaS_in_China.pdf (accessed on 31 December 2024).
26. International Energy Agency (IEA). 14th Five-Year Plan on Circular Economy. Available online: https://www.iea.org/policies/24989-14th-five-year-plan-on-circular-economy (accessed on 31 December 2024).
27. SMM Review of Chinese Steel Scrap Market: Supply and Demand and the Prospect. Available online: https://news.metal.com/newscontent/102053000/SMM-Review-of-Chinese-Steel-Scrap-Market-Supply-and-Demand-and-the-Prospect/ (accessed on 31 December 2024).
28. English.gov.cn. Policy Watch: 14th Five-Year Plan. Available online: https://english.www.gov.cn/policies/policywatch/202107/08/content_WS60e64b42c6d0df57f98dc94e.html (accessed on 31 December 2024).
29. Metal.com. Notice on the Promotion and Application of Financial Subsidy Policy for New Energy Vehicles in 2022. 13 December 2021. Available online: https://news.metal.com/newscontent/101714008/notice-on-the-promotion-and-application-of-financial-subsidy-policy-for-new-energy-vehicles-in-2022 (accessed on 2 January 2025).
30. International Energy Agency (IEA). Financial Subsidy Policy for the Promotion and Application of New-Energy-Vehicles. Available online: https://www.iea.org/policies/2609-financial-subsidy-policy-for-the-promotion-and-application-of-new-energy-vehicles (accessed on 2 September 2024).
31. China Briefing. China Extends NEV Tax Reduction and Exemption Policy to 2027. 21 June 2023. Available online: https://www.china-briefing.com/news/china-extends-nev-tax-reduction-and-exemption-policy-to-2027/ (accessed on 2 January 2025).
32. CnEVPost. Factbox: China NEV Purchase Tax Policy 2024. 1 January 2024. Available online: https://cnevpost.com/2024/01/01/factbox-china-nev-purchase-tax-policy-2024/ (accessed on 2 January 2025).
33. China Daily. China Sees Faster NEV Adoption Thanks to Favorable Policies. 21 June 2023. Available online: https://shorturl.at/9WNwt (accessed on 4 January 2025).
34. Zhang, L.; Wang, J.; Liu, H. Recent advances in battery management systems for electric vehicles: A comprehensive review. *Energy Rep.* **2024**, *10*, 123–145. Available online: https://www.sciencedirect.com/science/article/pii/S2949720524000298 (accessed on 4 January 2025).
35. *GB 31574-2015*; Technical Specifications for Lithium-Ion Battery Recovery and Recycling. Chinese Standard. 2015. Available online: https://www.chinesestandard.net/PDF/BOOK.aspx/GB31574-2015 (accessed on 4 January 2025).
36. Statista. Size of the EV Battery Recycling Market in China. Available online: https://www.statista.com/statistics/1534764/china-size-of-the-ev-battery-recycling-market/ (accessed on 4 January 2025).
37. Fastmarkets. China Needs Clearer Standards to Support Battery Recycling Industry. Available online: https://www.fastmarkets.com/insights/china-needs-clearer-standards-to-support-battery-recycling-industry/ (accessed on 4 January 2025).
38. GME Recycling. EV Waste Battery Recycling: The Leadership of China. Available online: https://www.gme-recycling.com/ev-waste-battery-recycling-the-leadership-of-china/ (accessed on 5 January 2025).

39. Asia Times. China's Surging Lead in the EV Battery Circular Economy. Available online: https://asiatimes.com/2024/11/chinas-surging-lead-in-the-ev-battery-circular-economy/ (accessed on 5 January 2025).
40. China's Growing Lead in EV Battery Recycling. Supply Chain Report. Available online: https://supplychainreport.org/chinas-growing-lead/ (accessed on 5 January 2025).
41. Beijing Municipal Government. Electric Vehicle Policies in Beijing. Available online: https://english.beijing.gov.cn/so/s?siteCode=englishbj&tab=all&qt=electric%20vehicle (accessed on 5 January 2025).
42. Shanghai Municipal Government. Shanghai Tops Global Cities in NEV Ownership. Available online: https://english.shanghai.gov.cn/en-Latest-WhatsNew/20240305/53790527f40b46b083010a9690009184.html (accessed on 5 January 2025).
43. Circle Economy. Shenzhen's Electric Mobility Plan. 2024. Available online: https://knowledge-hub.circle-economy.com/article/5218?n=Shenzhen (accessed on 5 January 2025).
44. Brunp Recycling. Official Website of Brunp Recycling. Available online: https://en.brunp.com.cn/ (accessed on 6 January 2025).
45. Wang, H. Public transport strategies in the context of urban sustainability: A comprehensive review. *Transp. Res. Part A Policy Pract.* **2023**, *175*, 123–134. Available online: https://www.sciencedirect.com/science/article/abs/pii/S0965856423002215 (accessed on 6 January 2025).
46. Metal.com. MIIT Issued a Lithium Battery Specification Document to Further Strengthen the Management of the Lithium-Ion Battery Industry. 13 September 2021. Available online: https://news.metal.com/newscontent/101672387/MIIT-Issued-a-Lithium-Battery-Specification-Document-to-Further-Strengthen-the-Management-of-the-Lithium-ion-Battery-Industry/ (accessed on 6 January 2025).
47. EnergyTrend. Analysis of the Global Lithium Battery Market in 2025. 11 May 2024. Available online: https://www.energytrend.com/news/20240511-46916.html (accessed on 6 January 2025).
48. China Briefing. China's Energy Storage Sector: Policies and Investment Opportunities. 24 August 2023. Available online: https://www.china-briefing.com/news/chinas-energy-storage-sector-policies-and-investment-opportunities/ (accessed on 6 January 2025).
49. S&P Global Commodity Insights. China Targets to Cut Battery Storage Costs by 30% by 2025. 22 March 2022. Available online: https://www.spglobal.com/commodityinsights/en/market-insights/latest-news/energy-transition/032222-china-targets-to-cut-battery-storage-costs-by-30-by-2025 (accessed on 8 January 2025).
50. Meegoda, J.N.; Malladi, S.; Zayas, I.C. End-of-life management of electric vehicle lithium-ion batteries in the United States. *Clean Technol.* **2022**, *4*, 1162–1174. [CrossRef]
51. U.S. Environmental Protection Agency. Hazardous Waste Recycling. 2024. Available online: https://www.epa.gov/hw/hazardous-waste-recycling (accessed on 11 January 2025).
52. California State Senate. Assembly Bill No. 2440. Judiciary Committee. June 2022. Available online: https://shorturl.at/34JDc (accessed on 11 January 2025).
53. New York Senate. Senate Bill S2906: Relates to Establishing a Program for the Collection, Transportation, and Recycling of Batteries. 2021. Available online: https://www.nysenate.gov/legislation/bills/2021/S2906 (accessed on 15 January 2025).
54. Washington State Department of Ecology. Electronics E-Cycle Program. Available online: https://ecology.wa.gov/waste-toxics/reducing-recycling-waste/our-recycling-programs/electronics-e-cycle (accessed on 15 January 2025).
55. U.S. Department of Energy. Battery Policies and Incentives Search. Office of Energy Efficiency & Renewable Energy. Available online: https://www.energy.gov/eere/vehicles/battery-policies-and-incentives-search (accessed on 8 January 2025).
56. U.S. Congress. Energy Independence and Security Act of 2007. Public Law 110-140. December 2007. Available online: https://www.congress.gov/110/plaws/publ140/PLAW-110publ140.pdf (accessed on 8 January 2025).
57. Infrastructure Investment and Jobs Act. Public Law No. 117-58, 117th Congress of the United States. 15 November 2021. Available online: https://www.congress.gov/117/plaws/publ58/PLAW-117publ58.pdf (accessed on 8 January 2025).
58. The White House. Fact Sheet: The Bipartisan Infrastructure Deal. November 2021. Available online: https://bidenwhitehouse.archives.gov/briefing-room/statements-releases/2021/08/05/fact-sheet-the-bipartisan-infrastructure-investment-and-jobs-act-advances-president-bidens-climate-agenda/#:~:text=The%20deal%20makes%20our%20communities,a%20major%20investment%20in%20weatherization. (accessed on 8 January 2025).
59. U.S. Department of Transportation. Bipartisan Infrastructure Law. November 2021. Available online: https://www.transportation.gov/bipartisan-infrastructure-law (accessed on 8 January 2025).
60. Capanna, S.; Devranoglu, S.; Loper, J. *A Review of Federal Agency Compliance with Energy-Efficient Procurement Law*; Alliance to Save Energy: Washington, DC, USA, 2008. Available online: https://www.ase.org/sites/ase.org/files/file_AgencyComplianceW_EE.pdf (accessed on 8 January 2025).
61. U.S. Congress. Energy Policy Act of 2005. Public Law 109-58. August 2005. Available online: https://www.govinfo.gov/content/pkg/PLAW-109publ58/pdf/PLAW-109publ58.pdf (accessed on 10 January 2025).

62. U.S. Government. Executive Order 14008: Tackling the Climate Crisis at Home and Abroad. January 2021. Available online: https://www.federalregister.gov/documents/2021/02/01/2021-02177/tackling-the-climate-crisis-at-home-and-abroad (accessed on 10 January 2025).
63. U.S. Congress. H.R. 3684—Infrastructure Investment and Jobs Act. 117th Congress. November 2021. Available online: https://www.congress.gov/bill/117th-congress/house-bill/3684/text (accessed on 10 January 2025).
64. U.S. Department of Energy. National Blueprint for Lithium Batteries 2021–2030. June 2021. Available online: https://www.energy.gov/sites/default/files/2021-06/FCAB%20National%20Blueprint%20Lithium%20Batteries%200621_0.pdf (accessed on 10 January 2025).
65. The Battery Show. Battery Recycling and Second-Life Applications. Available online: https://www.thebatteryshow.com/content/dam/Informa/amg/novi/2022/docs/09_45%20-%20Burrell.pdf (accessed on 10 January 2025).
66. GovDelivery. DOE Advances in Battery Recycling. Available online: https://content.govdelivery.com/accounts/USEERE/bulletins/35f91aa (accessed on 10 January 2025).
67. The National Law Review. New Opportunities for DOE Advanced Batteries. Available online: https://natlawreview.com/article/new-opportunities-doe-advanced-batteries (accessed on 10 January 2025).
68. Supply Chain Dive. How the U.S. Plans to Strengthen Lithium Supply Chain for EV Batteries. Available online: https://www.supplychaindive.com/news/us-strengthening-lithium-supply-processing-ev-batteries/634544/ (accessed on 10 January 2025).
69. Investopedia. Biden Administration Announces $3 Billion in Battery Supply Chain Investments. Available online: https://www.investopedia.com/biden-administration-announces-usd3-billion-in-battery-supply-chain-investments-8715780 (accessed on 11 January 2025).
70. The Verge. Stellantis and Samsung to Receive $7.54 Billion Federal Loan for EV Battery Factories. Available online: https://www.theverge.com/2024/12/3/24312175/stellantis-samsung-doe-atvm-ev-battery-loan (accessed on 11 January 2025).
71. U.S. Congress. Senate Bill 1918. 117th Congress. May 2021. Available online: https://www.congress.gov/bill/117th-congress/senate-bill/1918/text (accessed on 11 January 2025).
72. Redwood Materials. Redwood Materials: Building a Circular Supply Chain. 2024. Available online: https://www.redwoodmaterials.com (accessed on 11 January 2025).
73. Li-Cycle. Li-Cycle: A Leader in Lithium-Ion Battery Resource Recovery. 2024. Available online: https://li-cycle.com (accessed on 11 January 2025).
74. U.S. Department of Energy. Gaines 2019 Presentation: Life-Cycle Analysis of Batteries and Electrification of Transport. 2019. Available online: https://rb.gy/ckzzv8 (accessed on 11 January 2025).
75. U.S. Environmental Protection Agency. Regulations. 2024. Available online: https://www.epa.gov/laws-regulations/regulations (accessed on 11 January 2025).
76. Ascend Elements. Products. 2024. Available online: https://ascendelements.com/products/ (accessed on 2 September 2024).
77. Vermont General Assembly. Act No. 152. 2024. Available online: https://rb.gy/zn1td2 (accessed on 15 January 2025).
78. Ministry of Industry and Information Technology (MIIT). Recycling Efficiency of Lithium Ion Batteries in China. 2023. Available online: https://www.miit.gov.cn (accessed on 26 March 2025).
79. National Development and Reform Commission (NDRC). 14th Five-Year Plan. 2021. Available online: https://en.ndrc.gov.cn (accessed on 26 March 2025).
80. Wu, Q.M. The embrace and resistance of Chinese battery investments in Hungary: The case of CATL. *Asia Eur. J.* **2024**, *22*, 201–223. [CrossRef]
81. Chen, Y.; Su, X.; Huang, N. ESG Information Disclosure and Innovation in New Energy Enterprises: A Case Study of CATL. *Adv. Econ. Manag. Polit. Sci.* **2024**, *121*, 113–122. [CrossRef]
82. Leong, J.Y. Review on Circularity in the Electric Vehicle (EV) Industry. *World Electr. Veh. J.* **2024**, *15*, 426. [CrossRef]
83. Tian, T.; Zheng, C.; Yang, L.; Luo, X.; Lu, L. Optimal recycling channel selection of power battery closed-loop supply chain considering corporate social responsibility in China. *Sustainability* **2022**, *14*, 16712. [CrossRef]
84. Yang, J.; Jiang, Q.; Zhang, J. Bridging the regulatory gap: A policy review of extended producer responsibility for power battery recycling in China. *Energy Sustain. Dev.* **2025**, *86*, 101697. [CrossRef]
85. Dzienis, A.M.; Mccaleb, A. Digital and Green Transitions and Automotive Industry Reconfiguration: Evidence From Japan and China. *J. Contemp. Asia* **2024**, 1–27. [CrossRef]
86. Huang, J.; Dong, X.; Chen, J.; Zeng, A. The slow-release effect of recycling on rapid demand growth of critical metals from EV batteries up to 2050: Evidence from China. *Resour. Policy* **2023**, *82*, 103504. [CrossRef]
87. Zhang, H.; Liu, G.; Li, J.; Qiao, D.; Zhang, S.; Li, T.; Guo, X.; Liu, M. Modeling the impact of nickel recycling from batteries on nickel demand during vehicle electrification in China from 2010 to 2050. *Sci. Total Environ.* **2023**, *859*, 159964. [CrossRef]
88. Jiang, S.; Zhang, L.; Hua, H.; Liu, X.; Wu, H.; Yuan, Z. Assessment of end-of-life electric vehicle batteries in China: Future scenarios and economic benefits. *Waste Manag.* **2021**, *135*, 70–78. [CrossRef]

89. Jiang, R.; Wu, C.; Feng, W.; You, K.; Liu, J.; Zhou, G.; Liu, L.; Cheng, H.M. Impact of electric vehicle battery recycling on reducing raw material demand and battery life-cycle carbon emissions in China. *Sci. Rep.* **2025**, *15*, 2267. [CrossRef]
90. Redwood Materials. Recycling Innovations and Efficiency Reports. 2023. Available online: https://www.redwoodmaterials.com/news/sustainable-battery-materials-process/ (accessed on 26 March 2025).
91. Duong, J.T.T. Decolonizing EV Battery Recycling: Unveiling Impacts on Native Communities. 2024. Available online: https://escholarship.org/content/qt0wr9x3gj/qt0wr9x3gj.pdf (accessed on 26 March 2025).
92. Kendall, A.; Slattery, M.; Dunn, J. Lithium-ion car battery recycling advisory group. *Calif. Environ. Prot. Agency* **2022**. Available online: https://calepa.ca.gov/wp-content/uploads/2022/05/2022_AB-2832_Lithium-Ion-Car-Battery-Recycling-Advisory-Goup-Final-Report.pdf?utm_source=chatgpt.com (accessed on 26 March 2025).
93. Curtis, T.L.; Smith, L.; Buchanan, H.; Heath, G. *A Circular Economy for Lithium-Ion Batteries Used in Mobile and Stationary Energy Storage: Drivers, Barriers, Enablers, and US Policy Considerations*; National Renewable Energy Lab. (NREL): Golden, CO, USA, 2021.
94. Meegoda, J.; Charbel, G.; Watts, D. Second Life of Used Lithium-Ion Batteries from Electric Vehicles in the USA. *Environments* **2024**, *11*, 97. [CrossRef]
95. U.S. Department of Energy. Battery and Critical Mineral Recycling Act (S.1918). 2023. Available online: https://www.energy.gov (accessed on 26 March 2025).
96. U.S. Environmental Protection Agency (EPA). Critical Minerals Recovery. 2022. Available online: https://www.epa.gov (accessed on 26 March 2025).
97. Umicore. Annual Report: Advancements in Hydrometallurgy. 2023. Available online: https://www.umicore.com (accessed on 26 March 2025).
98. Deng, H.; Wang, B.; Xu, J.; Yang, G.; Shi, Z.; Zhu, H.; He, W.; Li, G. A comprehensive review of whole process typical hydrometallurgical technologies for recycling waste lithium-ion batteries. *Sep. Purif. Technol.* **2025**, *363*, 132234. [CrossRef]
99. Huifan; Chen, Y.; Cui, Z.; Su, Y.; Liu, Z.; Dong, X.; Wang, Z.L.; Tang, W. One-step green hydrometallurgical recycling of spent lithium-ion batteries' cathode. *J. Hazard. Mater.* **2025**, *484*, 136769.
100. Wang, J.; Wang, H.; Zhang, G.; Ma, X.; Liu, Y.; Hao, J.; Xie, W.; He, Y. A review of research progress on combined pyro-hydrometallurgical technology for spent lithium-ion batteries. *Chem. Eng. J.* **2025**, *505*, 159403. [CrossRef]
101. Tesla. Innovations in Direct Recycling for EV Batteries. 2023. Available online: https://www.tesla.com (accessed on 26 March 2025).
102. Lipson, A.L.; Macholz, J.D.; Dai, Q.; Melin, P.; Gallagher, S.M.; LeResche, M.; Polzin, B.J.; Spangenberger, J.S. Cost-Effective and Scalable Approach for the Separation and Direct Cathode Recovery from End-of-Life Li-Ion Batteries. *Adv. Energy Mater.* **2025**, 2405430. [CrossRef]
103. Liu, S.; Dolotko, O.; Bergfeldt, T.; Knapp, M.; Ehrenberg, H. Towards Sustainable Direct Recycling: Unraveling Structural Degradation Induced by Thermal Pretreatment of Lithium-Ion Battery Electrodes. *ChemSusChem* **2025**, *18*, e202400727. [CrossRef] [PubMed]
104. Gao, X.-s.; Meng, W.U.; Zhao, G.-J.; Gu, K.-H.; Wu, J.-J.; Zeng, H.-B.; Qin, W.-Q.; Han, J.-W. Recycling technologies of spent lithium-ion batteries and future directions: A review. *Trans. Nonferrous Met. Soc. China* **2025**, *35*, 271–295. [CrossRef]
105. A.I. Grid. A.I. in Advanced Battery Recycling and Resource Recovery. 20 September 2024. Available online: https://aigrid.wordpress.com/2024/09/20/ai-in-advanced-battery-recycling-and-resource-recovery/ (accessed on 15 January 2025).

MDPI AG
Grosspeteranlage 5
4052 Basel
Switzerland
Tel.: +41 61 683 77 34

Recycling Editorial Office
E-mail: recycling@mdpi.com
www.mdpi.com/journal/recycling

www.ingramcontent.com/pod-product-compliance
Lightning Source LLC
Chambersburg PA
CBHW040816250926
51850CB00007BA/1181
9783725862689